ANIMAL BEHAVIOUR

ECOLOGY AND EVOLUTION

C.J. Barnard

1807 1982

A WILEY—INTERSCIENCE PUBLICATION
JOHN WILEY & SONS
NEW YORK CHICHESTER BRISBANE TORONTO SINGAPORE

Published in the USA
by Wiley–Interscience, a Division of
John Wiley & Sons, Inc., New York.

ISBN 0–471–88929–6

Printed and bound in Great Britain

CONTENTS

To Siân, my Parents
and Ken and Pegi

PREFACE

The study of animal behaviour, particularly from evolutionary and ecological viewpoints, has been one of the major growing points in biology over the last 10 to 15 years. The degree of quantitative rigour in theoretical, observational and experimental approaches to behaviour has increased dramatically. As more of the rapidly growing research literature becomes a basic requirement for students reading animal behaviour at undergraduate level, there is a need for a readily comprehensible text, covering all major aspects of behaviour study, to accompany their courses. This book, based on my first, second and third year undergraduate lectures at the University of Nottingham, is designed to meet that requirement.

The book begins with a discussion of the physiological and anatomical bases of behaviour: the relationship between nervous system structure and function and behaviour; hormonal effects on behaviour; biological clocks; perceptual mechanisms; and stimulus filtering. This leads to a consideration in Chapter 2 of how the animal integrates internal and external stimuli in making decisions about its behaviour and the way natural selection has shaped decision-making processes and the organisation of motivation. The first two chapters therefore deal with the instigation or causation of behaviour within the animal.

Chapters 3 and 4 deal with developmental aspects of behaviour. Chapter 3 discusses behaviour genetics, including the relationship between specific genes and behaviour, the heritability of behaviour patterns, the site of gene action in the body and the evolutionary consequences of a genetic basis to behaviour. This is followed by an examination of the role of experience and learning and the interaction of genetic and environmental factors in ontogeny.

In Chapters 5 to 10, the book moves on to examine how an animal's decisions are modified by its ecology. Once the animal has decided to feed, for instance, an enormous number of environmental variables, such as the distribution, abundance and quality of different prey types, climatic factors, competition and the risk of predation influence the way it must organise its feeding behaviour. Models, experiments and observations concerning the ecology of habitat choice, migration, foraging behaviour, predator avoidance, reproduction, social behaviour and communication are discussed in detail with emphasis placed on the rationale behind theoretical predictions and the testing of these predictions in the field and laboratory. Throughout, an attempt has been made to put ideas and results across in simple, readily comprehensible language and to avoid couching them in their less accessible mathematical format. Nevertheless, interested students are encouraged to follow up particular points in their original form in the literature.

The book concludes with a discussion in Chapter 11 of some key evolutionary concepts mentioned in previous chapters: coevolution, arms races, levels of selection and evolutionarily stable strategies (ESSs). The last two are discussed in detail because there is still a good deal of confusion about them in the research literature; it is particularly important in an undergraduate textbook, therefore, to make their meaning clear. All chapters close with an enumerated summary to reinforce their important points.

A number of people have very kindly helped in the preparation of the book over the past twelve months by reading and criticising one or more of the chapters. I am particularly indebted to Patrick Green, David Parkin, Geoff Parker, Richard Cowie, John Lazarus, Ian Duce, Hilary Stephens and Des Thompson, and to Geoff Parker, Peter Davies and Colin Galbraith for permission to reproduce their excellent photographs. I should also like to take this opportunity to express a much longer-standing debt of gratitude to Rex Knight, Professor A.J. Cain, Dr R.G. Pearson, Geoff Parker and John Krebs, all of whom at one time or another, through discussion or advice, have enormously stimulated my interest in the study of evolution and behaviour. I am grateful to Marlies Rivers, Dawn Thompson, Katherine Lyon and Wendy Lister for typing the manuscript and to Tim Hardwick of Croom Helm Ltd for suggesting the book in the first place and for advice and encouragement during its production. Finally, my very special thanks go to my wife Siân for her immeasurable support and assistance as always and to my daughters Anna and Lucy for patience with an absentee father.

1 PHYSIOLOGICAL MECHANISMS AND BEHAVIOUR

Behaviour is the tool with which an animal uses its environment. Through behaviour the animal manoeuvres itself in an organised and directed way and manipulates objects in the environment to suit its requirements. In order to behave, the animal must act as an integrated and co-ordinated unit. It must juggle a bewildering array of stimuli from inside its body and from the external environment and organise the information into a series of commands to its muscles. In Chapter 2 we shall examine this process of integration and see how internal and external stimuli from the environment are translated by the animal into behaviour. Before we can do that, however, we need to know something about the sources and processing of information in the animal's body. Animals have evolved complex systems of cells and chemicals whose task it is to detect, transmit, integrate and store environmentally supplied information for later use in making decisions. They consist of (a) various types of sensory cell which pick up different changes in the environment, (b) a more or less complex system of nerve cells which transmits and integrates information from sensory receptors, (c) chemical messengers which transmit information on a more leisurely time scale than the nervous system and (d) muscle cells which transform information from the nervous system into actions. In this chapter, we shall examine these components to see how they interact to produce behaviour.

1.1 Neural Mechanisms and Behaviour

Although a complex nervous system is not essential for behaviour — protozoans get by quite nicely with only rudimentary sense cells — the scope and sophistication of behaviour within the animal kingdom is quite clearly linked with the evolution of neural complexity. The behavioural capacities of protozoans and earthworms are extremely limited compared with those of birds and mammals. What, then, are the properties of a nervous system which make complex behaviour possible? In the sections which follow, we shall look briefly at the structure and function of nervous systems. This is not the place, however, to discuss neural anatomy and physiology in detail. Readers wishing to find out more are referred to the excellent books by Katz[155] and Usherwood[320]. In addition, a good introduction to the relationship between nervous system organisation and behaviour is provided by Guthrie[123].

1.1.1 Nerve Cells and Synapses

Nerve Cell Structure. True nervous systems are only found in multicellular animals. Here they form a tissue of discrete, self-contained nerve cells or *neurons*.

Figure 1.1 (a) A Motor Neuron Connecting with Muscle Fibres. cb, cell body; d, dendrite; a, axon; nS, nucleus of Schwann cell; nR, node of Ranvier; S, Schwann cell; my, myelin sheath; m, muscle fibre; e, end plate. (b) The Relationship between Ion Flow and Membrane Potential during an Action Potential in a Non-myelinated Axon. rp, resting potential; mrp, minimum refractory period; ap, action potential; solid circles Na⁺ ions; open circles, K⁺ ions. (c) Net Ionic Movements during an Action Potential and the Recovery Phase. Ion flow during an action potential is energetically downhill and occurs spontaneously, while recovery requires a metabolic pump. Poisoning of the pump does not prevent propagation of action potentials, but eventually ion concentration gradients become dissipated and conduction fails.

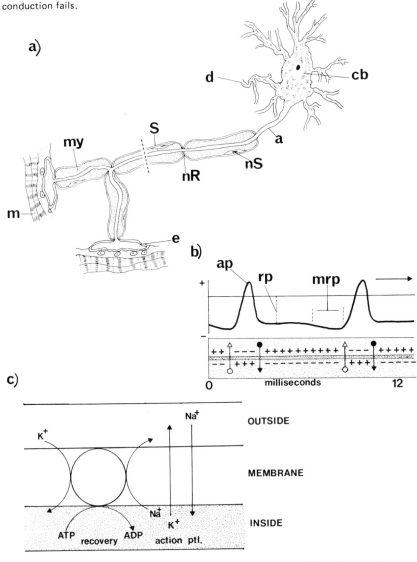

Source: Modified after Adrian, R.H. (1974). *The Nerve Impulse.* Oxford Biology Readers.

Like any other type of animal cell, neurons comprise an intricate system of cell organelles surrounded by a cell membrane (Figure 1.1a). Unlike other animal cells, however, they are specialised for transmitting electrical messages from one part of the body to another. This specialisation is reflected both in their structure and their physiology.

A neuron has three obvious structural components. The main body of the cell, the *soma*, is a broad, expanded structure housing the nucleus. Extending from the soma are two types of cytoplasm-filled processes called *axons* and *dendrites*. Axons carry electrical impulses away from the soma and pass them on to other neurons or to muscle fibres. Dendrites receive impulses from other neurons and transport them to the soma. All three components are usually surrounded by *glial cells*. Although glial cells are not derived from nerve tissue, they come to form a more or less complex sheath around the axon. In invertebrates, the glial cell membranes may form a loose, multilayered sheath in which there is still room for cytoplasm between the layers. In this case the arrangement is known as a *tunicated axon*. In vertebrates the sheath is bound more tightly so that no gaps are left. The glial cells are known as Schwann cells and are arranged along the axon in a characteristic way. Each Schwann cell covers about 2 mm of axon. Between neighbouring cells there is a small gap where the membrane of the axon is exposed to the extracellular medium. These gaps are known as the nodes of Ranvier. Axons with this interrupted Schwann cell sheath are called *myelinated* or *medullated* axons. The formation of the myelin sheath enhances enormously the speed and quality of impulse conduction.

Nerve Cell Function. The basis of the neuron's ability to conduct electrical impulses is the distribution of electrically charged atoms (ions) inside and outside the cell. In squid neurons, for instance, the concentration of potassium (K^+) ions is some 20 times greater inside the cell than outside. Sodium (Na^+) ions, on the other hand, are about 10 times more concentrated outside the cell. While both ions tend to diffuse towards equilibrium along their concentration gradients, diffusion is counteracted both by the low permeability of the cell membrane to Na^+ ions (which have large hydration shells) and by a metabolic pump which transfers the ions against their gradients. The net result of this ionic imbalance is a negative *resting potential* across the membrane of some 60-70 mV, depending on the type of neuron.

When the neuron is stimulated, the membrane suddenly becomes highly permeable to Na^+ ions at the site of stimulation (Figure 1.1b). As Na^+ ions flood into the cell, they reduce the membrane potential at that point. When the change in potential reaches a certain threshold, there is a massive influx of Na^+ ions which results in a sharp depolarisation to around +40 mV, known as an *action potential*. The formation of an action potential at one part of the membrane stimulates an increase in Na^+ permeability in the adjacent part and a wave of depolarisation courses down the axon. The rate of action potential conduction, of course, depends on how good a conductor the axon is. One way to increase its conductivity is to increase its diameter. This is precisely what has occurred in the 'giant' axons of some invertebrates (e.g. earthworms, squid). However, increasing the size of axons is costly both in materials and space. Insects and vertebrates have solved the conduction problem by insulating their axons instead. Their myelinated (see above) axons allow the flow of current only through certain areas (nodes of Ranvier). Consequently, action potentials are conducted from node to node extremely rapidly. As soon as an action potential has passed a given point on the membrane, there is a sudden increase in K^+ permeability and K^+ ions diffuse out along an electrical gradient. At the same time, Na^+ ions are actively pumped out of the cell and the negative resting potential is restored (Figure 1.1c).

In most cells, action potentials are 'all-or-nothing' events. Information is transmitted by their frequency of generation rather than their graded strength. In sensory receptors and certain other specialised cells, however, where short-distance communication is required, information is coded as graded potentials.

Synapses. The membrane boundaries of neurons are complete and the contact between cells which is essential to the transmission of impulses is accomplished by close juxtaposition instead of the formation of a continuous syncytium. The region of juxtaposition is known as a *synapse*, a term which is also applied to contacts between neurons and sensory receptors and neurons and muscle fibres. In most cases, transmission across the synapse occurs through the medium of transmitter substances, although electrical communication is known in cells where the juxtaposition is very close (e.g. in invertebrate giant axons). Various transmitter substances are known, including acetylcholine (an excitatory transmitter at vertebrate neuromuscular junctions and in invertebrate and vertebrate central nervous systems), noradrenaline, dopamine, glutamic acid, γamino butyric acid (GABA) and 5-hydroxytryptamine (serotonin). Some of the latter also appear to act as inhibitory transmitters at certain synapses. Synapses provide a powerful means of organising complex cross-connections between neural pathways and are thus a crucial feature in the evolution of sophisticated behaviour patterns and the ability to synthesise sensory information.

1.1.2 Evolutionary Trends in Nervous Systems and Behaviour

As we ascend the phylogenetic scale from simple unicellular organisms to vertebrates, the organisation and complexity of the nervous system changes in two major ways. The first is a trend towards greater *differentiation*. In unicellular organisms, the whole individual acts as a sensory receptor and motor effector rolled into one. It is not until the advent of simple multicellular animals that these two functions are performed by separate anatomical components and become spatially localised. The process of localisation occurs in response to the functional needs of the animal. In order for this to occur, however, there must be some means of maintaining communication between sensory and effector components. This means is provided by the extensive axons of the nerve cells and by the facility for linking the axons and dendrites of different neurons via synapses. In an advanced nervous system there are five major components linking stimulus perception and motor response. These are (a) a cell or group of cells acting as a sensory receptor, (b) an *afferent* or *sensory* neuron carrying impulses from the sense cells, (c) an *efferent* or *motor* neuron carrying impulses to effector cells, (d) an *internuncial* or *interneuron* linking b and c, and (e) effector organ.

Invertebrate Nervous Systems and Behaviour. Examples of this link-up system in its simplest form are found in the nerve nets of cnidarians (sea anemones, jellyfish, etc.) and echinoderms (starfish, sea urchins, etc.). The simplest kind of net is the type found in the sedentary freshwater cnidarian *Hydra*. In *Hydra*, the nerve net lies just under the epidermis and consists of synaptically-linked bipolar and tripolar cells. Impulse transmission is slow because impulses have to traverse large numbers of synapses and because their transmission lacks directionality. The synapses allow impulses to dissipate in several different directions. Trends in other groups which still possess nerve nets are towards the concentration of nervous tissue into tracts. Impulses can then be channelled in particular directions to bring effector organs into play more rapidly.

The behaviour of organisms relying on nerve nets is characteristically stereo-typed. Most exhibit simple reflexes (Section 1.1.3), stereotyped motor sequences (*fixed action patterns*) and rhythmic locomotory activities. Even the more elaborate behaviour patterns, like shell-climbing in epizooic anemones, consist of only three or four distinct elements. There is a low level of stimulus discrimination and such learning as exists consists of habituation and reflex facilitation rather than conditioning or other forms of associative learning (see Chapter 4). These simple responses function well in the animals' relatively stable aquatic environments with their ample food supplies and effective medium for passive dispersal.

Initially, through-conduction tracts are little more than local, directional thickenings in an otherwise diffuse net (e.g. *Actinia* spp.). In more advanced invertebrates, like flatworms (Platyhelminthes), the tracts are more pronounced and the nervous system begins to show signs of a second major evolutionary trend: *centralisation*. Flatworms are the first level of life to possess a *central nervous system* (*CNS*). Even within the Cnidaria there is a trend towards more deeply seated nerve rings and tracts in mobile species, but in the higher invertebrates, culminating in the metamerically segmented groups (annelids, arthropods) and the molluscs, the nervous system has become differentiated into concentrations of nervous tissue (*ganglia*) linked by nerve cords lying near the ventral surface of the body. While the initial aggregation of nervous tissue into tracts simply enhanced the conduction of impulses between specific parts of the body, the centralisation process has ultimately produced a kind of neural 'switchboard'. Afferent fibres from sensory receptors plug into the central switchboard where a mass of interneurons is ready to connect them with a variety of motor neurons.

Depending on the type of input from a sensory receptor, different effectors are brought into play so that the animal can respond appropriately.

CNS ganglia (other than cerebral ganglia) may contain anything from 400 (leeches) to 1,500 (the tectibranch mollusc *Aplysia*) cells. Well-defined tracts and glomeruli (aggregations of neuron terminals) can also be distinguished, especially in the cerebral ganglia which may form an elaborate brain-like structure. The primary function of each ganglion is the regulation of local reflex activities providing for local control of movement impossible with a diffuse nerve net. Ganglia also exercise longer-range control via long interneurons extending through the nerve cords thus controlling the co-ordinated operation of several parts of the body.

Along with the centralisation of nervous tissue and its organisation into a more or less linear system of cords and ganglia comes a number of behavioural changes. The most obvious are an increase in discriminatory abilities and responses to specific *key stimuli*. In addition, the elaboration of appendages and musculature makes subtle movements and the performance of complex tasks possible. Good examples are the elaborate courtship songs and postures of many insects and particularly the building of ornamented mud nests by certain hymenopteran species. However, while invertebrates certainly show learning ability, learned responses are seldom retained for long. This may be in part because of the small size of the 'brain' (cerebral ganglia) and in part because of the rapid turnover nature of many invertebrate life cycles. In short-generation life cycles where there is little time for learning, simple pre-programmed responses are likely to be more economical.

Among the invertebrates, there are trends towards enlargement of the brain, a coalescence of somatic ganglia and an increase in the brain's control over regional centres (collectively referred to as *cephalisation*). Despite the increasing importance of the brain, however, the somatic ganglia still retain considerable independence of control. Earthworms, for instance, can crawl normally, feed, copulate and burrow after removal of the cerebral ganglia, but they are hyperactive and their behaviour phrenetic. Nereid worms are able to learn certain tasks or persist with previously learned tasks after disconnection of their cerebral ganglia from the rest of the CNS. The cerebral ganglia therefore appear to be just one of many memory storage sites.

Independent control of behaviour by somatic ganglia is particularly well developed in arthropods. If still connected with its ganglion, the isolated leg of a cockroach will continue to show stepping movements when stimulated appropriately (e.g. by pressing on the trochanter). Indeed, even if the ventral nerve cord is completely severed, co-ordinated walking can be elicited by stimulating a single leg. Movement of an anterior leg, for instance, exerts a traction force on the leg behind, stimulating proprioceptors and eliciting a reflex response. Even in cephalopod molluscs (squids and octopuses) where cephalisation has reached its peak within the invertebrates, many responses are still under the control of somatic ganglia. A limitation in the octopus's (*Octopus vulgaris*)

Figure 1.2: Sagittal Section of an Octopus Brain Showing Component Ganglia. bg, brachial ganglion; cg, cerebral ganglion; pg, pedal ganglion; pvg, pallio-visceral ganglion; sbg, superior buccal ganglion; o, oesophagus.

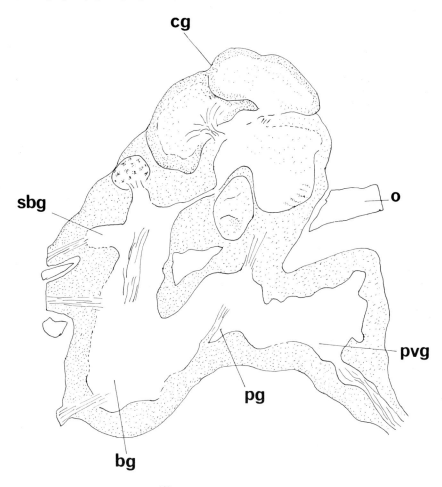

Source: Modified after Guthrie[123].

otherwise impressive object learning abilities is its inability to distinguish objects by weight. This is because the movements of each tentacle are regulated by local axial ganglia. Information from tentacle proprioceptors is processed in the ganglia rather than being passed on to the brain where it could become available for learning.

Invertebrate brains vary considerably in structure and complexity. At the lower end of the scale, flatworm brains contain some 2,000 cells. Insect brains are intermediate with around 340,000, whereas those of cephalopods contain up to 90 million, almost a tenth of the number found in the human brain.

Despite its outwardly advanced appearance, however, the anatomy of the cephalopod brain reveals its construction to be of several amalgamated somatic ganglia (Figure 1.2). The cephalopod CNS contrasts sharply with the loose string of ganglia (consisting in total of fewer than 50,000 cells) found in the slower moving gastropod and lamellibranch molluscs. There are separate visual and tactile learning centres within the lobes of the brain and centres for the integration of visual information from the high-resolution eye and tactile information from the mobile tentacles.

Among arthropods, brain structure is remarkably constant. In both annelids and arthropods, the brain develops as three main regions: (1) the *protocerebrum*, the main components of which in arthropods are the paired optic lobes, the median body and the *corpora pedunculata*, each functioning in the integration of information from the anterior sense organs and the control of subsequent behaviour; (2) the *deuterocerebrum*, which contains association centres for the first antennae; and (3) the *tritocerebrum*, nerves from which extend to the upper alimentary canal, second antennae (where they exist) and the upper 'lip'. The same functional zones can be recognised throughout most of the phylum and are derived from similar zones found in the errant polychaete annelids. Within broad taxonomic groups, the size of different centres is related to behaviour. In hunting spiders (Lycosidae), for instance, the optic centres in particular and the brain in general are better developed than in the web-spinners (Saltidae). Similarly, the largest single neuron in the semi-nocturnal cockroaches, which depend on tactile and olfactory senses, is the descending cell from the antennae to the thoracic ganglia (controlling activity levels). In the visually-hunting dipteran flies, it is the neurons receiving their input from the eyes which are best developed. In some species, the optic lobes may contain as many neurons as the remainder of the protocerebrum. Also, the *optic glomeruli* develop a laminated structure reminiscent of the stratified cortex of vertebrate brains.

Vertebrate Nervous Systems and Behaviour. In vertebrates, the nervous system develops from dorsal tissue and as a tube rather than solid primordia. Nevertheless, traces of the ancestral segmental pattern remain in the distribution of sensory and motor zones within the system. Centralisation, cephalisation and the functional differentiation of the nervous system reach their peak within the vertebrates. Indeed, while structural centralisation occurs quite low down on the phylogenetic scale, true centralisation of *function* is the exclusive property of vertebrate nervous systems.

CNS Structure and Function. The vertebrate CNS consists more clearly than that of invertebrates of two principle components: the brain and the spinal cord. The spinal cord consists of an outer region of mostly myelinated tracts (so-called 'white matter') connecting the brain with spinal control centres, and an inner region of neuron cell bodies ('grey matter') (see Figure 1.5). The *dorsal horns* of the inner region accept afferent, sensory neurons entering the CNS, while the *ventral horns* send out efferent motor fibres (Figure 1.5). The arrangement of axon and cell body material in the vertebrate spinal cord is the opposite of that in the invertebrate CNS where the open circulatory system requires the cell bodies to be exposed in the haemocoel.

The brain develops as three major regions: the forebrain (*prosencephalon*), the mid-brain (*mesencephalon*) and the hind brain (*rhombencephalon*). The prosencephalon is further divided into two parts: (1) the anterior *telencephalon* incorporating the olfactory lobes, *corpus striatum* and *cerebral cortex* (cerebral hemispheres) and (2) the *diencephalon*, comprising the *thalamus* and *hypothalamus* and connecting via the latter with the pituitary gland. The mesencephalon consists mainly of the *optic tectum* (including the *optic lobes* or, in mammals, the *corpora quadrigemina*) and the *tegmentum*. The rhombencephalon is also divided into two parts: the *metencephalon*, containing the anterior part of the *medulla oblongata*, the *cerebellum* and (in mammals) the *pons*, and the *myelencephalon* which contains the posterior medulla oblongata. The relatively simple arrangement of 'white' and 'grey matter' in the spinal cord is extensively modified in the brain. Here it is the central regions surrounding the fluid-filled *ventricles* which originate in the outer regions of the spinal cord. The outer areas of the brain are specialised,often stratified, tissues consisting of dense masses of cell bodies, homologous with the 'grey matter' of the spinal cord.

Vertebrate brains show two distinct evolutionary trends. One, exemplified by the bony fish (Actinopterygii), is an elaboration of the mesencephalon where the optic tectum becomes thickened and stratified with tracts from most other regions projecting into it. The mid-brain has thus become the major site of neural integration (Figure 1.3a). In mammals, the main thrust of evolution has been in the elaboration of the telencephalic lobes (cerebral hemispheres) which now become the major association centres (Figure 1.3b), although the tectum still retains its stratified appearance. The lobes consist of *neocortical* material and in man have extended to cover the rest of the brain.

The neocortex (referred to from now on as 'cortex') has several externally visible divisions which correspond with functional divisions. In advanced mammals, voluntary motor control areas lie in front of those for somatic sensory functions and are separated from them by a deep fissure. A proportion of the motor cortex cells communicate directly with the spinal cord via a large through-conduction pathway, the *pyramidal tract*. The two halves of the cortex are connected by another large tract, the *corpus callosum*. Severance of the corpus callosum has revealed pronounced differences in the functional dominance of the left and right hemispheres. In man, the dominant left hemisphere controls speech (and has been shown in birds to control singing), whilst the right appears to be involved in comprehension.

Further motor control occurs subcortically in various centres. The corpus striatum is one such centre which in birds is highly elaborated to form a major

Figure 1.3: Diagrams of the Brain of a Teleost Fish (a) and a Mammal (b) Showing the Main Connective Pathways. (a) dtn, dorsal thalamic nuclei; vtn, ventral thalamic nuclei; Mc, Mauthner cell; sc, sensory column; mc, motor column; cs, corpus striatum; ol, olfactory lobe; vc, valvula cerebelli; cc, cerebellum; m, medulla; ot, optic tectum. (b) tn, thalamic nuclei; sc, sensory column; t, tegmentum; ct, cerebral cortex; c, cerebellum; bg, basal ganglia; h, hippocampus; lg, lateral geniculate nucleus; mc, motor column; mrf, medullary reticular formation; mg, median geniculate nucleus; pt, pyramidal tract; ol, olfactory lobe; on, optic nerve.

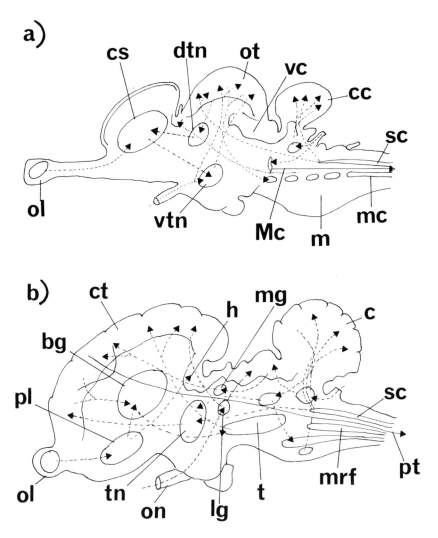

Source: Modified after Guthrie[123].

association centre important in the control of fixed action patterns (see Section 1.1.3). Another important subcortical region is the so-called *limbic system* which comprises the *hippocampus, cingulate gyrus* and the *amygdala*. The limbic system connects with the hypothalamus and the *reticular formation* (a system of branching cells traversing the three major brain regions which, among other things, appears to control motivational levels and specificity) and plays a role in conditioning (hippocampus), agonistic behaviour (amygdala) and decision-making.

The thalamus contains a number of important nuclei (well-defined groups of cells) which relay information to the cortex from the retina (*lateral geniculate nucleus*), the ear (*medial geniculate nucleus*), the cerebellum (*ventrolateral nucleus*) and the tectum (*posterolateral nucleus*) and appears to function in the appreciation of temperature, pain and pleasure. A complex array of nuclei in the hypothalamus regulate, via connections with the pituitary gland, the production of various hormones important in the control of behaviour (see Section 1.2) and also the activity of the autonomic (sympathetic and parasympathetic) nervous system. In addition, the hypothalamus controls complex behaviours like sleep, feeding (*ventromedial nucleus*) and aggression and hypothalamic *osmoreceptors* (cells sensitive to blood Na^+ ion concentration) appear to play a role in the initiation of drinking-orientated behaviour.

Crucial to the control and maintenance of posture and locomotion is the cerebellum. So-called *climbing* and *mossy* fibres ascend from peripheral proprioceptors and excite the cerebellar *Purkinje cells*. Excitation of the Purkinje cells in conjunction with the inhibitory effects of other specialised centres (e.g. *basket* and *Golgi cells*) produces zones of neuronal activity which control the sequential activity of functionally related groups of muscles. An important pathway for the transfer of information from the Purkinje cells to the cortex relays via the *red nucleus* in the tegmentum and the thalamus. Below the red nucleus is the *substantia nigra* which connects with the corpus striatum and is part of a dopaminergic pathway (a pathway in which dopamine acts as a transmitter substance) important in controlling locomotion.

Spanning the rhombencephalon is the medulla oblongata. The medulla contains a number of nuclei for the cranial nerves supplying the head and body (including the vagus). In addition, large numbers of fibres pass through the medulla on their way from the spinal cord to the cortex and vice versa. The medulla is the centre of control for certain vital functions including respiration and blood vessel tone. The vagus nerve which originates in the medulla is concerned with heart and gastro-intestinal reflexes. An animal can live if the brain is severed at the level of the mesencephalon, but serious damage to the medulla is invariably fatal.

Brain Morphology and Behaviour. As in invertebrates, elaboration of different parts of the vertebrate brain correlates well with the behavioural characteristics of the animal. Species of *Acanthopagrus*, for instance, which are nocturnally-

Figure 1.4: The Relationship between Brain and Body Weight (Encephalisation Quotient) in Higher (Solid Symbols: circles, mammals; squares, birds) and Lower (Open Symbols) Vertebrates. a, vampire bat; b, mole; c, rat; d, opossum; e, baboon; f, wolf; g, *Australopithecus*; h, chimpanzee; i, man; j, porpoise; k, male gorilla; l, lion; m, elephant; n, blue whale; o, crow; p, hummingbird; q, ostrich; r, goldfish; s, eel; t, coelacanth; u, alligator.

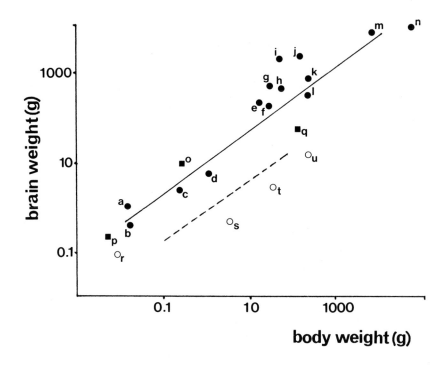

Source: Modified after Jerison, H.J. (1973). *Evolution of the Brain and Intelligence.* New York, Academic Press.

hunting fish, have large olfactory lobes. Other predators with a well-developed olfactory sense, like sharks, have pronounced diencephalic structures called *habenulae* which are associated with olfactory centres. Similarly, the optic lobes in visually-hunting fish like tunnies (*Thynnus*) and mackerel (*Scomber*) are relatively large. Actively-moving animals such as dolphins (*Coryphaenus*) and tunnies also have a well-developed and often elaborate cerebellum.

A graph of the relationship between brain and body size in vertebrates shows a pronounced discontinuity (Figure 1.4). In lower vertebrates, the brain accounts for less than 0.1 per cent of the body weight. In birds and mammals, however, it often exceeds 0.5 per cent with a high extreme in man (2.1 per cent) and a low extreme in ostriches (0.04 per cent). The discontinuity is due mainly to the dramatic increase in the forebrain cortex of higher vertebrates which correlates with superior discriminatory, association and learning abilities. Conditioning experiments with fish and rodents of different size suggest that small forms learn easy tasks more quickly but large forms have better retention. Such relationships, however, are far from conclusive because there are confounding effects of metabolic rate, activity levels and species-specific differences in brain structure.

Brain size alone is an unreliable guide to functional sophistication because of extensive specialisations within the brains of different species. Owls (Aves: Strigiformes) and parrots (Aves: Psittaciformes), for example, possess the largest and most specialised brains in birds. However, the major developments of the cortex have taken place in different areas. In owls, it is the dorsal cortex (dorsal and accessory hyperstriatum and intercalated cells) which has become elaborated, whereas in parrots it is the ventral hyperstriatum, neostriatum and palaeostriatum.

In many cases, however, regional specialisations do correlate with behaviour. In the raccoon (*Procyon*), there is a specialisation of part of the cortex which provides a sensory map of the palmar and digital areas of the 'hand'. The map has evolved in conjunction with the raccoon's manual skills in preparing and handling food. Similarly, among otters, the forelimb field in the cortex is much larger in the manually dextrous sea otter (*Enhydra*) than in the less dextrous *Lutra* and *Pteroneura* species.

1.1.3 Reflexes and Complex Behaviour

We have seen that the behaviour of an animal is related in a broad way to the organisation and degree of sophistication of its nervous system. However, this only provides a gross view of the relationship between neural organisation and behaviour. To what extent can we identify the precise neuronal pathways which control particular behaviours? Our first problem is to define behaviour.

Deciding which kinds of response do and which do not qualify as behaviour is not easy. However, we clearly need a definition of behaviour if we are to ask how it is controlled in the nervous system. For our purposes we shall adhere to Skinner and Hebb's working definition. This includes as behaviour '*all observable processes by which an animal responds to perceived changes in the internal state of its body or in the external world*'. Thus attacking prey, withdrawing a limb from a painful stimulus or 'freezing' at the sight of a predator, all count as behaviour, though of differing degrees of complexity. Although Skinner and Hebb's definition admits some physiological responses, like the secretion of a sweat gland, this is not as serious a drawback as it at first seems. Such responses may say a lot about an animal's motivational state. In terms of the co-ordinated movement of the whole animal or part of the animal, however, *reflexes* represent most people's idea of the simplest form of behaviour. A reflex is an automatic, stereotyped unit of behaviour, whose occurrence may vary with context and habituation (see Chapter 4), but whose *form* does not vary. Because reflexes are relatively simple, they provide a good opportunity to examine the relationship between behaviour and the functioning of the nervous system.

Reflexes. In vertebrates, the spinal cord performs two major roles. It acts as a conductor for nerve impulses travelling to and from the brain and it integrates reflex behaviour by the trunk and limbs (it is important to point out, however, that the brain also integrates some reflexes). Here the term 'reflex' describes the rapid automatic response of the body or part of the body to a simple stimulus. The classic human knee jerk in response to a tap on the patellar tendon and the withdrawal of a limb from a painful stimulus are well-known examples. In Figure 1.5, we can trace the neural pathways for these two reflexes.

The nerve path traced by a reflex is known as a *reflex arc*. It includes the five neural components discussed in Section 1.1.1. The reflex arc for the knee jerk is rather special. Instead of connecting within an Interneuron in the spine, the sensory neuron connects directly with a motor neuron. The knee jerk therefore forms a *monosynaptic arc* (it contains only a single synapse), a type more typical of invertebrate nervous systems. The reflex arc described when a limb is pulled away from a painful stimulus is more complex. The passage of impulses from sensory to motor neurons is mediated by a mass of interneurons (called *association neurons*) in the spinal cord. The reflex arc is thus *polysynaptic* (Figure 1.5). The apparent simplicity of the withdrawal reflex belies the complex pattern of muscle control involved. The animal must simultaneously contract the *flexor* muscles and relax the antagonistic *extensor* muscles of the stimulated limb so that the limb is pulled in towards the body. There is first of all, then, a *flexion reflex*. The neural co-ordination of antagonistic sets of muscles occurs by a process of reciprocal excitation and inhibition. As the affected limb is flexed, the animal may need to use its other limbs to steady itself. Similar controlled actions of flexor and extensor muscles in these limbs are effected by crossed-extension reflexes. These two reflexes combined together control not only rapid emergency actions but also organise limb movement for locomotion. In order to control such movements smoothly, *stretch reflexes* operate via stretch receptors in the muscles to grade the flexion and extension of antagonistic pairs. Basically, stimulation of the stretch receptors causes periodic contraction and extension of the extensor muscle so that limb movement occurs in controlled stages rather than in one violent action.

Reflexes also occur, of course, in invertebrates. Although invertebrates do not possess such an elaborate CNS as vertebrates, sensory receptors and effector organs still communicate through the switchboard of the central nerve cord and ganglia. One reflex which has been investigated is the rapid withdrawal of the siphon and gill in the tectibranch mollusc *Aplysia*. The gill is a delicate and sensitive respiratory organ which is retracted in response to mechanical stimulation. This shows the way in which the withdrawal reflex can be related directly to the properties of identifiable neural components (Figure 1.6). The stimulation thresholds of the sensory and motor neurons involved

Figure 1.5: Diagram of Adjacent Spinal Cord Segments Showing Connections between Sensory Neurons, Interneurons and Motor Neurons Making up Certain Reflex Arcs. The monosynaptic arc involved in the knee jerk reflex comprises A, B and C components; the polysynaptic arc of the withdrawal reflex involves D, E, F, G and H components. sg, sensory ganglion; sn, sensory neuron; mn, motor neuron; dh, dorsal horn; vh, ventral horn; cc, cerebro-spinal canal; sr, stretch receptor; ex, extensor muscle; f, flexor muscle; pt, patellar tendon; int, interneuron; sk, skin.

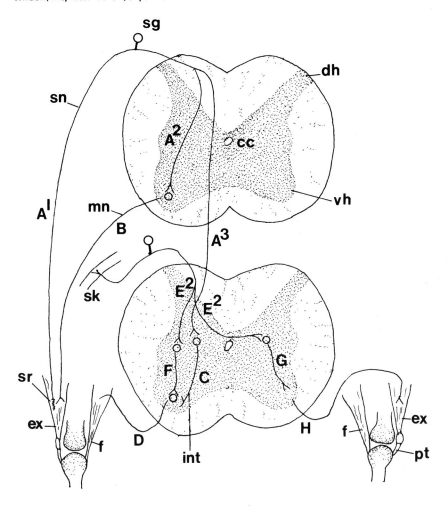

Source: Modified after Lloyd, D.P.C. (1955). Synaptic mechanisms. In (J.F. Fulton, ed.) *A Textbook of Physiology*. Philadelphia, Saunders.

Figure 1.6: Neural Circuit of the Gill Component of the Withdrawal Reflex of *Aplysia* in Response to Weak Stimulation of the Siphon. Input from siphon receptors is mediated by two excitatory interneurons (Exc L22, 23). The motor neurons (L7, LDg1, LDg2, L9g1, L9g2, RDg) and excitatory and inhibitory interneurons are identified cells. SN, sensory neurons; open squares, excitatory synapses; shaded triangles, inhibitory synapses.

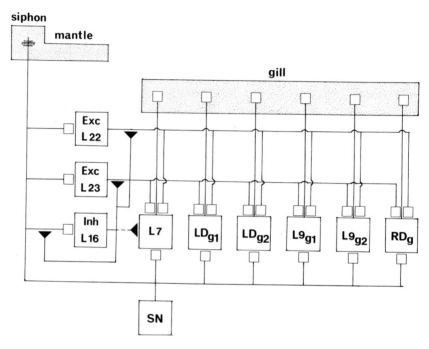

Source: modified after Kandel, E.R. (1976). *The Cellular Basis of Behaviour*. San Francisco, Freeman.

are similar to that for the reflex as a whole (about 0.25 g) and the firing of both types of neuron and the magnitude of gill retraction are linearly related to the intensity of the mechanical stimulus. Together these relationships make for a smooth withdrawal of the gill.

Reflex arcs have physiological characteristics which set them apart from other kinds of neuronal activity. First there is usually a *latency* in response. When a dog withdraws its paw from a painful stimulus, the response should occur within 27 ms if the only limiting factor is the speed of impulse conduction through appropriate neurons. In fact the response does not appear for 60–200 ms.[281] The reason for the delay is the number of synaptic junctions the impulse has to traverse. Unlike more complex behaviour, however, the latency of reflexes decreases with increased strengths of stimulus. This is important because reflexes are designed to act in emergencies.

The second characteristic of reflexes involves the pattern of muscle relaxation after stimulation. Normally the muscle relaxes a few seconds after stimulation ceases. If contraction is elicited through a reflex arc, however, it may persist for several seconds after stimulus cessation. The duration and strength of this *'after-contraction'* is positively related to the strength of the stimulus.[281]

Thirdly, reflexes show *summation*. Among its other integrating properties, the CNS is able to accumulate repeated stimuli over time (*temporal summation*) from different parts of the body (*spatial summation*). The scratch reflex of the dog is a good example. A dog shows a scratching response with its hind leg if an irritating stimulus is applied to certain areas on its back. If the stimulus is weak, scratching may not occur until after 20 or more applications.[281] The CNS summates the irritations over the time they are applied. In this way summation may lead to a fourth characteristic of reflexes, the *warm-up* effect. Reflexes often do not occur in full strength until a stimulus has been applied several times. Such warm-up effects are well known in various mobbing and courtship activities. In reflexes where the warm-up effect occurs, it seems that successive stimulation brings more motor neurons into play thus producing a stronger reaction. This is known as *motor recruitment*.

Finally reflexes show accentuated *fatigue*. Normally, a muscle stimulated to contract remains responsive for several hours. If the muscle is stimulated via a reflex arc, however, its response declines very rapidly. In some cases, like the scratch reflex in the dog, responses last only about 20 seconds. What appears to happen is that, with repeated stimulation, the interneurons in the CNS begin to block impulse transmission by increasing the resistance of their synaptic junctions. A stronger or qualitatively different stimulus, however, will soon re-establish a fatigued reflex.

With simple behaviours, like reflexes, it is usually not too difficult to trace their origins within the nervous system. With more complex behaviour, it may be almost impossible. Nevertheless, in some cases, complex behaviour patterns have been traced to specific neural components.

Complex Behaviour. The distinction between complex behaviour and simple reflexes hinges on the number of different factors involved in the expression of behaviour. In reflexes, there is little more than a simple neural pathway and a clear-cut, often momentary, response. Most other behaviour involves several complicating additions: the animal's motivational state, hormone levels, physiological rhythms, cognitive processes, subtle changes in the behaviour of other animals and so on. However, this does not mean that complex behaviour cannot be stereotyped. Many behaviours, like the classic egg-retrieval response of nesting greylag geese (*Anser anser*), consist of different functional elements which occur in an inflexible and predictable sequence. Such stereotyped sequences have been called *fixed action patterns* (*FAPs*). The elicitation of a FAP is almost solely dependent on the presentation of a specific stimulus, variously called a *key stimulus*, *sign stimulus* or *releaser*. Key stimuli are seen as releasing a specific behavioural mechanism or *innate releasing mechanism*

Figure 1.7: Model of a Neural Network Controlling Escape Swimming in *Tritonia*. G, general excitor neuron; dfn, dorsal flexor neuron; vfn, ventral flexor neuron; DM, dorsal muscle; VM, ventral muscle; T, trigger neurons; SN, sensory neurons; shaded triangles, inhibitory synapses; open triangles, excitatory synapses.

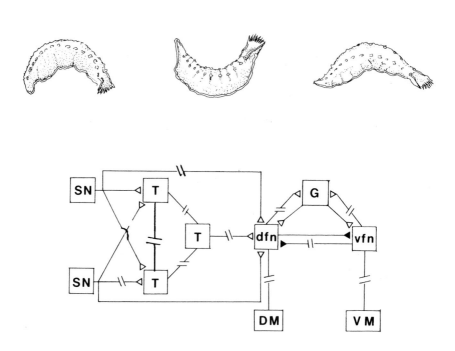

Source: Modified from Willows *et al*[333].

(*IRM*) which is a neural mechanism distinct from the receptor that initially detects the stimulus. While the concept of an IRM has been criticised for its over-simplicity, there are nevertheless neurons, like feature detectors (see Section 1.3) which appear to perform precisely that function. Because of their relative simplicity, FAPs are the most promising form of complex behaviour for studies of neuronal circuitry. Indeed, they are almost the only form for which we are likely to obtain the complete 'wiring diagram' kind of detail of reflex arcs.

One of the most extensively studied FAPs is the defensive swimming response of *Tritonia*, an opisthobranch mollusc.[333] If *Tritonia* touches a starfish (a predator), it pushes off from the substrate and swims away. Swimming is accomplished by alternate flexing of dorsal and ventral longitudinal muscles. The whole response has four components: local withdrawal from the predator, postural preparation for swimming, swimming and termination (the animal returns to the substrate). Willows and coworkers[333] have managed to elucidate the nerve pathways controlling the swimming response (Figure 1.7). These pathways are focused on various neurons with the CNS. Excited sensory cells in the epithelium stimulate 'trigger' neurons (TNs) on each side of the CNS and 'dorsal-flexor' motor neurons (DFs) dorsally. These control withdrawal from the predator. Strong excitation of the TNs stimulates the DFs (controlling dorsal muscle contraction). Since only the dorsal muscles can respond, the animal bends into a U-shape. When impulses from the TNs decay, inhibition of the ventral flexors (VFs) is removed. These are then free to respond to impulses from 'general excitor' neurons (GEs) and the animal bends in the opposite direction. Convulsive swimming movements are thus maintained by reciprocal inhibition between the antagonistic DFs and VFs. Activity is finally halted by impulses from special 'terminator' neurons and the animal settles on the substrate. The terminators are normally inhibited jointly by the GEs, DFs and VFs. A number of other studies of mollusc and insect species are slowly yielding neural circuits for particular behaviours, particularly with respect to feeding and visual perception, but the pictures are by no means as complete as the *Tritonia* example.

In vertebrates, the complexity of behaviour is enormously increased and it is singularly unlikely that complete 'wiring diagrams' will ever be obtained for most behaviours. Instead attention has focused on broad control centres and pathways within the CNS using lesion (cutting through selected parts of the CNS), ablation (selectively destroying parts of the CNS) or electrical stimulation techniques to elucidate particular pathways of control. Several behaviours have been successfully investigated.

Using localised lesions, Nottebohm and coworkers[228] have described the CNS centres controlling singing in the canary (*Serinus canaria*) (Figure 1.8a). Central to the control system is the *hyperstriatum ventrale pars caudale* (*HVC*) nucleus in the ventral hyperstriatum. The HVC has two major projections: one to the parolfactory (X in Figure 1.8a) lobe which does not play a role in singing, and one to a nucleus in the archistriatum, the *robustus archistriatalis* (*RA*). Neurons from the RA project into the hindbrain motor nucleus controlling the syrinx (vocal apparatus) (nXIIts in Figure 1.8a). Ablation of the HVC prevents the production of song but not the postures associated with singing. The decision to sing appears to be taken in a different area to that controlling song production. Work in other species suggests that the *nucleus intercollicularis* (*ICo*) may be the centre controlling the elicitation of singing, while song construction appears to depend on an auditory projection area underlying the HVC (L in Figure 1.8a).

The neural control of agonistic behaviour has been studied in a number of vertebrate species. In the cat, stimulation of the mid-hypothalamus and the *ventromedial nucleus* (*VMN*) (Figure 1.8b) produces clawing actions and postures associated with defence, while stimulation of the *mediodorsal nucleus* in the thalamus produces crouching and escape behaviour. Stimulation of the *latero-dorsal nucleus*, however, produces offensive biting and attack responses. Via the *path of Nauta*, connecting with the hypothalamus, the amygdala in the limbic

Figure 1.8: (a) Structures in the Canary Brain Connected with Song Production. hvc, hyperstriatum ventrale pars caudale; ra, nucleus robustus archistriatalis; ico, intercollicular nucleus; nXIIts, motor nucleus controlling syrinx; mld, mesencephalic lateral nucleus; s, pathway to syrinx; C, cortex; see text for explanations of X and L. (b) Some Behavioural Control Centres in the Diencephalon and Neighbouring Areas of the Mammalian (Rat) Brain. amt, anterior medial thalamic nucleus; cc, corpus callosum; sco, superior colliculus; t, tegmentum; zi zona incerta; p, pituitary; ma, mammilary nuclei; lh, lateral hypothalamic nucleus; vmh, ventromedian hypothalamic nucleus; o, optic nerve; so, supraoptic nucleus; po, preoptic nucleus; mi, massa intermedia; mt, mediodorsal thalamic nucleus; f, fornix.

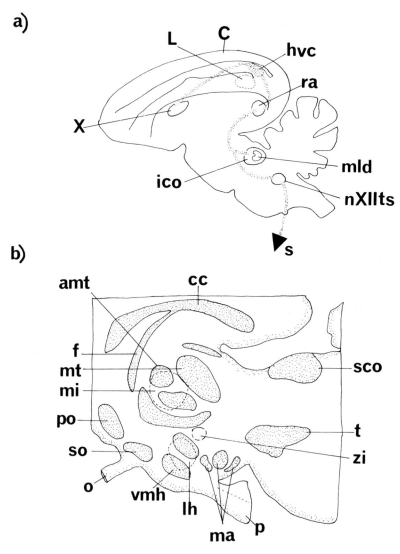

Source: (a) Modified after Nottebohm[228]. (b) Modified after Guthrie[123].

system also play a role in defence and fear behaviour. Lesion and stimulation investigations of monkey brains have shown that the amygdala and hypothalamus may control long-term dominant/subordinate relationships within social groups (see Chapter 9).

The VMN in mammals is also involved in the control of feeding behaviour. In conjunction with the *lateral nucleus (LN)*, the VMN appears to monitor glucose levels in the blood, although there is now evidence that monitoring occurs outside the brain as well. When the glucose level drops, the VMN releases its inhibitory control over the LN which then stimulates food searching and ingestion. Once the animal is satiated, sensory feedback re-establishes the VMN's control and feeding ceases. If the VMN is ablated, the animal will gorge itself to distension. If the LN is ablated, it will starve to death even in the presence of preferred food. Conversely stimulation of the VMN inhibits feeding, while stimulation of the LN causes it to be elicited.

While many such control centres have been found within the vertebrate brain, not all complex behaviour sequences in vertebrates are controlled from the brain. In a series of studies on 'spinal' male dogs (animals in which spinal connections with the brain had been severed) Hart[134] found that copulatory behaviour was controlled largely by centres in the spine. Tactile stimulation of the spinal dog's penis resulted in four major components of copulatory behaviour: erection of the penis, shallow pelvic thrusts, intense thrusts and ejaculation. Similar results were found in male rats. In spinal female mammals, mating activities and postures can be produced by tactile stimulation of the genital region or by a mounting male. The spinal cord therefore seems to house a number of control centres for complex behaviour.

Electroencephalograms. Another way of investigating the relationship of CNS activity with behaviour, is to monitor changes in gross electrical potentials as different behaviours are performed. Recordings can be taken either by implanting electrodes into the brain or by attaching them to the scalp. At least in man, such electroencephalographic (EEG) recordings have revealed interesting relationships between changes in behaviour and changes in electrical activity in the brain.

The EEG of a sleeping person is a more or less continuous potential made up of a basic slow rhythm of large phasic changes (several a second) over which are superimposed several smaller-scale changes of higher frequency. In an alert person, the basic slow rhythm ceases and several small amplitude frequencies take over. EEG patterns also change during conditioning. A weak sound, for instance, may initially produce little change in the EEG, while a strong light both depresses large, slow rhythm frequency and evokes rapid small ones. After repeated pairings of sound and light presentations, the sound alone causes a depression in slow EEG components. While EEGs which change with behaviour have been recorded in other animals, they have so far proved too complex and their origins too obscure to provide conclusive explanations for particular behaviours.

1.1.4 Neural and Behavioural Hierarchies

A point that emerges clearly from both the vertebrate and invertebrate examples discussed above is that there tend to be 'chains of command' within nervous systems. In the example of feeding control in mammals, the VMN controls the activity of the LN and the latter controls, via several different neuronal pathways, the muscular movements leading to feeding. The control of muscular activity in swimming *Tritonia* is similarly co-ordinated by a chain of neural command centres. Indeed, such chains or *hierarchies* of command are an ubiquitous feature of nervous systems. Why have they evolved with this inherent pattern of organisation? Three interesting possibilities are discussed by Dawkins[78] in a thought-provoking paper on the organisation of behaviour:

(1) The Evolutionary Rate Advantage. This is best illustrated by a simple story. Imagine two watchmakers A and B. A's watches are as good as B's except that he takes 4,000 times as long making them. The reason for the time difference is that B first assembles the 1,000 components of his watches into 100 sub-assemblies of 10 components each. He then assembles these into 10 larger sub-assemblies and finally puts the whole watch together. In this way, he need only examine the current sub-assembly if anything goes wrong. A, on the other hand, puts the 1,000 pieces of his watches together in one go. If anything goes wrong, he has to dismantle the whole watch to correct it. What this story illustrates is that complex and thermodynamically unlikely systems like the nervous system might evolve more rapidly if they go through a series of intermediate but functional stages. We should therefore *expect* complex biological systems to have hierarchial structures rather than be surprised by them.

(2) The Local Administration Advantage. The analogy we can draw here is with central and local modes of government. Clearly centralised government is essential for all-round co-ordination of the different regions within a country. Depending on the size of the country, however, such a government may be very inefficient in its policy-making for individual regions. It is more effective to have local administrations managing the day-to-day problems with the central administration being available for major decisions. Similarly we might expect various motor activities (e.g. feeding) to be controlled by different centres in the CNS (e.g. the LN) but these centres to be controlled in turn by a higher-level co-ordinator (e.g. the VMN) which oversees a number of such centres.

(3) The Redundancy Reduction Advantage. If we imagine a complex physiological system within an animal, the number of possible states of that system is likely to be enormous. For instance, the human retina contains some 4 million light-sensitive cells. If we assume that each cell either signals the presence of light or not, the number of possible states of the retina is an unimaginable $2^{4,000,000}$. For the brain to cope with this number of possibilities its cubic capacity would have to be astronomical. Furthermore, a large proportion of states will

be *redundant* as far as the animal is concerned because they are biologically meaningless or duplicate the information of other states. There will thus be intense pressure to reduce the redundancy of the retinal system. This has occurred in the form of lateral inhibition within the retina so that maximal firing rates occur in cells bearing most information (for example, those whose fields cover the edges of a scene). We can erect similar arguments for redundancy reduction in the motor system. Instead of operating independently, muscle fibres contract and relax in co-ordinated groups. The correlation is not so good for all the fibres of one muscle, but graded contraction would not be possible if it was. As we have seen, there are also correlations of action between different muscles so that limb flexion and locomotion are possible. Such a hierarchal organisation of the motor system benefits the animal in the same way as subroutines benefit the computer programmer.[78] Whole ranges of ready co-ordinated low-level units are available for command by high-level control centres.

So far, we have considered neural hierarchies with respect to one set of functional behaviours (e.g. feeding) at a time. Animals, however, appear to have behavioural priorities (for a full discussion see Chapter 2). Certain behaviours take precedence over others in their probability of expression. We can thus envisage hierarchies of behavioural organisation which determine priority. In some cases it has been possible to trace the neural components of such hierarchies. Attempts in this direction go back historically to the classic work on chickens (*Gallus gallus*) by von Holst and von St Paul (see the excellent account in Eibl-Eibesfeldt[95]). More recently, however, attention has focused on invertebrates.

Investigation of a number of behaviours in the mollusc *Pleurobranchaea* has shown a distinct hierarchy of priority in their expression. The main aspects of the hierarchy are shown in Figure 1.9a. Feeding behaviour dominates mating, withdrawal from a tactile stimulus and righting behaviour, but is itself dominated by egg-laying. Not surprisingly, the emergency escape swimming response (like that of *Tritonia*) dominates all other activities. By means of carefully controlled behavioural experiments, Davis and co-workers have elucidated some of the neural components of the hierarchy (Figure 1.9b). They suggest that feeding dominates righting because chemoreceptors on the oral veil (a fleshy structure around the mouth) act to suppress the neurons controlling righting behaviour. Suppression is thus brought about by chemicals in the food. Withdrawal behaviour, however, is inhibited directly by feeding behaviour. When they are active, the complex set of neurons controlling feeding behaviour inhibit those controlling withdrawal. Withdrawal can only occur when the feeding neurons are not stimulated. The suppression of feeding by egg-laying appears to be hormonally based. A single hormone produced by egg-producing animals both induces egg-laying and suppresses feeding, perhaps by competing for messenger receptor sites on the feeding neurons.

We have come a long way in our efforts to trace behaviour to its neural components within the body. Nevertheless, we are left with the inevitable con-clusion that our ability to do so is limited by the complexity of the behaviour

Figure 1.9: (a) Proposed Behavioural Hierarchy of Escape, Mating, Egg Laying, Feeding, Righting and Withdrawal Behaviours in *Pleurobranchaea*. (b) Proposed Wiring Diagram for Part of the Hierarchy. TR, tactile receptors; CR, chemoreceptors; S, statocysts; CNW, CNF, CNR, central neurons controlling withdrawal, feeding and righting respectively; SR, satiation receptors; open triangles, excitatory synapses; closed triangles, inhibitory synapses.

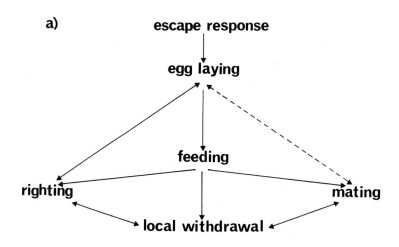

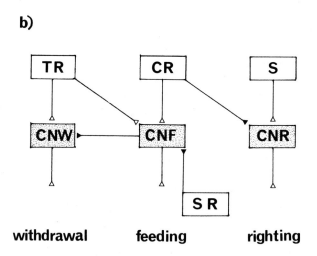

Source: Modified after Davis, W.J., Mpitsos, G.J., Pinneo, J.M. and Ram, J.L. (1977). Modification of the behavioural hierarchy in *Pleurobranchaea*. I. Satiation and feeding motivation. *J. Comp. Physiol.*, *117*: 99–125.

involved. Moreover, even if we could explain highly complex behaviours in terms of their neural components, the explanations would be very convoluted and unwieldy. They would not help us *understand* behaviour. Dawkins[78] draws the analogy between explanations of computer analyses and explanations of behaviour. If we wish to know how a computer carries out an analysis, it does not help to be told about integrated circuits and logic gates. We prefer an answer in terms of a high-level programming language. The 'software' program sets out the logic of the analysis in a readily assimilable form. The wealth of detail involved in a 'hardware' explanation is more than anyone can assimilate. In the same way, 'hardware', neural explanations of behaviour, are unlikely to help us understand its functional significance. To do that we need 'software' *models* of behaviour. In the chapters which follow we shall explore a number of such models and show how they have been useful in quantifying the principles underlying behaviour.

1.2 Hormones and Behaviour

Operating alongside the nervous system is a specialised group of organs called *endocrine glands*. These provide another means of communication within the animal's body, this time via *hormones*. Hormones are secreted by the glands into the bloodstream in response to specific stimuli. Because they operate via the circulatory system, hormonal messages are much slower than the electrical messages of the nervous system. They also have longer-lasting effects on their *target* organs, some effects persisting for months. In vertebrates the most important endocrine glands affecting behaviour are the pituitary gland, the gonads and placenta and the adrenal gland. Hormones of the thyroid and parathyroid glands, pancreas and gastrointestinal mucosa have no direct behavioural effects.

The pituitary gland is situated under the hypothalamus on the floor of the brain. It plays an important role as the central controller of the endocrine system. Between them, the pituitary's anterior and posterior lobes secrete hormones which indirectly affect blood pressure and water absorption (vasopressin), gamete production and the secretion of sex hormones (folliclestimulating hormones, luteinising hormone). In response to luteinising hormone (LH), testes secrete male hormones or androgens. One of these, testosterone, influences the development and maintenance of the male reproductive tract, the formation of secondary sexual characteristics and various aspects of behaviour, particularly aggression. In females, LH stimulates the ovaries to secrete oestrogens and progesterones. Between them they perform analogous functions to the androgens in males. The paired adrenal glands are situated next to the kidneys and have two anatomical components, the medulla and the cortex. The medulla secretes adrenalin and noradrenalin which play important roles in producing 'fight or flight' reactions in emergencies. These are mediated through the hormones' effects on smooth muscle contraction and the transmission of nerve impulses. In response to adrenocorticotrophic hormone (ACTH) from the pituitary, the adrenal cortex secretes a range of 28 steroid hormones which aid metabolism and resistance to infection.

1.2.1 Hormonal Actions Affecting Behaviour

The effects of hormones on behaviour can be traced to three major sites of action. These are: (1) the nervous system, (2) sensory perception and (3) effector organs and structures.

Hormonal Effects on the Nervous System. Hormones affect many aspects of the nervous system including its anatomy, biochemistry and transmission capabilities. Because of this, they are important agents of change in the state of the system. In hamsters primed with oestrogen, for example, behaviour typical of oestrus can be induced by the introduction of small doses of progesterone into the brain ventricles. In the absence of oestrogen, doses are inadequate to induce the behaviour. Similar evidence for hormonal 'priming' of the CNS comes from rats. Electrodes implanted in certain areas of a rat's brain can be connected in such a way that the animal can stimulate its own brain by pressing a pedal. Stimulation of certain areas results in more frequent and persistent pedal-pushing than others. Olds[230], however, found that by treating given areas of the brain with androgen he could enhance the rat's response to a standard stimulation of those areas.

In some cases, hormones may be directly responsible for basic structural and functional changes within the CNS. Sexual dimorphism in the anatomy of certain neurons in the rat hypothalamus appears to be mediated[251] and the maturation of reflex connections within the CNS is accelerated by high levels of thyroxin. Corticosteroid and sex hormones, by virtue of their effect on calcium metabolism, indirectly affect nerve conduction in which calcium ions play an important role. By altering the excitatory state of the CNS in these anatomical and physiological ways, hormones may influence learning processes.

Numerous, although sometimes conflicting, observations have shown a marked effect of adrenalectomy and corticosteroid treatment on conditioning. In monkeys, stimulation of the adrenal glands with ACTH results in more rapid extinction of conditioned 'fear' responses, while the amount of cortico-steroid hormone secreted by the adrenals in dogs influences the intensity of 'passive avoidance' behaviour.

Sometimes, hormones may not so much exert a positive effect on the nervous system as remove inhibition. In neonatal guinea pigs the lordosis posture is an integral part of excretory behaviour. At first excretion is stimulated by the mother, but, as the young mature, micturation and defaecation become internally controlled. Now the spinal centres which control these activities come under inhibitory control from the brain. In adults, ovarian hormones act to remove this inhibition. Alternatively, hormones may actually induce inhibition.

Oestrogens inhibit aggressive behaviour in female hamsters. Among invertebrates, the sexual receptivity of female grasshoppers (Orthoptera) is inhibited by hormones produced as a result of the spermatheca filling with sperm. Hormonal effects on the nervous system may also be multiple. Full sexual receptivity in female cats, for instance, appears to require different hormone levels to other sexual behaviours.

Hormonal Effects on Sensory Perception. Many studies suggest that hormones affect an animal's sensory capabilities. In doing so, they alter the animal's perception of its environment and the way it responds to certain stimuli. During the spring, male three-spined sticklebacks (*Gasterosteus aculeatus*) migrate from the sea to their freshwater breeding grounds. Migration is brought about by changes in hormonal output by the pituitary and thyroid glands. In particular, thyroxin alters the fishes' salinity preference from salt water to freshwater, thus providing the impetus for migration.[5]

In many female mammals, sensory perception is influenced by the oestrus cycle. Female rats fluctuate in their ability to detect certain odours according to the levels of oestrogen and progesterone. Similarly, visual sensitivity in female humans varies with the stage of their menstrual cycle. Visual sensitivity, as determined by ability to detect a small light, is most acute at around ovulation and least acute during menstruation. Sensory perception in males is also influenced by hormones. Experienced male rats prefer the odour of urine produced by oestrous females to that produced by females in dioestrous. This preference disappears in castrated males. Androgens produced by the testes therefore modify the animals' response to standard olfactory stimuli, although the effect is partly confounded with those of experience.

In female rats, oestrogen has the effect of extending the sensory field of the perineal nerve (the region of the body whose stimulation excites the nerve) which innervates the genital tract. During oestrus she is thus more responsive to the tactile stimulation of intromission and orientates her body to facilitate penetration. Hormones also affect genital sensitivity in male rats. Testosterone causes the skin of the glans penis to become thinner so that underlying sensory cells receive more stimulation. In this way the male can respond more effectively to the copulatory movements of the female. In female birds, oestrogen causes the formation of a 'brood patch'. Feathers are lost from part of the ventral body surface and the exposed skin becomes more vascularised. The brood patch increases the sensitivity of the female to the nest cup and influences her nest maintenance and incubatory behaviour.

Hormonal Effects on Effector Organs and Structures. Animals use a variety of their structural components in the execution of a behaviour. Hormones may affect these structures and hence the efficacy of the behaviour in a number of ways. Good examples are the secondary sexual adornments of male birds. These may be just bright plumage or bill colorations or they may be elaborate and gaudy structures like combs and wattles. Secondary sexual adornments appear to serve a variety of functions including the attraction of mates, the deterrence of sexual rivals and the declaration of reproductive condition. The production and maintenance of these adornments depend on androgen levels. For example, comb and wattle size in newly-hatched male and female domestic chicks can be increased to proportions normally found in sexually mature males by injections of testosterone. Conversely, castration of mature males results in a pronounced reduction of comb and wattle size. Bill pigmentation in many species fluctuates in intensity with the time of year. In both sexes of house sparrow (*Passer domesticus*), the bill changes from pale brown to black at the beginning of the breeding season. Removal of the gonads in either sex inhibits the deposition of melanin and results in a cream-coloured bill. Androgens are also important in determining seasonal changes in plumage patterns. Castrated male ruffs (*Philomachus pugnax*) and blackheaded gulls (*Larus ridibundus*) fail to assume their characteristic breeding plumage. In red-necked and Wilson's phalaropes (*Phalaropus* spp.), in which females instead of males acquire brightly coloured breeding plumage, pattern changes are also controlled by androgens.[153] Similar breeding and competitive status changes are found in mammals. Horn and antler size in certain male artiodactyls is used as a deciding factor in sexual disputes and mate choice and appears to be at least partly under the control of androgens. In these cases, the effects of hormones on sexual and aggressive behaviour may be mediated through their effects on relevant bodily structures. This is true in an even more direct way of the effects of androgens on the sexual behaviour of male rats. In intact males, the surface of the penis is covered with tiny papillae. These appear to enhance tactile sensitivity. If animals are castrated, the papillae disappear and there is a corresponding decrease in copulatory behaviour. Hormonally maintained sexual activity seems to be partly a function of the hormone's anatomical effects on the penis.

1.2.2 Developmental Effects of Hormones

So far we have dealt mostly with the behavioural effects of hormones in adult animals. Hormones, however, have profound effects on the development of young animals and impart some characteristic features to their behaviour.

At a very direct level, hormones affect embryonic development. Hormone-mediated anatomical and physiological changes are important in the development of sexual behaviour. In guinea pigs, testosterone levels influence the development of the genitalia. Treatment of pregnant females with testosterone proprionate results in female offspring with male-like genitals.[343] In rats, analogous effects occur after birth rather than *in utero*.[132] If female rats are treated with testosterone when about four days old, their oestrous cycle and sexual behaviour as adults are suppressed. The CNS in neonatal rats seems to be relatively undifferentiated as far as sexual behaviour is concerned, although there is a tendency towards characteristically female patterns. It is only by the direct action of testosterone that male behaviour develops.

The rat and guinea pig examples illustrate an important characteristic of developmental hormonal effects on behaviour. The effects are exerted during species-specific *'critical periods'*. The presence of an appropriate hormone during the critical period determines anatomical and behavioural characteristics for life. No amount of therapy with oestrogen and progesterone will revive female sexual behaviour in female rats treated with testosterone during their post-natal critical period. Developmental hormonal effects are therefore irreversible. There is an intriguing similarity here with critical periods in learning which we shall discuss in Chapter 4. Another important feature of hormonal effects on development, well-illustrated by sexual behaviour, is the temporal separation between the action of a hormone and the manifestation of its behavioural effects. This too is very similar to the imprinting process. Some early hormonal effects are due to changes in peripheral organs like the genitals, while others can be traced to sites within the CNS. Perinatal hormone levels bring about the neural sexual dimorphism found in the preoptic brain of rats and neonatal castration of male rats depresses the sexual reflexes mediated through the spinal cord.

Hormones then have diverse and important effects on behaviour. Because their effects are so varied, it is not surprising that they are dependent on a number of internal and external factors. One of the most obvious modifying factors is *individual genotype*. Several experiments have indicated strain-specificity in the effects of given hormones. Strains of domestic fowl which differ in their ability to complete mating sequences, showed very different mating responses to standard androgen treatment.[209] Similarly the increase in aggression shown by female mice after treatment with androgens depends on whether the males are aggressive. Differences in the effects of hormones are even more marked *between species*. Progesterone induces sexual receptivity in female rodents but inhibits receptivity in some primates. In Japanese quail (*Coturnix c. japonica*), testosterone enhances male sexual behaviour in males but has hardly any effect on females. On the other hand, oestrogen produces normal female behaviour in females but both male and female behaviour in males. The contrast between birds and mammals in this respect is interesting. It appears to depend on which sex is homogametic (i.e. has two XX chromosomes instead of XY). In mammals this is the female, in birds the male. In contrast to mammals, the development of sexuality without hormonal influence in birds is thus masculine.

Seasonal effects are also important. In red deer (*Cervus elephus*), the administration of testosterone in winter brings about full rutting behaviour. A similar administration in late spring has no effect at all.[190] Even the animal's *past experience* can influence hormonal effects. Copulatory actions in castrated male cats tend to be more protracted if they have copulated in the past. Inexperienced animals show reduced copulatory vigour. Clearly, hormonal influences on behaviour are not simple and readily predictable. Some impression of their complexity can be readily gleaned from Figure 1.10. This shows a much

Figure 1.10: The Relationship between Internal Hormonal and External Factors in the Control of Breeding Behaviour in the Canary. Solid lines, confirmed relationships; dashed lines, suggested relationships.

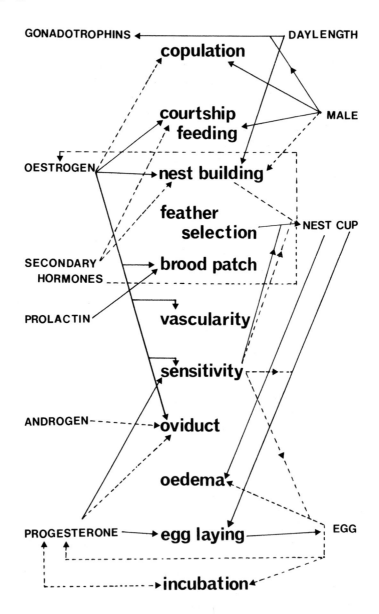

Source: Modified after Hinde, R.A. (1965). Interaction of internal and external factors in integration of canary reproduction. In (F.A. Beach, ed.) *Sex and Behavior*. New York, Wiley.

simplified breakdown of the relationships between hormones and various stimuli on the reproductive behaviour of the canary.

1.3 Biological Clocks

The behavioural roles of the nervous and endocrine systems we have discussed so far, have dealt largely with responses to instantaneously applied stimuli or with developmental effects. In many cases, however, behaviour exhibits a marked periodicity in its pattern of performance, a periodicity which involves complex interactions between the two systems. During their evolution, animals have acquired a variety of endogenous rhythms whose periods are matched with those of rhythmic events in the environment. Some rhythms are matched to the 24-hour cycle of light and dark (*circadian*), others to the 12.4- or 24.8-hour tide cycle (*circatidal*), the 29-day lunar cycle (*circalunar*), the yearly seasons (*circannual*) or the time (14.7 days) between successive spring low waters (*semilunar* or *circasyzygic*). The remarkable thing about these rhythms is that they persist even when the environmental cycles (*Zeitgebers*) to which they are *entrained* are artificially excluded. They therefore seem to form a time base for behaviour or '*biological clock*' within the animal.

Circadian Clocks. The best understood of the various clocks are those operating on an approximately 24-hour, circadian (from the Latin *circa* (about) *dies* (a day)) basis. Examples of circadian rhythms can be found in almost all the major taxonomic groups. In their simplest form they are reflected in the alternating periods of activity and sleep which correlate with the light/dark cycle. Various indices of this activity pattern can be used to illustrate the precision of the underlying clock. Figure 1.11 shows the periodicity of wheel-running activity by a flying squirrel (*Glaucomys* sp.).[87] This animal had been entrained to a 24-hour light/dark cycle and was then kept in continuous darkness. The figure shows the remarkable precision in the onset and cessation of wheel-running in the complete absence of the environmental cue to which it is related. It also shows that, in common with most circadian clocks, the squirrel's clock is not calibrated to exactly 24 hours. Wheel-running started slightly earlier each day.

Circadian clocks sometimes provide the underlying mechanism for apparent longer-term behavioural rhythms. The southward autumn feeding and northward spring breeding migrations of the white-crowned sparrow (*Zonotrichia leucophrys*) are good examples.

In simplified terms, the clock works as follows.[105] Birds show cyclical changes in light sensitivity over the 24-hour period. This sensitivity, however, is a property of cells in the brain itself, not of the eyes. The cycle is set each day at first light after which the bird remains insensitive to light for about 16 hours. After 16 to 20 hours, sensitivity increases and then declines again to a minimum after 24 hours. If daylength is short (11–12 hours), no light reaches the bird during its

Figure 1.11: Twenty-four-hour Rhythm of Wheel-running Activity in a Flying Squirrel under Conditions of Constant Darkness.

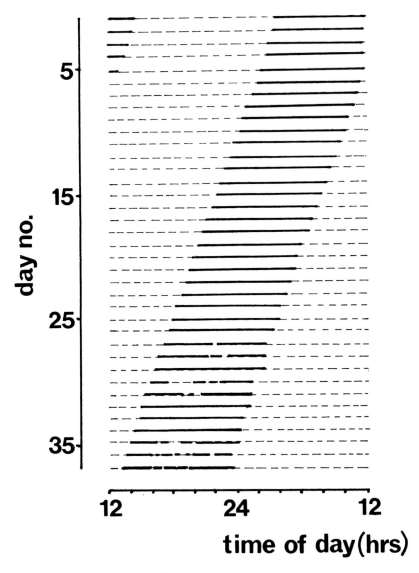

Source: Modified after de Coursey[87].

photosensitive phase. When days get longer (14–15 hours), hours of light extend into the sensitive phase and a number of light-activated systems are brought into play. Hormones are then released from the hypothalamus which stimulate the pituitary gland to secrete prolactin and gonadotrophic hormones. In response

Figure 1.12: Neural and Hormonal Control System, Including a Circadian Clock, which Regulates Reproductive Behaviour in the Male White-crowned Sparrow.

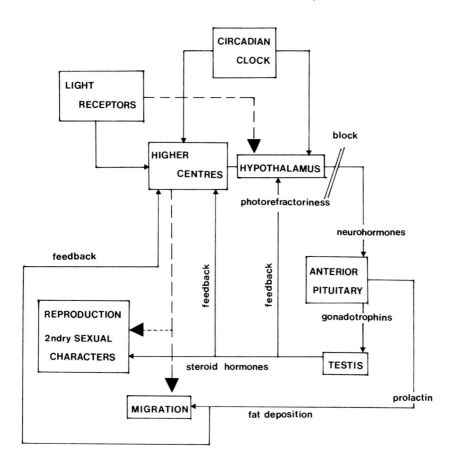

Source: Modified after Farner and Lewis[105].

to these, the gonads rapidly increase in size (*recrudesce*) from their winter resting state (at 1/300 to 1/400 their reproductive size). At the same time birds start to eat more, increase their fat deposition and become more active, particularly at night (migratory restlessness' or *Zugunruhe*). After migration and breeding, the gonads atrophy again and birds enter a photorefractory period. It is not until the following spring that they once again become sensitive to changing daylength. The interaction of internal and external factors in regulating the timing of migration is summarised in Figure 1.12.

Circadian clocks are also of crucial importance in the use of celestial cues (sun, stars) for navigation. These will be discussed in detail in Chapter 5. Among invertebrates, circadian rhythms occur in behaviours ranging from vertical migrations by marine plankton and the periodicity of pupal eclosion in dipterans to the sophisticated dance 'language' of foraging bees.

Circannual Clocks. Although the apparent long-term rhythmicity in white-crowned sparrow behaviour was a product of a circadian rhythm of light sensitivity, real circannual rhythms also exist. Among birds, some of the best examples are shown by warblers. Willow warblers (*Phylloscopus trochilus*), wood warblers (*P. sibilatrix*), and chiffchaffs (*P. collybita*) show similar circannual rhythms of body weight and Zugunruhe to white-crowned sparrows. However, if birds are kept in constant laboratory conditions of 12 hours light/12 hours dark, the annual cycles still continue and may persist even after 27 months of the constant laboratory regime.[125] Furthermore, there are differences in the phase-setting effects of photoperiod which are predictable from the assumption of a circannual rhythm.

Endogenous circannual rhythms also appear to control certain behaviours in mammals. Activity/hibernation cycles in hedgehogs (*Erinaceus europaeus*) and some rodents are clear examples. In ground squirrels (*Spermophilus* spp.) and chipmunks (*Eutamias* spp.) the activity/hibernation cycle persists under constant laboratory photoperiods.[138, 243] Like circadian clocks, however, the circannual rhythms seldom show periodicities of exactly one year. In ground squirrels, for instance, they range from 229 days to 445 days.

Circatidal and Circasyzygic Clocks. Any organism living in the intertidal zone of the seashore are alternately submerged by water and exposed to air. The number of environmental factors this alters is enormous: pressure, salinity, food supply, temperature, predation risks — the list is endless. It is therefore hardly surprising that many such organisms show behavioural periodicities associated with the tides. Straightforward examples of tide-entrained cycles which persist under laboratory conditions are the vertical migration cycles of sand-dwelling platyhelminths, polychaetes and diatoms, expansion and contraction rhythms in sea anemones, filtration rates in bivalve molluscs and swimming activity in fish. In some cases, however, a circatidal clock may act in subtle conjunction with a circadian clock. An example comes from the shore crab (*Carcinus maenas*). Crabs show peaks of activity at daytime high tides and at night.[224] In the laboratory, the tidal peaks persist for a few weeks but then decay leaving only the circadian, nocturnal peak. Interestingly, crabs which live along shores with little tidal fluctuation, like those in Mediterranean areas, exhibit only a circadian activity rhythm.

Some seashore organisms show semilunar or circasyzygic rhythms which are synchronised with the fortnightly cycle of spring and neap tides. The periwinkle, *Littorina rudis*, for example, shows a marked 15-day periodicity in its locomotory activity. The species lives high up on the shore and is only covered by the high water of spring tides. Even species which live intertidally and normally show circatidal periodicity exhibit marked increases in activity during spring tides. In some crabs, the fortnightly activity peaks may be the result of inter-action between circatidal and circadian rhythms.[224]

Circalunar Clocks. In some animals behaviour may be entrained to the 29-day lunar month. In these cases behaviour is said to show circalunar or *circasynodic* periodicity. Ciralunar cycles are particularly common among marine invertebrates. The Mediterranean polychaete *Platynereis dumerlii* transforms into its sexual phase (heteronereis) and swarms at the surface in synchrony with the full moon. The swarming worms release their gametes and then die. The periodicity of this cycle can be maintained under constant laboratory conditions although it appears to be labile to several Zeitgebers, particularly light. A very similar endogenous rhythm brings about the famous reproductive swarms of Palolo worms (*Eunice viridis*) in the Pacific.

Some terrestrial animals also show circalunar rhythms. Pit-building activity in antlions (*Myrmeleon obscurus*) reaches a peak at around full moon. Rather like the circannual rhythm in the white-crowned sparrows, however, this circa-lunar pattern may be the product of a shorter-term clock. Youthed and Moran have demonstrated a lunar day (24.8 hour) rhythm with antlions being most active about four hours after moonrise. They suggest that the overall circalunar rhythm of activity is really a product of interaction between the lunar day rhythm and a circadian rhythm.

The Physiological Basis of Biological Clocks

In the cases which have been investigated, the mechanisms underlying behavioural periodicity have turned out to be mainly neural, but with hormones playing an important role in their maintenance. In cockroaches (*Periplaneta americana*) and some other insects, the circadian clock appears to be regulated by large ocellar fibres extending from cells in the optic lobes to thoracic neurons and neurosecretory cells in the thoracic nerves. Both hormones and nerve impulses, however, are important in crustacean clocks. In shore crabs and crayfish (*Cambarus* spp.), neural components of the eyestalk provide the fundamental rhythmicity of activity, while hormones from the X-organ determine its persistence. Some neural clocks seem to be partially autonomous. The R15 ('parabolic burster') cell in the parieto-visceral ganglion of *Aplysia* emits a rhythmic 12–13 impulses per 30 s. Superimposed on this regular rhythm are diurnal fluctuations with a marked increase in spike frequency an hour before dawn. This rhythm persists in a modified form even when the ganglion is isolated. Recently, a seasonal variation in the R15 spiking pattern has been discovered, so the same mechanism may provide for both long- and short-term behavioural rhythmicity.

In the lugworm *Arenicola*, there appears to be an independent rhythmic pacemaker in the muscles of the anterior 'extrovert' which functions during burrowing. The rhythmicity persists after isolation of the muscle from the nervous system and is very reminiscent of the myogenic rhythm of vertebrate heart muscle. In some molluscs, seasonal rhythmicity correlates with changes in the ionic content of the blood and neurons. After hibernation, the neuron resting potential is reduced from 60 mV to 40 or 50 mV, perhaps because of increased sodium and calcium ion permeability, which brings the potential nearer to the impulse threshold and may even elicit autorhythmic spiking. Ionic concentrations are reversed in the summer when activity levels are lower and animals aestivate. In these species, therefore, the chemical environment of neurons appears to correlate with the animal's activity levels.

The pacemaker of the clock and the physiological mechanism for its entrainment appear to be one and the same thing in invertebrates. In vertebrates, however, the picture is more complicated. In birds, the pineal organ in the brain may be the pacemaker for the photoperiodicity clock, but it is not the sensory area governing entrainment. This lies elsewhere on the surface of the brain. While in birds the eyes are not the site of photoreception for the clock, in mammals they are. Light detected by the eyes suppresses melatonin formation in the pineal organ. Since melatonin suppresses the development and maintenance of the gonads, increasing daylength stimulates reproduction by inhibiting its formation.

We have discussed biological clocks at some length because they illustrate an important point. They show that natural selection may impose temporal patterning on behaviour when the timing of risk or benefit in the animal's internal and external environment is predictable. The pre-programming of appropriate rhythmic responses which match environmental events reduces the potential for costly errors which might arise in the vicissitudes of decision-making. Animals cannot afford to mis-time their breeding activity or be exposed to desiccation and predators on a low-tide sea shore. The evolution of appropriate physiological and behavioural rhythms makes sure they are not caught out.

1.4 Perceptual Mechanisms and Behaviour

Different animals perceive an environment in different ways. Stimuli from the environment are filtered through an animal's sense organs and nervous systems so that it obtains a very selective picture of the world around it. Exactly what this picture contains depends on the results of natural selection. The environment is likely to contain an enormous amount of information, only a small proportion of which is important to the animal. Getting rid of this redundancy while at the same time maintaining adequate detailed sensory input is the task of the specialised sense organs and the CNS. In the first part of this section, we shall examine the various sensory modalities employed by animals and the way sense organs filter environmental information. A detailed discussion of the structure and physiology of sense organs, however, is beyond the scope of this book. What we are concerned with is the information these organs allow through for use by the nervous and endocrine systems in producing behaviour.

1.4.1 Sensory Mechanisms and 'Peripheral Filtering'

For an animal to behave in an appropriate way, its nervous system and endocrine glands must receive the right kind of information. The animal needs sensors which pick up relevant changes in the environment. Several types of change might be important. Variations in light intensity, temperature, sound, tactile experience, barometric pressure, smells and other factors impinge on the animal's activities and need careful monitoring. What sort of organs have evolved to cope with this monitoring task?

Visual Perception. A major potential source of information about objects in the environment is the range of electromagnetic waves which bombard the earth from space. These waves take different forms which vary in both their wavelength and intensity. This variation is critical to their usefulness in providing sensory information. Forms with very short wavelengths, like X-rays and gamma rays, have a high energy content. These are damaging to biological material and are therefore difficult to put to use. At the other end of the scale, long wavelength forms, like infra-red, have such a low energy content that only a narrow range of sensitive materials can detect them. In between is a biologically useful range of wavelengths which man and many other animals perceive as light. Animals have evolved a variety of light-sensitive organs.

Visual perception in single-celled organisms and cnidarians is limited to an appreciation of overall light intensity. Their simple, dermal light-sensitive cells do not allow discrimination of form, texture or colour. This, however, is sufficient to maintain them at depths in the water where temperatures, food availability and currents are most suitable. As we ascend the phylogenetic scale, visual organs become more sophisticated. Platyhelminthes and some annelids possess simple forms of eye consisting of sensory cells screened with pigment. Although these form definite eyespots and may be inset into a cup, they still only allow the perception of gross changes in illumination. It is not until the polychaetes and arthropods that more complex eyes with lenses occur. In arthropods many eye units or *ommatidia* are massed together to form a *compound eye*. Arthropods are particularly interesting from a behavioural point of view because they show impressive discriminatory abilities and behavioural sophistication, despite a relatively simple nervous system. While the compound eye is not well-suited to fine image resolution, it is supremely adapted to perceiving moving objects. Crabs, for instance, respond to a light spot moving at $0.002°/s$ (about the speed of apparent movement of the sun across the sky). Some insect eyes are able to perceive alternating light/dark patterns whose alternation rates are as high as 300 Hz. In man, alternation rates of only 50 Hz produce a continuous blur. The insects' alternating light/dark perceptual abilities are clearly advantageous in fast-flying species which need to discriminate between particular food sources while on the wing.

Vertebrate eyes operate in a similar fashion to a simple box camera. The chamber of the eyeball corresponds to the dark interior of the camera box.

Within the eye, the lens (and in land vertebrates the cornea as well) functions like the camera lens in focusing light on the *retina*, an heterogeneous layer of light-sensitive cells at the back of the eye. The retina thus receives an image in the same way as the camera film. The iris (comparable with its diaphragm analogue in a camera) regulates the size of the pupil and hence the amount of light entering the eye.

The most important component of the vertebrate eye is, of course, the retina. The remaining eye structures are more or less geared to training, arranging and focusing light rays on the retina which then transmits the information to the brain. Embryologically, the retina develops as two layers. Its complex sensory and nervous mechanisms, however, all develop from the inner of these two layers, the outer becoming little more than a layer of pigment cells. The inner layer consists of four broad sub-layers: (1) a striated zone containing the elongated tips of the light-sensitive *rod* and *cone* cells, (2) an outer nuclear zone containing the rod and cone cell bodies, (3) a nuclear zone containing the *bipolar cells* which transmit impulses inward from the rods, cones and accessory retinal cells and (4) a *ganglion cell* layer which picks up stimuli from the bipolar cells and relays them via the optic nerve to the brain.

Rods and cones differ markedly in their structure and function. Rods are sensitive to low-intensity illumination but provide poor image resolution and no colour discrimination, while cones are stimulated only by high light levels and provide good image and colour resolution. Cones are particularly concentrated in a central region of the retina (the *area centralis* or *macula lutaea*) where detail perception is most acute. In many groups, a *fovea* (a depression which is clear of blood vessels and other cell types) is also present in the area centralis and appears to function in the focusing of visual attention. In many birds there are two foveae, the second situated at the back of the eye and probably allowing acute visual discrimination ahead of the animal as it flies (the eyes of most birds are situated in an extreme lateral position).

The ganglion cells produce long fibres extending into the forebrain. From here they enter the *optic chiasma* where those from the right optic nerve cross to the left side of the brain and vice versa. In lower vertebrates the fibres project into the optic tectum, while in mammals they are relayed to the cerebral hemispheres (see Section 1.2). At least in mammals, the regions of the brain receiving information from different parts of the retina are arranged topographically so as to reproduce the arrangement of visual stimuli falling on the retina. A series of excellent studies on the leopard frog (*Rana pipiens*) shows how a complex visual system works.[186]

At one time the eye was thought of as little more than a means of translating an environmental image into electrical impulses. It was only in the brain that the information was sorted and interpreted. Early studies on the leopard frog had supported this view. By concentrating on retinal responses to points of light and dark, Hartline and coworkers concluded that the frog's eye simply registered changes in the tone of a viewed object. Lettvin and coworkers[186],

Figure 1.13: Diagram of a Frog Retina. rc, zone of rods and cones; Int, intermediate layer; Pl, plexiform layer; bi, bipolar cells; a, amacrine cells; g, ganglion cells. The ganglion cells spread out with varied branching patterns at different levels of the plexiform layer. The pronounced variation in the bipolar and ganglion cells may reflect the differences in the complex operation of the latter discussed in the text.

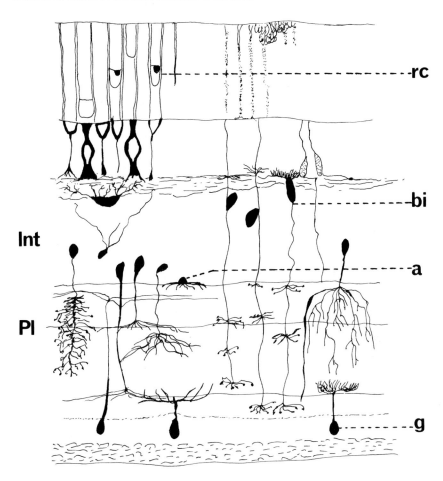

Source: Modified after Lettvin *et al*[186].

however, had the inspired idea of stimulating the retina with natural images. Instead of just using points of light, they used flies, twigs, bugs etc., objects the frog was likely to come across in its natural habitat. By this means they discovered a remarkable functional diversity among the ganglion cells of the retina (Figure 1.13). There were five main types of cell:

(1) *Sustained-edge detectors (SEDs)*: These showed the greatest response when a small, moving edge entered and remained in their receptive field. Immobile or long edges did not evoke a response.

(2) *Convex-edge detectors (CEDs)*: These were stimulated mainly by small, dark objects with a convex outline.

(3) *Moving-edge detectors (MEDs)*: These were most responsive to edges moving in and out of their receptive field.

(4) *Dimming detectors (DDs)*: These responded most to decreases in light intensity.

(5) *Light-intensity detectors (LIDs)*: The responsiveness of these cells was inversely proportional to light intensity. They were most responsive in dim light.

The frog's retina is thus responsive to a very selective range of stimuli. Furthermore, the stimuli to which it does show a response are those which are important to the frog. CEDs, for example, respond only when small, dark objects move haltingly into view. Such objects are often small beetles or bugs and the cells are known colloquially as 'bug detectors'. Similarly, SEDs, MEDs and DDs probably aid detection of predators. The first two respond to moving edges and the third to shadows cast suddenly over the frog. SEDs also detect edges which move into view and then stop. Many important predators of the frog, like herons and pike, often 'freeze' just before striking.

A lot of redundant information is thus weeded out in the retina, so that the brain only receives information of biological importance. The frog simply does not see inconsequential objects. If they are not food, predators or mates etc., there is no point cluttering up the nervous system with them. Similar retinal cell specificity is found in other species. In the retina of the rabbit (*Oryctolagus cuniculus*), for example, there are at least eight types of discriminating cell. These respond to a wide variety of specific stimuli including speed of movement, direction of movement, stimulus orientation, contrasting borders and even the absence of a stimulus ('uniformity detectors').[187] Pike possess retinal cells which respond preferentially to small vertical displacements of horizontal edges and others which respond most strongly to slow movement by small objects. The goldfish retina contains two populations of cells which respond to different size ranges of object. One population responds to objects 2.5-7° across, the other to objects 8-12.5° across.

Certain cells in the retina have evolved the ability to distinguish between different wavelengths of light. That is, they detect colour. In the leopard frog this is a property of the LIDs. Honey bees (*Apis mellifera*) have a trichromatic system of retinal receptors with absorbance peaks at 345 nm (ultra-violet), 440 nm (blue-violet) and 550 nm (green-yellow). Just as with object detection, an animal's colour vision is adapted to its specific needs. Thus humans can perceive red light and distinguish between yellow, orange and green whereas honey bees can do neither. Bees, on the other hand, can see ultra-violet light and put it to good use in choosing which flowers to visit. In herring gull (*Larus argentatus*) chicks, the classic pecking response to the red spot on the parents' bill[305] seems to be aided by colour filtering in the retina. Incoming light has to pass through a variety of red, orange and yellow oil droplets in the retinal cells before it reaches the light-sensitive pigment. These droplets tend to filter out light at the blue end of the spectrum while allowing red light to pass through. The birds are not blind to blue, but blue colours are reduced in intensity relative to red.

Auditory Perception. Sounds are relatively high frequency vibrations which are emitted from a variety of sources in the environment. Sound waves have sufficient energy to cause pressure changes and, in some cases, displace objects. Sense organs for perceiving sound have evolved to capitalise on this energy.

In many insects, these consist of tympanic organs, which are usually located on the legs, thorax or abdomen and form a diaphragm ('ear drum' or *tympanum*) connected to a series of sensory neurons. When sound waves cause the tympanum to vibrate, the sensory neurons are stimulated and send impulses down the auditory nerve to the CNS. In locusts (Orthoptera), the 70 or so sensory neurons of the tympanum are organised into four groups. Each group responds to a limited range of sound wave frequencies (from 1.5 Hz to 19 Hz). In this way, the locust can obtain a variety of information from the sound reaching its 'ears'. Nevertheless, in comparison with vertebrate hearing systems, insect ears are very limited in the kind of auditory stimuli they can recognise.

During its evolution, the vertebrate ear has become an organ of elaborate structural complexity. In mammals, the ear can be divided into three parts: the external ear (*pinna*), the middle ear and the inner ear. The cartilaginous pinnae serve to channel sound waves into the ear aperture (*auditory meatus*). At the end of the auditory meatus is the tympanic membrane (ear drum) which separates the external and middle ears. The membrane connects with the first (*malleus*) of three bony ossicles which are strung across the air-filled cavity of the middle ear. Vibration of the membrane causes a chain of articulations between the malleus and the other two ossicles (the *incus* and *stapes*). In this way vibrations from the membrane are transmitted to the *fenestra ovalis* of the fluid-filled *cochlea*. Inward bulging of the fenestra ovalis causes movement of the cochlea fluid (*endolymph*) so that the *fenestra rotunda* on the lower surface of the cochlea bulges outwards. Thus, as sound vibrations are transmitted through the ear, the cochlea oscillates back and forth. This causes movement of a membrane (basilar membrane) within the cochlea which stimulates special hair cells. Impulses then pass down the neurons leading from each hair cell to the temporal lobes of the brain where they are translated into what the animal recognises as sound.

Like vision, auditory systems are specialised in what they respond to. If animals responded to all the sounds in their environment, they would be subjected to a meaningless cacophony. In predators which hunt by sound, for

Figure 1.14: Barn Owls Match the Sound Input to Each Ear to Locate and Catch Prey in the Dark. A mouse at A will stimulate ear 1 but not ear 2 and vice versa at C. A mouse at B will stimulate both ears.

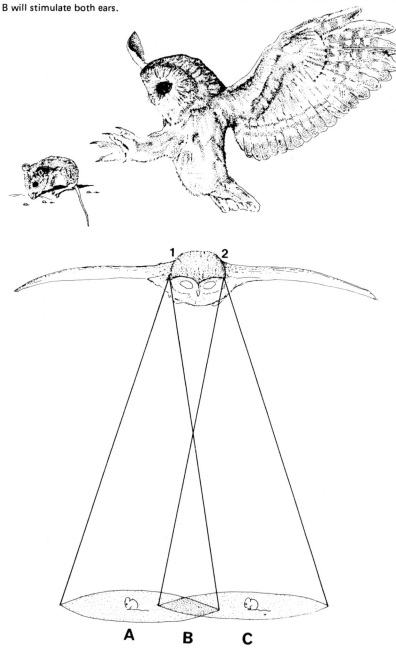

Source: In part modified from Alcock[1].

instance, auditory stimulation needs to be tightly controlled. Night-hunting owls like the barn owl (*Tyto alba*) are a good example.

Although barn owls do not possess the mobile pinnae of a mammal, they are uncannily good at locating a source of sound. Even barely audible sounds like a leaf being pulled across a floor can be pin-pointed.[241] The owl appears to achieve this accuracy by means of the orientation of the auditory meatus and the arrangement of feathers around the face ('facial disc'). Together these design features block all sounds except those arriving along the line of vision. All the owl has to do to locate a sound source is move its head about until the sound is loudest in both ears. It will then be facing the source. To aid the location process, the owl's ears are placed asymmetrically on its head. This narrows down the range of locations within the environments from which sounds can stimulate the two ears equally (Figure 1.14).

Auditory stimulus filtering in the owl mainly involves the removal of directional redundancy. Among insects, however, there are many examples of ears which respond solely to a limited range of sound frequencies. Male mosquitos, for instance, respond selectively to the sound of the female's wing-beat. The beat frequency of females is different from that of males. In this case it appears to be hair-like processes on the male's antennae which detect the sound, perhaps by vibrating in sympathy like a tuning-fork. An even more extreme example of selective 'listening' occurs in noctuid moths.[257] These moths are heavily preyed upon by bats. Their simple ears are geared almost entirely to detecting approaching bats. The moths possess a pair of ears, one either side of the thorax. Each ear consists of just two sensory neurons connected to the tympanum. One neuron is sensitive to low-intensity sound, the other to high-intensity sound. The 'low-intensity' (LI) neuron also responds more to intermittent rather than continuous sound. The two neurons, however, do not discriminate between different frequencies of sound, merely its intensity and pattern of emission. The LI neuron responds to the faint ultra-sonic emissions of bats up to ten metres away and fires as the emissions increase in intensity. The moth can therefore tell whether the bat is getting closer. By 'weighing-up' the relative degree of stimulation of the LI neurons in its two ears, the moth can also pin-point the position (above, below, beside, etc.) of the bat relative to itself and move away. The 'high-intensity' (HI) neuron only seems to come into play if the bat flies very close to the moth. Roeder[257] suggests that high-intensity emissions from a nearby bat stimulates the HI neuron to fire. Impulses are transmitted to the cerebral ganglia and inhibit the neural centre controlling the activity of the thoracic ganglion (Figure 1.15). The thoracic ganglion controls the wing beat pattern, so, when it is inhibited, the wings beat asynchronously or stop beating altogether. The moth's flight thus becomes erratic and it may drop like a stone out of the flight path of the bat.

Auditory systems, like visual systems, are tuned to the precise needs of the animal. In the extreme case of the moth, the only sounds to which it needs to respond are those emitted by hunting bats. These are therefore the only sounds

Figure 1.15: The Tympanum of a Noctuid Moth Showing the LI and HI Neurons (see text), Air Sacs (as) and Tympanic Membrane (tm), and a Diagram to Show the Spatial Relationship between the Ears (ea), Cerebral Ganglia (cg) and Thoracic Ganglia (tg).

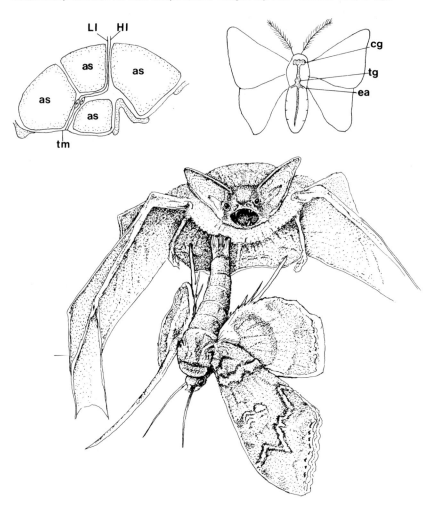

Source: In part modified after Alcock[1]. Drawing after an illustration by W. Linsenmaier.

it hears. In the owl, selection has geared the auditory system to distinguish a wide range of frequencies but to be highly sensitive to their direction of origin. Some animals like bats rely almost totally on sound for hunting and navigation. In many cases, there has been an elaborate coevolution of sound emission and detection systems so that the animal creates its own sonic topography within which to operate.

Tactile Perception. Auditory perception depends on specialised *mechanoreceptors* detecting changes in high frequency mechanical stimuli. Many other types of mechanical stimuli are also important to animals which have thus evolved a wide range of mechanoreceptors to detect them. Most of these are low-frequency stimuli which the animal senses as touch or pressure.

A wide range of species, including insects and mammals, possess touch-sensitive hairs over the body surface. The intricate communicatory dance of the honey bee depends on such hairs. Bees usually dance on the vertical surface of the honeycomb. In order to get the directional components of the dance right, they need to know their orientation on the comb. Touch-sensitive hairs between the head and thorax provide the necessary information. Because the top half of the bee's head (2.5 mg) weighs less than the bottom half (6.5 mg), the bee's orientation on a vertical surface causes the head to press on the hairs in different ways. If the bee is orientated with the head vertically up, gravity presses the bottom of the head against the ventral set of hairs. If it moves to the left or right, the left or right ventral hairs are stimulated. If it orientates vertically down, the lower part of the head is pushed forward and the top presses on the dorsal set of hairs. Central processing of this tactile information provides the bee with information about its orientation. If the sensitive hairs are denervated, the animal wanders chaotically over the comb. If weights are glued to different parts of the head, the bee's movements can be affected predictably on the basis of the head's pattern of pressure on the hairs.[191] Such pressure-sensitive hairs are also important for manoeuvrability in flying insects. Many mammals also use specialised sensitive hairs, called *vibrissae*, for assessing objects and distances in their environment. Shrews (Soricidae) and moles (Talpidae) depend almost entirely on their vibrissae to find food.

In a more fundamental way, mechanoreceptors in the form of stretch receptors affect behaviour by controlling posture and locomotion. The same is true of statocysts in crustaceans and the endolymph-filled semicircular canals of the vertebrate inner ear. Here gravitational displacement of a mineral statolith or fluid endolymph causes pressure changes on sets of sensitive hairs. In vertebrates, the sophisticated semicircular canal system provides information not only about orientation, but also about the speed and direction of acceleration for olfactory discrimination. It is thus able to make fine distinctions between different potential food plants.

Many animals have evolved chemical means of communication. *Pheromones* are special chemicals which convey information between conspecifics. They are thus distinguished from *allomones* which facilitate interspecific communication and *hormones* which operate within an individual. One of the best known pheromones is bombykol, the sex-attractant of the silk moth. A receptive female simply emits a small quantity of bombykol molecules into the air and sits tight. The molecules drift downwind where they are detected by a wandering male. Because of the extraordinary sensitivity of the male's antennal sense cells, he needs only 200 molecules to strike the antenna within a second to be able to orientate and home in on the female. Unlike the male, the female silk moth is insensitive to bombykol. Her antennae are small and thin compared with the male's elaborate, feathery structures. However, only the female is sensitive to the subtle olfactory cues which characterise good oviposition sites. Olfactory capabilities, like other senses, are designed for particular purposes rather than conferring blanket sensitivity. Other well-known pheromones include the complex scented secretions of mammals. Urine and faeces, as well as specialised scent glands, may be used to mark objects, conspecific territory boundaries or even to scent the air. Species, sex, age, motivational state, may all be coded in various secretions.

Olfaction is, of course, closely linked with the other major chemosensory modality, gustation (taste). In higher animals, however, the two are physiologically distinct and have separate excitatory properties. In insects, the gustation system consists of sensory hairs distributed over the body. Blowflies (*Phormia* spp.), for example, possess about 300 such hairs on their mouthparts and 3,120 more on their legs. Each hair has five afferent sensory neurons, one of which registers mechanical stimuli, and the other four, various gustatory stimuli (in this case one responds to water, one to sugars and two to salt). In vertebrates, gustatory sense cells are restricted to surfaces within the mouth. Well-differentiated 'taste' cells are distributed in distinct zones over the surface of the tongue in man. In general, 'bitter tasters' are located at the back of the tongue, sour tasters at the side and sweet and salt tasters at the front. The selective pressures underlying the evolution of sensitive olfactory and gustatory modalities are obvious. The chemical characteristics of objects in the environment provide invaluable information about their usefulness to the animal. Distinguishing subtle chemical differences in food may mean the difference between a nutritious meal and being poisoned. The scope for social interaction and avoidance is immeasurably enhanced by chemical communication, particularly in nocturnal or crepuscular animals.

The perceptual mechanisms discussed above are not, of course, the only ones animals have evolved. Most animals, for example, have some kind of temperature sensor on their outer surface and a few have evolved 'active' electrical means of perception (for example mormyrid fish).[39] However, they serve to illustrate the diversity of perceptual mechanisms and more importantly, their direct role in interpreting the environment in a way that is useful to the animal. Nevertheless, peripheral sensors are only one stage of the information filtering process. To understand the next, we need to examine the CNS.

1.4.2 Central Stimulus-filtering

Although peripheral sense organs perform much of the stimulus sorting process, they do not perform it all. In the leopard frog retina, for example, the LID cells can certainly distinguish colours but the frog only recognises the colour blue. This is because cells in the thalamus of its brain, which receive information from the LIDs, only respond to blue. The selective advantage of this extreme central filtering of colour stimuli appears to be that the frog tends to jump towards blue objects when alarmed. This presumably increases the chances of landing in the safety of water.[222] Colour-specific responses in some bird species also seem to rely on central filtering. Whereas the preference for red in herring gull chicks could be traced to colour-filtering oil droplets in the retina, similar preferences for black in some tern species (*Sterna* spp.) appear to be based in the CNS. The young of many precocial birds often show marked colour preferences which have no experiential basis. Mallard (*Anas platyrhynchos*) ducklings, for instance, have a strong preference for green, as do the young of some phasianids. There are no retinal mechanisms which could result in a bias towards green, so the preference seems to be mediated centrally.

In mammals, the central processing of stimuli may be very complex. In cats, cells in the lateral geniculate nucleus (LGN) of the brain enhance the differences in illumination striking the retina.[150] Most of these cells respond to impulses from only one of the eyes. Some LGN cells respond temporarily to white objects entering or black objects leaving their receptive fields. Others fire when black or white objects do either. The visual cortex itself contains two main types of cell. 'Simple' cells respond to lines and edges in particular orientations and locations. Their receptive fields are characterised by 'on' and 'off' zones in which stimulation with light results in an increase and decrease in firing rate respectively. 'Complex' cells also respond to lines, slits and edges orientated in a particular direction, but their receptive fields are not divided into 'on' and 'off' zones. Appropriate stimulation causes the whole unit to increase or decrease its firing rate. A 'complex' cell seems to have inputs from several 'simple' cells whose receptive fields occur within its own. Visual information is also passed on from the visual cortex to other parts of the brain. The recognition of complex visual stimuli by the cat therefore depends on neural activity at many stages along the visual pathway.

Although the cat relies more on central rather than peripheral filtering processes, the contrast with the leopard frog is not simply phylogenetic. Pigeons show even more retinal cell differentiation than the frog. Conversely, some lower vertebrates and invertebrates exhibit quite complex central filtering. In goldfish, for example, the optic chiasma and tectum of the brain possess a number of cell types which respond to distinct classes of stimuli. The optic nerve of the crab *Podophthalmus vigil* also contains differentially sensitive cells. Here cells differ in terms of the area of the visual field, the type of stimulus and the degree of contrast needed to elicit a response. In locusts (*Locusta migratoria*), 'novel movement detection' cells have been found in the tritocerebrum. These selectively respond to stimuli moving in a new direction.

Central Filtering in Auditory Perception. Central stimulus-filtering also occurs during auditory perception. The noctuid moths and their bat predators discussed in Section 1.4.1 are good examples.

While much auditory discrimination by moths takes place in the ear, Roeder

Figure 1.16: Generalised Mammalian Auditory Pathway. ac, auditory cortex; mg, medial geniculate nucleus; ic, inferior collicular nucleus; lln, lateral lemniscal nucleus; on, olivary nucleus; cn, cochlear nucleus; an, auditory nerve.

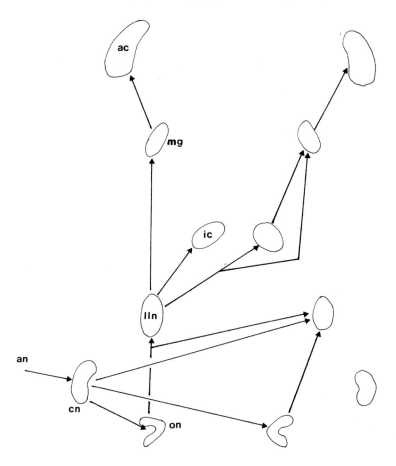

Source: Modified after Guthrie[123].

discovered at least four types of CNS neurons which also process auditory information. 'Repeater' interneurons in the thoracic ganglia appear to produce prolonged impulses in response to LI neuron discharge and allow the moth to respond to lower intensity stimuli. Further signal enhancement seems to be provided by so-called *binaural units* which sum and perhaps enhance inputs from each ear. Finally, two types of 'marker' interneuron have been discovered. Sparsely-distributed *train markers* appear to mark the beginning and the end of LI neuron activity with low frequency, non-adapting responses, while *pulse markers* are associated with three kinds of electrical event which have the net effect of producing a fast potential in response to each sound emission from the bat.

In mammals, the central processing of auditory information is mediated by a series of nuclei in the mid-brain (Figure 1.16). The basic mammalian auditory pathway shown in Figure 1.16 is considerably modified in insectivorous bats. The mid-brain auditory centres are some seven times larger than in other mammals, with the main size increase being in the *inferior colliculi, cochlear nuclei, olivary nuclei* and *lateral lemnisci*. Extensive connections between the collicular neurons and the reticular formation appear to provide the physical basis for a high degree of selective attention and rapid motor control. The reticular formation is especially well-developed in bats and may allow the integration of proprioceptive and vestibular (inner ear) information with auditory data.

In this chapter we have examined some of the important internal mechanisms which mediate behaviour. By processes of stimulus perception and long- and short-term transmission of information around the body, these mechanisms *cause* behaviour. They are internal *causal factors*. Via the sense organs, internal causal factors are mixed with others in the external environment (external causal factors). In the next chapter we shall discuss some other internal causal factors and examine the precise way in which they are blended with external factors to produce behaviour.

Summary

(1) Behavioural sophistication and specialisation is to some degree related to the complexity, functional organisation and activity of the nervous system.
(2) Some behaviour patterns, both 'simple' and 'complex', can be traced to specific neural pathways within the animal. However, owing to their complexity, neural explanations of behaviour and the organisation of behaviour sequences are unlikely to be very informative.
(3) Hormones transmit information around the body more slowly than nerve cells. They influence behaviour mainly through their effects on the CNS, sensory-perception and effector organs. However, these effects often involve complex interactions with a number of other factors.
(4) In some cases, behaviour may be programmed to occur on a predictable schedule. These schedules are correlated with a variety of long- and short-term rhythmic events in the environment and provide the animal with an adaptive internal 'clock'. Neural, endocrine and perceptual mechanisms interact in a complex way to regulate the clock.
(5) Neural and endocrine systems receive information from the animal's environment via specialised sense organs. Sense organs, however, do not necessarily transmit a faithful image of the environment but 'filter' their input so that internal mechanisms receive a highly selective picture. Both peripheral and central stimulus-filtering processes can be related to the functional needs of the animal.

2   MOTIVATION AND DECISION-MAKING

It is a matter of common observation that an animal does not respond to a stimulus in the same way every time that stimulus is encountered. We only have to think of the lion moving through the bush and coming upon a herd of wildebeest. On some occasions such an encounter results in the lion stalking and perhaps killing one of the wildebeest. On others, the lion walks casually past apparently ignoring the presence of potential food. A male chaffinch, during winter, vigorously defends his territory against conspecific intruders of either sex. In spring, however, he selectively admits and courts females while continuing to drive off other males. Something about the lion and the chaffinch in these two examples has changed between one encounter with food or a female and another. But what has changed? Since there is no difference in the stimulus itself, we are left with the possibility of some internal change in the animal. That internal change we can conveniently label *motivation*.

'Motivation' is a useful descriptive term which in itself does not help us to understand the change, but rather defines some kind of internal variable which influences the relationship, or *intervenes*, between stimulus and response. Historically, such *intervening variables* were sometimes referred to as *drives*, which built up within the animal as time passed since the last performance of a behaviour. As a drive built up, the animal's threshold of response to functionally related groups of environmental stimuli was lowered. Thus the build-up of a feeding drive resulted in an increased responsiveness to stimuli connected with food (e.g. the sight or smell of prey), a build-up of sexual drive to stimuli associated with courtship and mating and so on. The fact that an animal's thresholds of response to stimuli within such functionally related groups do appear to rise and fall together makes it tempting to think of motivational changes as being highly specific, linked to one aspect of the animal's requirements.

In some ways, however, it may be misleading to think of motivation as being specific to particular requirements or *goals*. At least in mammals, a stimulus can evoke two types of response within the brain. One, directly related to the stimulus, operates via what can be called *specific sensory pathways*. Visual stimuli evoke responses in the visual centres of the brain, sounds in the auditory centres and so on. The second type is less specific because incoming pathways also give off branches to the reticular formation (see Section 1.2) where they connect with non-specific pathways to other brain centres which then become aroused. This means that a stimulus may not only evoke a response relevant to its own modality (e.g. vision), but may also increase the animal's responsiveness to other, unrelated stimuli (e.g. auditory, olfactory). The relationship between hunger and thirst is a good example. Rats trained to run down an alley for water will run down faster and drink more the thirstier they are.

However, if they are then tested when not thirsty, but hungry, they show similar increased tendencies to search for and drink water. The irrelevant state of hunger therefore appears to facilitate appetitive and consummatory behaviour appropriate to thirst. Similarly, mild disturbance or shock, which usually disrupts the current behaviour, may increase the responsiveness of rats to food or water.

While the problem of *general* versus *specific* arousal is an interesting one, it may not be too serious from a functional point of view. As we shall see later, it is possible for more than one motivational system to be aroused at any one time but for responses specific to only one of them to be performed. For all practical purposes it is usually reasonable to talk in terms of a specific motivation or 'drive'. A hungry animal can then be said to have a motivational drive to find food, a thirsty animal to find water and so on.

2.1 The Drive Concept and Energy Models of Motivation

An animal motivated to search out a particular requirement like food or water behaves in a characteristic way. Ethologists and comparative psychologists have often divided up the expression of a drive into three components:

(1) A phase of searching for the particular requirement or goal ('appetitive behaviour').
(2) Behaviour orientated around and/or to do with processing the goal ('consummatory behaviour').
(3) A quiescent period following achievement of the goal ('quiescent' or 'refractory period').

Animals can rarely just reach out and grab food, for instance. They have to spend time and energy foraging for and processing suitable items. Following consummatory behaviour, the animal may enter a temporary period of quiescence with respect to that particular goal, although it may still show appetitive behaviour for other goals.

Early attempts at modelling the way motivation is controlled relied heavily on the drive concept. Since we are going to meet many different kinds of behavioural model in this book, it is worth spending a little time explaining what they are and what they aim to do. As Manning[199] succinctly put it, *'the purpose of a behaviour model is to devise a system of hypothetical components to which particular properties are given and which are so connected together that their "behaviour" reproduces that which we observe'*. A model is therefore an educated guess at how an unknown system works based on an analogous and well-understood system. Analogous systems may be familiar pieces of technology, or perhaps mathematical equations. We stretch the analogy as far as it will go by experimenting with the behaviour and comparing the results with what we know would happen in the case of our analogue. If the results are similar, the model

Figure 2.1a: Lorenz's Hydraulic Model of Motivation.

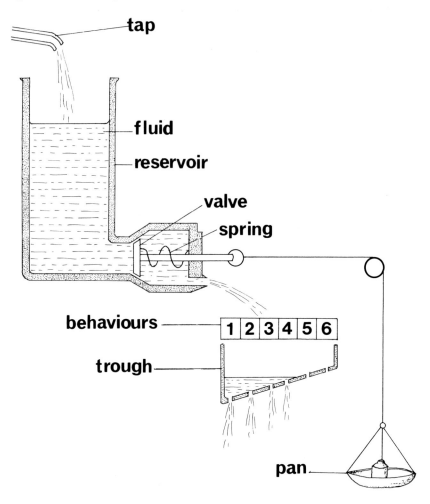

Source: Modified after Lorenz, K.Z. (1950). The comparative method of studying innate behaviour patterns. *Symp. Soc. Exp. Biol., 4*: 221–68.

is retained and tested further; if they are different we suspect something is wrong with our analogy and seek to improve it. Later, we shall see how such empirical testing and improvement can achieve a close fit between a model and observed behaviour.

2.1.1 Early Models of Motivation

Lorenz's Hydraulic Model. One of the earliest attempts to model motivation used the analogy of an hydraulic flow system (Figure 2.1a). The fluid in the

Figure 2.1b: Tinbergen's Hierarchical Model of the Reproductive Instinct of the Male Three-spined Stickleback. Motivational impulses (arrows) 'load' behavioural centres. Impulses may come from the external environment, superordinated centres or spontaneously from within the centre itself. IRMs (open squares) inhibit the discharge of centres until released by an appropriate stimulus.

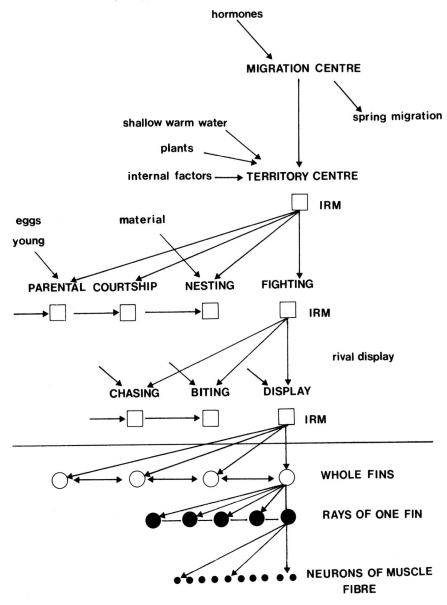

Source: Modified after Tinbergen, N. (1951). *The Study of Instinct*. Oxford University Press.

system is analogous to *action-specific energy* (*ASE*) which accumulates spontaneously over time. The longer an animal has not performed a behaviour, the more energy specific to that behaviour accumulates in the animal's system and the more likely the behaviour is to be performed. The animal's drive for that behaviour is increased. Energy is released through the valve when the animal encounters an appropriate stimulus. The strength of the stimulus is related to the weight on the spring pan. The ease with which the valve opens is a joint function of the characteristics of the stimulus (weights) and the amount of energy (fluid) in the reservoir. Depending on how far the fluid shoots out of the valve, different numbers of outlets in the trough come into operation. In behavioural terms, each outlet represents one stage of expression of the animal's drive.

Alcock[1] uses the example of a cat which has not eaten for some time. The cat's feeding drive has been building up like the fluid in the reservoir, when suddenly a mouse appears. The mouse provides a strong stimulus to feed (it represents a heavy weight on the spring pan) because it is a preferred prey of cats. On seeing the mouse, the cat may (1) stalk it, (2) catch it, (3) kill it, (4) eat it. It is as if the mouse released all the ASE available for feeding which poured into the trough and flowed out as the sequential performance of acts 1–4, making up the cat's feeding behaviour. But what would happen if the cat had not been without food very long? Now it is as if only a little fluid has built up in the reservoir. When the fluid is released it shoots out only a short way and only brings into operation the first and second outlets in the trough. The cat is thus observed to stalk and catch, rather than follow through to kill and eat. On the other hand, if the cat had gone without food for a considerable period, a very mild stimulus (a light weight on the pan), like a bird high up in a tree, might trigger all four acts. Usually the bird would be ignored because the cat is unlikely to catch it, but even the remotest possibilities are worth a try if the animal is very hungry.

Tinbergen's Hierarchical Model. Tinbergen proposed a more complex model of motivation designed to account for the organisation of behaviour over longer periods of time. In this model energy is not completely specific to individual behaviours but is held in higher level behavioural centres such as those for reproductive behaviour or feeding behaviour. Tinbergen envisaged motivational energy as flowing down through various inhibitory blocks to progressively finer levels of appetitive behaviour or consummatory acts through to specific muscular activities and finally indivisible motor units (Figure 2.1b).

In Tinbergen's model the observed patterns of behaviour reflect an order of functional organisation (an *instinct*) within the central nervous system. In the example shown in Figure 2.1b, hormones are assumed to affect the highest centre – the migratory centre – controlling reproduction in the stickleback (*Gasterosteus aculeatus*). This results in appetitive behaviour in the form of migration. Appetitive migration ceases when the fish encounters a particular type of habitat. Stimuli from the appropriate habitat excite a specific IRM (see Chapter 1). In its resting state, the IRM blocks the next centre in the hierarchy and inhibits the performance of behaviours beyond that point. When it is excited, the block is removed and the centre is freed for propagation. Impulses can now travel down to lower centres controlling, for example, brood care, resting and fighting, but each of these centres is blocked until their appropriate or *key* stimulus appears. In the case of fighting, the key stimulus would be a rival. The rival must then provide a range of further key stimuli to elicit particular fighting behaviours like biting or chasing.

2.1.2 Usefulness and Limitations of 'Drive'

The chief convenience of the drive concept is its economy. Drive is a convenient shorthand way of recognising the relationship between a number of environmental conditions and behavioural responses (Figure 2.2). Terms like 'thirst drive' also acknowledge that the animal's physiology is geared to help it meet biological needs. They indicate that environmental conditions do something to the animal's internal state which it then acts to put right. At first sight, it might seem better to abandon motivational terms altogether and describe behaviour purely in terms of physiological changes. An object lesson in why it is not is provided by Dethier's classic study of feeding in the blowfly (an excellent summary can be found in Alcock[1]).

Economical though 'drive' undoubtedly is, however, it has many serious drawbacks. To begin with, calling something a 'drive' merely labels it. It explains nothing. One criticism of the early energy models of motivation is their implicit suggestion that drives activate behaviour in the same way that energy activates a physical system. 'Drive' also results in complex and heterogeneous mechanisms being subsumed under one label. Thus Miller found that the strength of the drinking response in rats following a standard saline injection, depended on how it was measured. Bar-pressing, quinine tolerance and drinking rate all gave different results. Since different internal mechanisms seem to be involved, a single term 'thirst' seems inappropriate.

Another problem with drives is that they can be subdivided endlessly. We might start by saying that a blackbird's behaviour during spring and summer is governed by a reproductive drive but, before long, we find we have to distinguish several different 'sub-drives'. Nest-building, for example, is only part of the bird's total reproductive activity, but it can clearly be considered as a complete act which is independent of other reproductive acts like courtship, copulation and incubation. Should we postulate a nesting drive? If we do, we immediately admit a mass of further subdivisions like stick-finding drive, nest-construction drive, nest-lining drive and so on. 'Drive' as a concept rapidly becomes valueless. For these and other reasons (see Hinde[143] for a full discussion) the drive concept as a model for motivation has been more or less abandoned and in its place has arisen a different and quantitatively more rigorous approach.

Figure 2.2: Drive Terminology Provides an Economical Way of Describing the Relationship between Antecedent Conditions (a–c) and Behaviour (1–3).

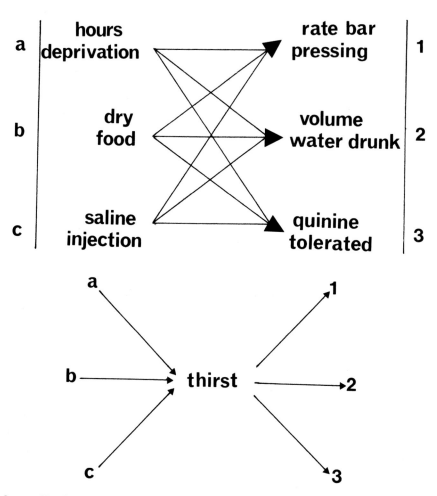

Source: Modified after Miller, N.E. (1957). Experiments on motivation. Studies combining psychological, physiological and pharmacological techniques. *Science, 126*: 1271–8.

2.2 Feedback Models of Motivation

2.2.1 Deutsch's Model

An early attempt at the kind of motivational model which has proved very powerful in recent years is the loop system proposed by Deutsch (Figure 2.3). Although Deutsch's model was designed to cover much wider aspects of behaviour, only the part representing motivation is reproduced here. The model operates in the following way. A deficit or imbalance in the animal's physiology

Figure 2.3: Deutsch's Feedback Model of Motivation.

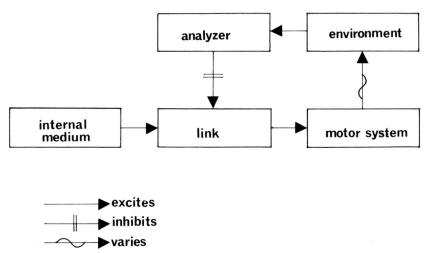

Source: Modified after Deutsch, J.A. (1960). *The Structural Basis of Behaviour*. Cambridge University Press.

is detected by and excites a 'central structure' or 'link'. The persistence and strength of this excitation depends on the magnitude of the imbalance. The excited link then activates an appropriate motor system which sets corrective behaviour in motion. Although in its original form this portion of Deutsch's model does not represent external stimuli which obviously induce behaviour, we can easily modify it to do so. As a result of the animal's behaviour some aspect of its internal and/or external environment changes. For example, it has eaten and its stomach is now full of food, or it has escaped from a predator which is now no longer present. The change in the environment is registered by the 'analyzer' which then inhibits the link so that it no longer responds to input from the internal/external environment. This inhibition slowly decays until the link is once again sensitive to excitation.

This is a very different type of model to those discussed earlier and, because it uses rather more abstract analogies, it is perhaps not so immediately appealing. However, the important question is whether it is any better as a model for behaviour.

A good test of the two types of model is the series of experiments carried out by Janowitz and Grossman[152]. Oesophageal fistulas were placed in a number of dogs so that when they ate, food passed out through the fistula instead of into the stomach (so-called 'sham-eating'). On the other hand, food could be placed directly in the stomach through the body wall. If a fistulated dog is deprived of food but then has food placed directly in its stomach, Lorenz's model predicts that it will still eat because the ASE for eating has not been used up. Deutsch's model, however, predicts that the dog will not eat. Although the 'link' has been excited (by deprivation), it is inhibited by the 'analyzer' which has been activated by the insertion of food. In fact the dogs in the experiments did *not* eat, so Deutsch's model appears to be the better analogue of their behaviour. What appears to happen is that stretch receptors in the stomach are stimulated by distention, send impulses to the central nervous system, and inhibit feeding activity. Receptors in the throat may also play a small part but fistulated dogs continue sham-eating long after they would normally have stopped.

The reason that Deutsch's model is more satisfactory than Lorenz's is that it incorporates sensory feedback from the animal's environment. There is a *negative* feedback loop between the behaviour and its consequences. Clearly such feedback is important if the animal is to avoid behaving in a way that is no longer relevant to its requirements. Deutsch's model is also more successful in another way in that we can find neural and physiological counterparts to the components of the model (for example, in the hormonal control of drinking via the hypothalamus in vertebrates).

Deutsch's model represents an early application of the terminology of control theory — a branch of engineering mathematics — to the problem of motivation. Control theory substitutes the concept of motivational *state* for the older, problematical concept of drive. For the moment this distinction is not important and it will be explored more fully later. Since Deutsch's model, control theory models of motivation have been developed much further and have become more sophisticated in their attempts to describe behaviour. The main pioneers in this field have been D.J. McFarland and his coworkers, and the reader is referred for detailed information to three excellent books by McFarland[210] and Toates[306,307]. While the present book is not the place to discuss the intricacies of the control theory approach, it is worth taking a brief look at the development of such a model and how, by a process of testing and modification, the model comes to simulate behaviour very accurately.

2.2.2 Control Models and Newt Courtship

Houston *et al.*[148] have published one of the few examples of how a behavioural model has been progressively developed for a clear-cut behaviour. The behaviour they chose to model is the courtship sequence of the male smooth newt (*Triturus vulgaris*).

Fertilisation in the newt is internal but, in the absence of a male intromittent organ, sperm is transferred to the female in the form of a spermatophore. At the end of an elaborate courtship sequence between male and female, the male deposits one or, more usually, several spermatophores which may or may not be picked up by the female.[128]

Houston *et al.* were concerned primarily with modelling the temporal patterning of the later phases of courtship (retreat display, creep and spermatophore transfer) (Figure 2.4), which is largely under the control of the male. The duration of each phase is very variable and there is also considerable variation in

Figure 2.4: Later Stages of Courtship in the Smooth Newt. The shaded individual is the female.

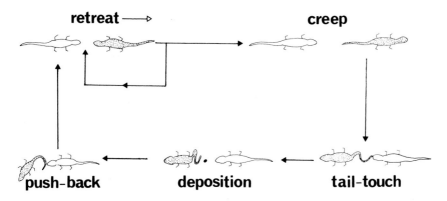

Source: Modified after Halliday, T.R. (1974). The sexual behaviour of the smooth newt, *Triturus vulgaris* (Urodela, Salamandridae). *J. Herpetol., 8*: 277–92.

the way transitions from one phase to another occur. In some sequences the male will make a simple transition from, for example, retreat display to creep; in others he will vacillate a number of times between the two before proceeding with the spermatophore transfer phase.

The Newtsex Model. As a basis for their model of newt courtship, Houston *et al.* used a similar set of assumptions to Deutsch. They assumed that internal and/or external factors in the newt's environment induce a change in its motivational state causing it to behave in an appropriate way. The changes wrought in the newt's internal/external environment as a result of the behaviour are registered by an integrator (analyzer in Deutsch's terminology) which in turn induces a change in the animal's motivational state.

Four versions of the newt courtship model (Newtsex Mks I-IV) are presented by Houston *et al.* which resulted from simulating each successive version on a computer, comparing the output with observed behaviour and adjusting the model to cater for discrepancies. The final version of the model (Figure 2.5a) is worth describing in some detail. The variable 'hope' represents the male's assessment of the readiness of the female to accept his spermatophore. Hope, therefore, depends on the state of readiness of the female, F, and the sperm supply of the male, S. At the beginning of a courtship sequence the male shows retreat display. During retreat display, the value of hope rises (via boxes 4 and 1) until it reaches threshold T_2 and there is a switch to creep. The threshold boxes, T_1 to T_4, operate in such a way that when the variable represented by the arrow leading into the box (in this case hope) exceeds a certain value, then the arrow leading *out* of the box is activated. During creep, hope decreases (by

Figure 2.5: (a) Newtsex Mk IV Control Model of Newt Courtship (see text). i, integrator; inr, increase in oxygen debt with time; inv, invertor. (b) The Lower Figure Shows the Sequence of Retreat (r), Creep (c) and Spermatophore Deposition (d) Predicted by Computer Simulation of the Model in (a); the Upper Figure Shows the Sequence Observed during Experiments.

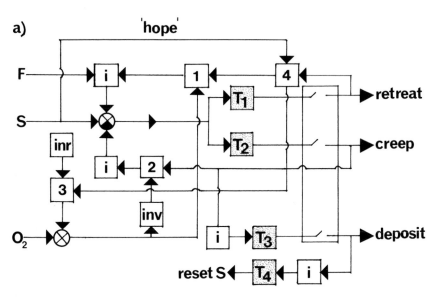

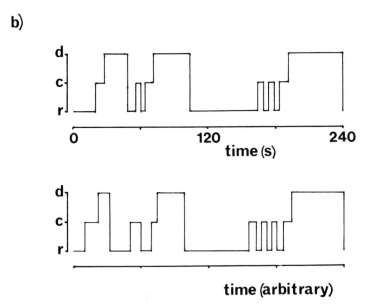

Source: Modified after Houston *et al.*[148]

negative feedback via box 2 — negative feedback is indicated by the circular symbol with a shaded segment) because the male can no longer see the female and thus does not know whether she is responding. When a certain amount of creep has been performed (T_3) the male proceeds to spermatophore deposition. In real life, however, he must receive a tail-touch from the female. If hope decays too quickly during creep so that it drops below T_1 before the tail-touch, the male reverts back to retreat display.

The other important factor is oxygen debt (O_2). Although newts are able to respire to a certain extent through their skin and buccal cavity, they are dependent during their aquatic life on air obtained at the water surface. The frequency with which newts have to ascend to the surface depends on their level of activity, the water temperature, the gas content of the water and the relative concentration of atmospheric gases above the water. This requirement for air places a serious constraint on the duration of sexual behaviour. Breathing behaviour and sexual behaviour are in competition. On the basis of Halliday and coworkers' studies, oxygen requirement is built into the simulation as a behaviour competing with courtship. In Newtsex Mk IV, the magnitude of the oxygen debt affects the rate at which hope changes as a result of retreat display (via box 1) and creep (via box 2). As O_2 will increase through the sequence, the male will be more likely to get through to spermatophore deposition as time goes on because he behaves at a higher rate.

Figure 2.5b shows the sequencing of retreat display, creep and spermatophore deposition which emerges from simulation of Newtsex Mk IV and the same sequencing as actually observed in newts. The remarkable similarity between the output of the model and the newt's behaviour illustrates the power of this type of modelling. However, while we have described a model of newt courtship and seen that it can be made to simulate behaviour quite accurately, what does such a model really represent? What are the biological analogues of the mathematical symbols? In fact, what the model simulates is a series of decisions the male newt makes during courtship. These decisions are made on the basis of the animal's motivational state and the feedback it obtains from pertinent environmental stimuli (in this case a female conspecific). We must emphasise here, however, that these decisions are not seen as intelligent calculations by the newt (although it would not make any difference to the model if they were), but as naturally selected stimulus-response relationships built in to the animal's nervous system. The 'decision' is therefore taken by evolution. This is an important point and one to which we shall return when we discuss decision-making in this and other chapters.

Apart from their more accurate and detailed simulation of behaviour, feedback models of motivation have two other important features. First, in substituting the concept of *motivational state* for the older concept of drive, they dispense with the difficulties of the earlier approach, and secondly, they assume that behaviour acts to maintain the animal's internal milieu at some preferred or *optimal* state. These two features together have provided the basis for a new and extremely fruitful approach to the problem of motivation which has become known for obvious reasons as the *state space approach*.

2.3 The State Space Approach

2.3.1 Physiological and Regulatory Space

McFarland pointed out that the internal environment of an animal can be viewed as a system of interacting variables (consisting, for example, of hormones, nerve impulses, temperature, sugar levels, water levels and so on). The animal's internal *state* at any given time then depends on the instantaneous values of all these variables. The state of any biological system can be characterised by its *state variables*, the minimum number of variables needed to describe the system completely. The state of an animal's internal environment can therefore be described by a finite number of *physiological state variables*, the minimum number of factors like hormones and nerve impulses which defines its physiology.

Each of these physiological state variables can be represented as an independent axis of an n-dimensional hyperspace which we can call *physiological space*. All this means is that we can imagine being able to measure changes in any given variable in terms of the direction in which it changes the animal's internal environment. Within this *physiological space*, there are boundaries determined (a) by what is biologically possible (for example, negative hormone levels are impossible) and (b) by the values of state variables beyond which the animal cannot live (for example, certain temperature ranges are lethal). A convenient notion for the origin of this physiological space is the optimal point on each axis; that is, the value of each state variable which is ideal in a biological or physiological sense, and which the animal would do best to achieve. While this assumption is convenient from a theoretical standpoint, the ideal may not be achieved in real life because of constant perturbations in the animal's internal and external environment and because the optimum for one set of axes may not be the optimum for others. The idea of physiological space is illustrated in two-dimensional form in Figure 2.6.

Let us now imagine that the animal's physiological state is displaced from the optimum (O) for some reason. Perhaps it has been deprived of water and now has a water deficit. We can represent this in Figure 2.6 as a displacement to point P. In order to correct for this displacement, physiological regulatory and/or acclimatisation processes come into operation. It may be, of course, that these processes are unable to reverse the displacement which may continue beyond P to cross the boundary and result in the death of the animal. Even if they can compensate for the shift, there are a variety of ways in which compensation might occur. The arrows a and b in Figure 2.6 represent two of these: 'b' represents direct compensation whereas 'a' represents 'compensatory' displacement along another axis. A good example of 'a' is the interaction between hunger and thirst in Barbary doves (*Streptopelia risoria*). Thirsty doves conserve water during

Figure 2.6: A Two-dimensional Physiological Space Defined by the Axes x
(e.g. Temperature) and y (e.g. Hormone Levels). O, the optimal physiological state; L, the
lethal boundary for conditions of x and y. The shaded area represents regulatory space; P, a
point of displacement of physiological state and a, b, trajectories of recovery from P to O
(see text).

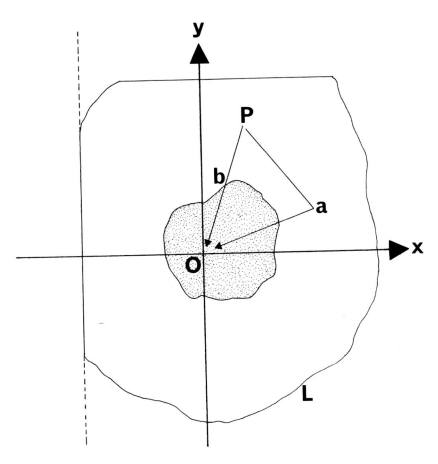

Source: Modified after Sibly and McFarland[284].

deprivation by cutting down on their food intake.[284] As a result they become
increasingly hungry even though food is available.

The ways in which an animal can adapt to physiological displacement vary
from short-term regulation (like breathing faster at high altitude) which occurs
within a day and may be a reflex reaction or complex behaviour, to long-term
acclimatisation which occurs over several days and is biochemical (involving,
for example, changes in enzyme properties). Short-term regulation compensating
for losses of sugars, water, heat, etc., tends to maintain the animal's physiological

state within a sub-space of physiological space. This sub-space Sibly and McFarland[284] call *regulatory space*. For convenience, the origin of regulatory space in Figure 2.6 is the same as that for physiological space, but it is possible that an animal placed in an extreme environment will acclimatise so that a new physiological optimum is created. In this case, regulatory mechanisms will act to maintain the animal's internal state at this new optimum and the origin of regulatory space will be displaced from its position in Figure 2.6. We can therefore regard the origin of regulatory space as an acclimatised optimum for the animal which will vary between environments.

2.3.2 Motivational Space

Behaviour can be viewed as a parallel process to physiological regulation. The 'goal' of much behaviour is to return a disturbed internal environment back to its optimum state. Animals drink to replace water, eat to replace blood sugars and so on. This optimum state is equivalent to the acclimatised origin of regulatory space. When *behavioural* regulation is involved, we can take this to be the origin of *motivational space*. Motivational space is therefore regulatory space where regulation is achieved by behavioural instead of physiological means. Just as displacement of the animal's internal state from the origin of regulatory space results in compensatory physiological mechanisms coming into play, so displacement from the origin of motivational space elicits compensatory behaviour. Such displacement in motivational space are called *commands*, because they set behaviour in motion.

Figure 2.7 shows an imaginary motivational space for drinking behaviour. The axes represent two physiological variables, blood osmolarity and the water content of the gut. These tell the animal something about its requirement for water by determining its degree of thirst and by telling it how much it has drunk. Let us assume that the animal is deprived of water for 24 hours. Its physiological, and hence motivational, state is moved away from the origin (optimum state) because blood osmolarity is increased. When the animal is allowed (or chooses) to drink, its motivational state will recover in the manner indicated by the curve. First, the water content of the gut increases sharply as water is imbibed, then, as water is absorbed through the gut wall, both the gut content and blood osmolarity drop until the animal's motivational state returns to its optimum. The axes for water content of the gut and blood osmolarity in this hypothetical example represent *causal factors* for drinking behaviour. We can picture such causal factor axes for many other motivational states scattered throughout the physiological space. Some axes might represent blood sugar levels and the food content of the gut and thus determine hunger state, others might represent testosterone levels and testicular sperm counts and determine sexual state. In reality, of course, each motivational space would have many axes. Drinking behaviour would not depend on the amount of water in the gut and blood osmolarity alone, but on these plus the distance to a drinking place, the risk of predation while drinking and many other causal factors.

Figure 2.7: A Hypothetical Trajectory of Recovery to the Optimal State in Motivational Space after Water Deprivation.

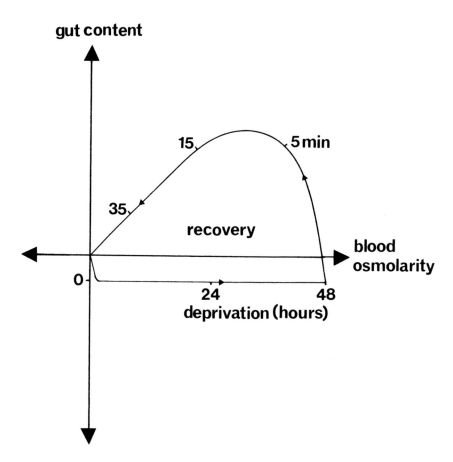

Source: Modified after McFarland[210].

2.3.3 Costs and Benefits in Decision-making

When an animal regulates its physiological state by behavioural means, it must 'decide' (a) which behaviour is appropriate and (b) how to perform that behaviour. There are an infinite number of ways in which a displaced physiological state could be returned to its optimum (we could have drawn an infinite number of curves in Figure 2.7), but which should the animal choose?

To find out how an animal weighs up various factors in coming to a decision, we need to ask it to choose between alternative different combinations of the same two factors. For instance, how would it choose between an easily accessible small amount of food and a large amount of food which could only be obtained

Figure 2.8: Indifference Curve for Feeding Behaviour Trading Off Predation Risk and Food Availability.

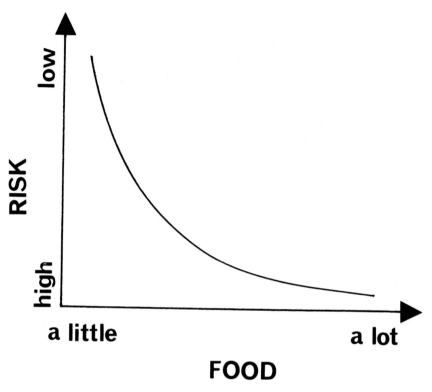

Source: Modified after McCleery[208].

with difficulty or by exposing itself to predation? This weighing-up process can be represented by an *indifference curve* (Figure 2.8). The curve in Figure 2.8 joins up combinations of danger and the amount of food which the animal is equally likely to choose. Thus, a lot of food in a very dangerous place is equivalent to only a little food in a safe place. The *utility* of choosing to feed in either place is the same. Indifference curves can be measured experimentally. Logan presented rats in a runway with a choice between food rewards of different sizes which were delayed for different times in their availability. She found that the rats were prepared to wait a long time for food as long as the amount they eventually obtained was large. The utility of choosing this option was exactly the same as choosing a small amount of food but having to wait only a short time to get it. The concept of utility is important because it reflects an animal's assessment of the costs and benefits of its actions. It is also on the utility of behaviour that natural selection might be expected to act.

2.3.4 Objective Functions and Cost Functions

When we ask how an animal's decision-making machinery operates, we are really asking how natural selection has shaped the animal to arrive at a correct solution to a behavioural problem. To answer that question we need some quantitative yardstick with which to compare the utilities of different solutions to the problem. An appropriate yardstick must be one related to the ultimate evolutionary advantage or disadvantage of indulging in different behaviours. The assessment of the relative advantages of different behaviours ought, therefore, to be made with reference to the effects of those behaviours on the individual's inclusive fitness (the reproductive potential of the individual and its close kin – see Chapter 3). Each behaviour performed can be thought of as incurring a certain probability of death (which includes not just the risk attached to the behaviour in question but also the costs of not performing any of the other activities the animal could have chosen), and a certain probability of successful reproduction. Jointly, these two factors can be lumped together and called the *cost* of a behaviour.[208] The cost of behaviour is therefore the chances of dying while performing the behaviour minus the chances of successful reproduction as a result of the behaviour.

This relationship between utility and the different options open to the animal – the cost of choosing each option under specified conditions – is referred to as an *objective function*. We therefore expect natural selection to shape the animal's decision-making process so that the choice it makes at any given time minimises the objective function for the problem in hand. Any other choice would result in a less efficient solution in evolutionary terms.

In practical terms, however, it is likely to be extremely difficult to calculate the objective function for each of a range of choices. Take a seemingly simple decision like a bird deciding to leave the eggs it is incubating to look for food. The problems involved in assessing how all the internal and external factors affecting the bird's long-term inclusive fitness should be combined together to make the decision are enormous. This does not mean that the notion of an objective function is worthless; it means we have to find a less direct way of measuring it. To do this we make an assumption. We assume that, if an animal is to minimise the overall long-term cost of its behaviour (the objective function), it must minimise the cost of each individual act it performs. That is, it should perform each behaviour in the most cost-effective way. The task of measuring the short-term costs and benefits of an individual behaviour is considerably easier. In our nesting bird, for instance, it would be feasible to measure (a) the short-term cost of leaving the eggs in terms of heat loss and slowed development or death, (b) the cost of staying too long on the nest as weight loss and (c) the net benefit of flying to a food source in terms of the distance to the source and the availability/quality of the food. The net cost of performing a particular behaviour (e.g. leaving the eggs to look for food) in a particular environment is called the *cost function* of that behaviour. If the animal is to minimise its hypothetical objective function, it should minimise its measurable cost function for each behaviour.

2.3.5 *The Relationship between Internal and External Causal Factors*

Before we can test the relationship between the cost of behaviour and its mode of performance, we need to know how the animal integrates the internal and external factors which influence its behaviour. Older models of motivation, like Lorenz's hydraulic model, made qualitative predictions about the relationship between internal and external causal factors. Can we describe quantitatively how they interact to produce a given behaviour?

One way we can do it is to examine the shape of the *motivational isoclines* for any given set of internal and external factors. Motivational isoclines are analogous to the indifference curves described earlier except, instead of joining points of equal choice utility, they join combinations of strengths of internal and external causal factors which are equally likely to produce a given behaviour. Motivational isoclines can be represented in a two-dimensional motivational space as a function of two causal factor axes (Figure 2.9). In Figure 2.9, one axis represents the strength of an internal causal factor like food deficit and the other an external causal factor such as the amount of food the animal can see. This, of course, is only one set of causal factors within motivational space and Figure 2.9 could be said to represent *causal factor space* for eating behaviour. Causal factor space is simply a sub-space of motivational space which is specific to particular causal factors.

The shape of the motivational isocline depends on the mathematical nature of the interaction between internal and external factors. One type describes a smooth curve and can be shown to result from a *multiplicative* relationship between the strengths of internal and external factors. The behaviour the animal performs is a function of the product 'strength of internal factor x strength of external factor'. Another type describes a straight line and results from an *additive* relationship. Behaviour is therefore a consequence of 'strength of internal factor + strength of external factor'. Can we deduce the mathematical relationship between internal and external factors experimentally?

Baerends *et al.*[4] investigated the courtship behaviour of the male guppy (*Lebistes reticulatus*). In their study, the most important *external* causal factor determining courtship behaviour in the male was the size of the female; males preferred large females. Baerends *et al.* categorised the male's *internal* sexual state according to various marking patterns. They drew up a calibration curve relating the male's marking pattern to the relative frequency with which he performed a number of different courtship activities and were thus able to construct a rank order of male internal states. Their experiment consisted of establishing the size of female needed to elicit a given display from males in known sexual states (as determined by their markings). Motivational isoclines were then constructed representing combinations of female size and male

Figure 2.9: Motivational Isoclines Join Combinations of Internal (e.g. Hunger) and External (e.g. Food Availability) Causal Factors Which Are Equally Likely to Elicit a Given Behaviour. The shape of the isocline is defined by the mathematical relationship between internal and external factors.

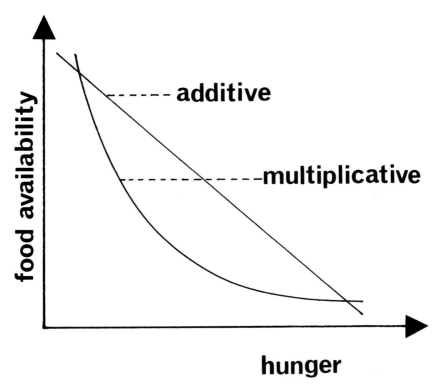

marking pattern which were equally likely to result in a given courtship display by the male (Figure 2.10a). The isoclines obtained by Baerends *et al.* described smooth curves suggesting that male sexual state and female size interacted in a multiplicative way to produce behaviour.

An example of what may be an additive relationship between internal and external causal factors is provided by Heiligenberg's[136] study of attack readiness in the cichlid fish *Pelmatochromis subocellatus*. Heiligenberg measured the effects of presenting a dummy male fish on the aggressive behaviour of a real male towards young fish sharing its tank. Rates of attack with and without dummy presentation are shown in Figure 2.10b. Lines represent the relationship between observed rates and rates expected on the basis of control counts. The slopes of the lines are the same but the line for 'with dummy' tests is shifted up. The external stimulus of the dummy is therefore added on to male's internal state to produce a higher level of aggressive behaviour. Figure 2.10c shows a similar additive relationship between the rate at which a male cricket (*Acheta*

Figure 2.10: (a) Titration Curve for Male Motivational State (Lower Figure) and Apparently Multiplicative Motivational Isoclines for Courtship in the Male Guppy. (b) Apparently Additive Isoclines for Aggression in Male Cichlids. dum, with dummy presented; con, without dummy (see text). (c) Apparently Additive Isoclines for Singing in Male Crickets. Chirp rate after stimulation with recorded cricket song increases in an apparently additive fashion relative to chirp rate immediately prior to stimulation.

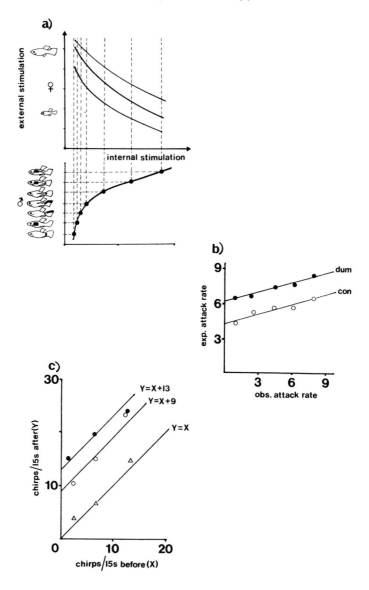

Source: (a) Modified after Baerends *et al.*[4] (b) Modified after Heiligenberg[136]. (c) Modified after Heiligenberg[137].

domesticus) chirped, before and after being stimulated by experimental chirps of different rates.[137]

Having suggested how animals should weigh up the choices open to them and how internal and external causal factors might interact to produce behaviour, let's see how theory works in practice.

2.3.6 The Control of Feeding and Drinking

Sibly and McFarland[285] attempted to quantify the choices made by an animal by investigating the choice between feeding and drinking behaviour in Barbary doves. The birds had been deprived of both food and water and were allowed to obtain them by pecking at different keys in a Skinner box. By examining the problem in the laboratory, the workers were able to control the number of causal factors likely to impinge on the doves' behaviour.

The Cost Function. Although we have just claimed that the short-term costs of a behaviour could and should be measured directly, Sibly and McFarland decided this was not practicable in their case. Instead they made an educated guess that the cost function for each of the two behaviours would be convex (Figure 2.11). This makes the reasonable assumption that an increase in food or water deficit when the deficit is already large is more costly than the same increase when the deficit is small. That is, the risk to the dove of a unit increase in deficit is much higher if it is already very hungry or very thirsty. Although the shape of this cost function is a guess, we shall see later that it can be verified indirectly.

The Performance of Feeding or Drinking. Intuitively it is reasonable to expect that, if the cost of having an internal state far removed from the optimum is high, then it is worth incurring a high cost by behaving at a high rate in order to move the internal state out of danger. As the optimum is approached, however, incurring such a cost is less worthwhile because the risk attached to a small displacement of internal state is likely to be negligible. What we should expect to see, therefore, as the animal eats food or drinks water, is a steady decrease in the rate of food or water intake as satiation is approached.

It can be shown mathematically that, if the cost function is of the form suggested by Sibly and McFarland, the rate of food or water intake should describe a negative exponential curve. Experimental evidence from Barbary doves and other species suggests that this is the case for feeding and drinking.[283,285] In the most stringent test so far of the match between the form of cost function and the rate of performance of behaviour, McCleery[208] found that alternative decelerating curves predicted by other forms of cost function did not fit as well as the negative exponential. The original assumptions about the form of the cost function are therefore supported.

The Sequencing of More Than One Behaviour. Animals are often faced with a

Figure 2.11: A Proposed Cost Function for Food or Water Deprivation in a Dove. The cost of prolonged deprivation is disproportionately higher than the cost of short deprivation.

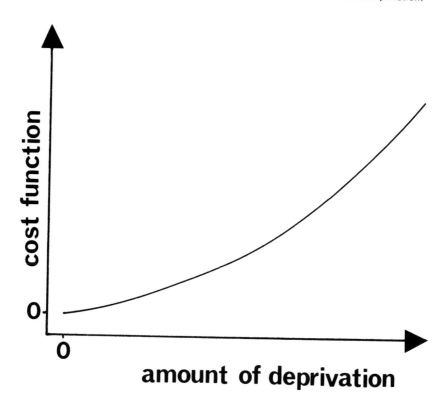

Source: Modified after Sibly[283].

choice between two or more behaviours, each of which needs to be performed in order to return a displaced internal state to the optimum. The problem is, they cannot all be performed simultaneously. In what sequence should the animal perform them in order to optimise the rate of return to its optimum internal state?

As Sibly and McFarland point out, which behaviour is chosen at any one time depends on three things: (1) the position of internal state relative to the optimum (the degree of deficit), (2) the maximum rate at which the behaviour can be performed (because this imposes a limit on the usefulness of the behaviour in reducing the deficit) and (3) the availability of the relevant commodity (e.g. food or water). Sibly and McFarland were able to show that, in doves, the interrelationship between these internal and external causal factors is multiplicative. This amounts to saying 'choose the behaviour for which the product deficit X commodity availability (called *incentive*) X maximum rate of behaviour is the greatest'. In Sibly and McFarland's dove experiment, birds are faced with the problem of reducing both food and water deficits but are only able to perform one behaviour (eating or drinking) at a time. The way in which the deficit X incentive rule would move the doves' internal state is shown in Figure 2.12a. The diagram represents motivational space with causal factor axes for increasing food and water deficits. Internal state will move towards the origin of the axes (optimum state) by first approaching the line which divides states where feeding is most advantageous from those where drinking is most advantageous (phase 1). This *switching line* should then be tracked back to the origin (phase 2). If the dove is designed optimally, we shall expect the tendency to perform a given activity to reflect the needs of the animal. States in which one behaviour is most advantageous should therefore be those in which it is most likely to be performed. States along the switching line are those in which feeding and drinking are equally advantageous and therefore equally likely to be performed. Is this predicted path of movement of internal state borne out by experimental evidence?

When a dove is working for food and water in a Skinner box, a graph of its food and water intake plotted against each other represents the changes in motivational state produced by its behaviour. At any one time the bird is said to be choosing the behaviour which is *dominant*. Experimentally, the dominant behaviour can be defined as that which is resumed after an interruption during which no feeding or drinking is possible. In the dove's case, the lights were turned off. If a bird feeds (pecks the food key) after an interruption, h (hunger) is plotted on the graph (Figure 2.12b). If it drinks (pecks the water key), t (thirst) is plotted. Changes in the motivational state of the bird are therefore inferred from its behaviour. A straight line (the *dominance boundary*) can now be drawn dividing points where feeding is dominant (h) from those where drinking is dominant (t). The assumption is that points along the dominance boundary are those where food and water deficits are equal and therefore where feeding and drinking tendency are equal. The experimentally-obtained dominance boundary should be the equivalent of the theoretically-predicted switching line. Sibly[283] tested this equivalence.

If one of the factors in the equation 'tendency to perform a behaviour = internal state X commodity availability X maximum rate of behaviour' is changed, the dominance boundary, if it is equivalent to the switching line, ought to rotate in a predictable way towards one of the deficit axes. To begin with, Sibly changed the availability of food by increasing the reward rate for pecking the food key. We would now predict that the dominance boundary will rotate towards the water deficit axis because feeding tendency ought to be greater than drinking tendency for lower values of food deficit than was the case before the change. Figure 2.12c shows that this was precisely what happened. The dominance boundary rotated by 33° towards the water deficit axis after the change in food availability. When water availability was increased, the opposite happened (Figure 2.12d).

Figure 2.12: (a) Predicted Trajectory for Returning Physiological State to the Optimum after Food and Water Deprivation. P, point of displacement; S, switching line = dominance boundary. (b) Sequences of Feeding and Drinking Following Interruptions Conform to the Predictions in (a). (Deficits in grams to satiation.) (c) When Food Availability is Increased (Change), the Dominance Boundary Rotates towards the Water Deficit Axis. (d) When Water Availability is Increased, the Boundary Rotates towards the Food Deficit Axis. (e) The Change in Slope of the Boundary Relative to a Change in Food or Water Reward Rate (Solid Line) is Consistent with the Assumption that Doves are Following the 'Incentive X Deficit' Rule in Deciding What To Do Next (Dashed Line).

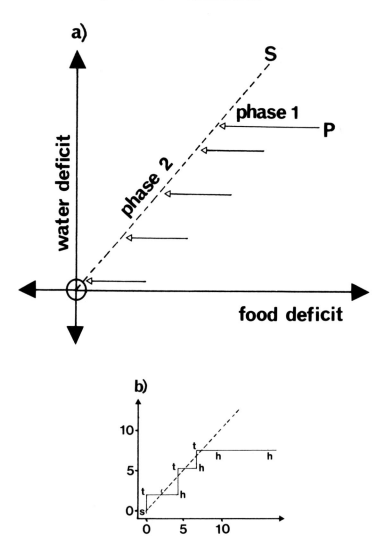

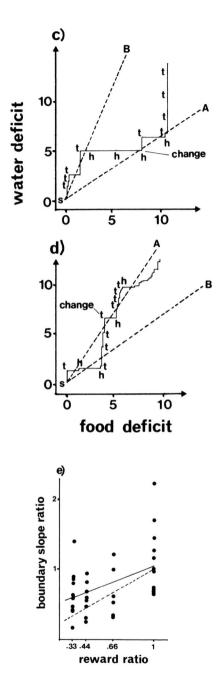

Source: (a) Modified after Sibly and McFarland[285]. (b)–(e) Modified after Sibly[283].

As a final test of the deficit X incentive rule, Sibly examined the relationship between the ratio of the dominance boundary slopes before and after reward rate changes and the ratio of the reward rates. If the rule is being followed, the two ratios should be directly proportional. Figure 2.12e shows the relationship obtained in Sibly's experiments plotted with the relationship expected if the deficit X incentive rule was being followed. The slopes of the solid and dashed lines do not differ significantly so that the birds appeared to behave as predicted.

2.3.7 *The Cost of Changing Behaviour*

When an animal changes from one behaviour to another, like the doves in the above example switching from feeding to drinking, there may be a cost involved. The act of changing to another behaviour may itself cost time and energy. A bird eating seeds in a dry field will build up a considerable thirst but may have to fly half a kilometre to find water. In addition to the energetic cost of the journey, the bird will lose time in the transition from feeding to drinking. It will not be reducing its food and water deficits during the transition and it may be incurring other risks, like predation. If changing behaviour is costly, we might expect the cost to affect an animal's decision to switch between behaviours. The question is, to which behaviour does the animal attach the cost of changing – the behaviour it is changing *from* or the behaviour it is changing *to*? Larkin and McFarland[181] tried to find out.

Using Barbary doves, Larkin and McFarland observed birds switching between feeding and drinking in Skinner boxes and in an experimental room. The cost of switching from feeding to drinking and vice versa was varied by making the birds jump up a step or negotiate a partition in order to change. To quantify the birds' allocation of switching cost, Larkin and McFarland used a similar technique to Sibly. They examined the direction and degree of rotation of the dominance boundary for switches between feeding and drinking when the cost of switching was varied. They predicted that the cost would be incorporated into the behaviour *to* which the bird was changing. If a bird switched from feeding to drinking and water was placed on top of a high step so that the bird had to spend a lot of energy before it could drink, the boundary should rotate towards the food deficit axis (see Figure 2.12d). The effect of adding the switching cost to drinking behaviour should be similar to that of reducing the availability of water. Conversely, for switches in the other direction when changing to feeding was expensive, boundary rotation should be towards the water deficit axis. These predictions were largely borne out.

2.4 Interactions between Motivational Systems

So far we have seen how we can model motivational state in order to make predictions about what sort of behaviour should be performed and how it should be performed. We have also seen how an animal copes when it is motivated to behave in several different ways at the same time. In this section we shall examine interactions between motivational states more closely.

2.4.1 Motivational Systems and Priorities

When we think of the motivational control of different behaviours, it is convenient to imagine *motivational systems* which control functionally related behaviours. The animal's total behavioural repertoire is composed of distinct behaviour categories, each of which is controlled by one of a number of motivational systems. Thus we can envisage a 'feeding' system, a 'drinking' system, a 'sexual' system and so on.

At any particular time, many motivational systems will be active (the animal will be hungry, thirsty, sexually aroused, etc., simultaneously) and must compete for access to the *behavioural final common path* (BFCP). BFCP is a term coined by Tinbergen to describe the last link in his chain of motivational centres (Figure 2.1b) which must be overcome before one or other behaviour can be performed. The term is still useful for expressing the competitive nature of motivational interactions. However, at any one time, the level of causal factors (internal, external or both) for some systems will be greater than that for others. These systems will therefore win access to the BFCP and the animal can be said to have *motivational priorities*. Feeding might be top, grooming next and sleeping third, etc. Systems which have top priority at any given time can be thought of as *dominant* systems, and those which do not, as *subordinate* systems. In general, dominant systems *inhibit* subordinate systems and behaviours appropriate to the dominant system are expressed. If animals do possess such motivational priorities, we should be able to find some evidence of them in the organisation and sequencing of behaviour.

Motivational Systems in Incubating Gulls. On the basis of an enormous amount of field data, Baerends[3] has constructed a model of the control of incubation in the herring gull (*Larus argentatus*). His model (Figure 2.13) shows four motivational systems: a preening system, a predator escape system, a resting system and an incubation system. Each system has under its control a number of functionally related behaviours. The preening system thus controls behaviours related to feather care, like combing and stroking, the escape system controls various locomotory actions and so on. Each system is also responsive to a variety of external stimuli (external causal factors for their respective behaviours) and has its own endogenous characteristics that determine its response to those stimuli.

As well as organising the gulls' behaviour into interrelating but self-contained systems, Baerends postulated a variety of excitatory and inhibitory relations between each system and (a) external stimuli, (b) various behavioural outputs and (c) all the other systems. There is one major system and three minor systems. The nest system controls nest-building, nest-maintenance and incubatory

Figure 2.13: Proposed Motivational Systems Interacting in the Control of Incubation in the Herring Gull. Solid lines, excitatory relationships; dashed lines, inhibitory relationships.

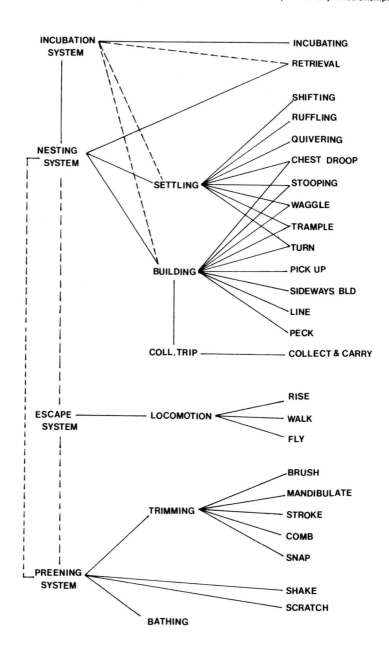

Source: Modified after Baerends[3].

behaviours and, at the same time, inhibits the escape and preening systems. However, if an alarm stimulus like an approaching predator should arise in the environment, the escape system is activated. Because of the inhibitory relationship between the escape and incubation systems (via the nesting system), a strong input to the escape system can block nest-and-egg orientated behaviour and allow the bird to flee. In the absence of alarm stimuli, the nesting (and hence incubation) system inhibits the escape system, just as it inhibits all the other activities which conflict with incubation (settling, building, trimming, etc.). As a result of these excitatory/inhibitory relationships between motivational systems, the bird can spend long, uninterrupted periods keeping the eggs at the optimum temperature. Also, because of the reciprocal inhibitory relationships between the different systems, the bird can usually avoid awkward motivational conflicts (see Section 2.4.2).

There are also variations in the degree of inhibition imposed by one system on another. A female who has laid five eggs and has been off the nest some time to find food is highly motivated to incubate. Inhibition of the escape and preening systems by the nesting and incubation systems is therefore strong. On the other hand, a female with only one egg who has been on the nest for several hours and may need food is only weakly motivated to incubate. Inhibition is thus correspondingly weak. However, even a highly motivated female will leave the nest if a predator ventures near. It is not unreasonable to suppose that the vital escape system has evolved in such a way that it can block any other system should an emergency arise. Less urgent behaviours, like preening, are allocated to subordinate systems which are expressed as and when the opportunity arises. In other words, the gulls have motivational priorities.

2.4.2 Conflict Behaviour

It is conceivable that two or more motivational systems might become equally aroused. What happens then? In the absence of any resolution involving the conflicting systems themselves, behaviour appropriate to a completely different system may be performed. Using an example from Baerend's model, a gull may be equally stimulated to incubate and to flee (it has only just returned to a large clutch of eggs, but it can see a predator approaching on the horizon) but, instead of doing either, it preens. The two conflicting motivational systems appear to cancel each other out and 'allow' the performance of a behaviour appropriate to another motivational system. Such behaviour is known as *conflict behaviour*.

Conflict behaviour has been important in developing an understanding of motivational systems and can be divided into a number of classes:

Displacement Activities. Displacement activity is a general term for seemingly irrelevant behaviours which are performed when two or more motivational systems conflict. In many aggressive or sexual contexts, passerine birds may wipe their beaks and often preen or perform drinking and feeding actions. Clearly none of these is functionally related to fighting or courting. We must, however, recognise that 'irrelevant' is a relative term and may sometimes be misleading. In cichlid mouth-brooders, for example, many motor patterns associated with fighting are not only similar to those used in feeding, they are motivationally related. Thus stimuli eliciting fighting also elicit feeding and vice versa. Activities occurring apparently out of context may not be causally irrelevant at all. However, such coincidences seem to be rare and can only account for a small fraction of observed displacement activities, so what about the rest?

In some cases displacement activities appear to stem from autonomic responses. The autonomic nervous system brings into play several physiological responses to alarm stimuli and controls homeostatic reflexes, like panting and sweating. These sometimes occur apparently out of context during other activities. A good example is the cooling behaviour shown by male buntings (Emberizidae) during sexual chases.

As part of courtship, the male chases rapidly after the female. During the chase the male and female undergo similar degrees of muscular exertion but, after it is over, only the male shows cooling responses (panting, feather-sleeking and wing-raising). Also if, instead of courting, the male attacks the female, he again performs cooling activities. The interesting point is that, during both types of interaction between the sexes, the male is in a conflict between approaching the female and retreating. Instead of doing either, he shows cooling responses. Andrew suggests that the occurrence of such responses during conflict may be due to changes in the peripheral (skin) circulation as a result of autonomic nervous activity. These result in a warming or cooling of the surface temperature receptors.

Another hypothesis, and one we shall discuss in more detail, is that, when mutual incompatability prevents the appearance of behaviours which would otherwise have highest priority, patterns which would have been suppressed are 'permitted' to appear. Such patterns are said to appear by *disinhibition*. A nicely controlled example of disinhibited displacement preening in terns (*Sterna* spp.) is given by van Iersel and Bol[151].

Van Iersel and Bol showed that incubation behaviour in terns inhibited preening (as in Baerends' model of incubation in gulls). Preening during incubation only occurred when the terns' motivation to incubate was low. However, preening also occurred as a displacement activity when birds were fighting or showing escape responses. Van Iersel and Bol assessed the strengths of various conflicting motivational tendencies in the terns and found that dis-placement preening was most likely when they had a certain relationship to each other.

For example, when a bird returns to incubate after being alarmed, the distance at which it lands from the nest can be regarded as a function of both its moti-vation to incubate and its motivation to flee. The latter can be divided into qualitative 'grades' which van Iersel and Bol labelled OAP, ½ AP, thick 1/1 AP and thin 1/1 AP (their precise meaning is not important here). When birds in van Iersel and Bol's study were only weakly alarmed, displacement preening was

Figure 2.14: The Relationship between Landing Distance from the Nest and Degree of Alarm in Producing Displacement Preening in Terns. (a) Thick 1/1 AP (medium/strong alarm), (b) OAP (weak alarm), (c) ½ AP (medium/weak alarm), (d) thin 1/1 AP (strong alarm) (see text).

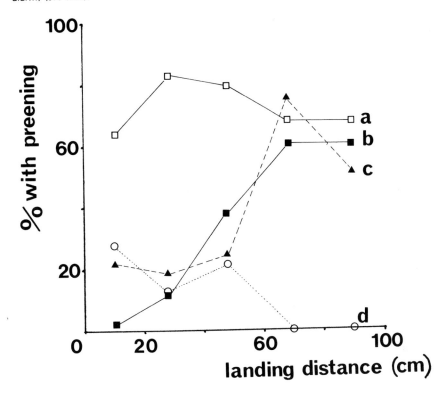

Source: Modified after van Iersel and Bol[151].

most likely when the bird landed far from the nest. This indicated that it was only weakly motivated to incubate (OAP and ½ AP lines in Figure 2.14). Strongly alarmed birds, however, were most likely to preen when they landed near the nest, indicating they were strongly motivated to incubate (thin 1/1 AP line). With an intermediate degree of alarm, preening was most likely at an intermediate distance, when birds were moderately motivated to incubate (thick 1/1 AP line). Displacement preening was therefore most likely when the conflicting systems for incubation and escape were both strongly, both weakly or both intermediately aroused. Under these three conditions they were mutually inhibited and their inhibitory effect on the preening system was removed.

Redirected Behaviour. Sometimes motor patterns appropriate to one of the conflicting motivational systems are shown but are directed to an irrelevant object. A male herring gull confronted on the boundary of its territory by an aggressive neighbour may suddenly tear at the grass. Its aggressive behaviour is thus redirected from the object by which it was elicited. In a similar way, stone-chats (*Saxicola torquata*) observed attacking a stuffed cuckoo's (*Cuculus canorus*) head frequently pecked at conspecifics instead of the head.[1]

Again there may be problems with the term 'irrelevance' when referring to the object to which behaviour is redirected. For instance, jungle fowl (*Gallus gallus*) show similar redirected pecking to the herring gull. However, redirected 'aggressive' ground pecks in the fowl sometimes result in food being taken so that feeding and redirected aggression occur hand-in-hand. Moreover, when a bird becomes aggressive to a coloured stick, it pecks more at food particles which are similar in colour to the stick than at particles which are different.

Intention Movements. It may be that inhibition of one motivational system by another is incomplete so that behaviour appropriate to the inhibited system is not suppressed altogether but is merely reduced in intensity or frequency or occurs in an incomplete form. In some cases this may consist of the initial phases of some movement or sequence of movements. Such truncated movement patterns are then referred to as *intention movements*. For example, the take-off leap of a bird consists of two phases. First, the bird crouches and withdraws the head and tail, then it reverses these movements as it springs off. During conflict, the first phase may be repeated several times before the bird finally becomes airborne.

If the bird is in motivational conflict, such intention movements may be repeated for a considerable time. Gulls and gannets (*Sula bassanus*) which are torn between remaining with their clutch of eggs and flying off to avoid predation or find food, show repeated 'wing-lifting'. Wing-lifting is normally a prelude to flight, but here it is repeated several times without the bird taking off.

Ambivalence. Sometimes intention movements appropriate to two conflicting motivational systems are combined into a single motor pattern containing elements of both. A half-tame moorhen (*Gallinula chloropus*) which is offered food may make abortive pecks towards it and even perform swallowing movements. At the same time it keeps its distance from the proffered food and may even edge away. Thus components of both feeding and fleeing are shown simultaneously but in an incipient form.

Conflict behaviours are interesting because they indicate the nature of inter-actions between different motivational systems within the animal and suggest how behaviours can be suppressed or released. One functional explanation for conflict behaviours, which was first put forward by ethologists in the 1950s and which may still explain some aspects of conflict, is that the arrangement between motivational systems permits the resolution of conflict through conflict behaviours rather than resulting in a paralysed indecision. In Section 2.4.3, we shall discuss another explanation.

2.4.3 Motivational Competition and Time-sharing

When we ask 'what factors determine which motivational system is to have priority and gain overt expression?', the initial answer seems to be 'that system with the highest level of causal factors gains priority by virtue of *competition* with other systems'. Other systems are then said to be subject to behavioural inhibition. Taken at face value, this suggests that causal factors for a particular behaviour may build up to a level sufficient to oust outgoing behaviour. If the continued performance of the behaviour produces consequences that feed back negatively on the system controlling it, the level of its causal factors may be reduced to a point where those relevant to another behaviour become strong enough (relatively) to take over. The fact that animals rarely indulge in one activity for long without interrupting it to perform some other activity has long been interpreted by ethologists and psychologists as the product of motivational competition. McFarland,[211] however, has defined the concept of motivational competition more precisely: 'a change in behaviour due to competition can in practice be recognised when a change in the level of causal factors for a second-in-priority behaviour results in an alteration in the temporal position of the occurrence of that activity.' That is, a change from behaviour A (top priority) to behaviour B should occur earlier when the strength of the causal factors for B is high than when it is low. Competition is thus characterised by the causal factors for a particular behaviour being ultimately responsible for the removal of its own inhibition.

Alternatively, that inhibition may be removed for other reasons so that the strength of the causal factors for the released behaviour plays no role in its removal. In this case the behaviour is said to occur by disinhibition. As McFarland puts it: 'the time of occurrence of a disinhibited activity is independent of the level of causal factors relevant to that activity.'

A simple experiment illustrates the distinction between competition and disinhibition. McFarland deprived Barbary doves of food and allowed them to work for food in a Skinner box. After 5–10 minutes, feeding was interrupted by preening and then resumed. To see whether the timing of preening was determined by competition or disinhibition, the experiment was repeated under the same conditions except that birds had a paper clip fastened to the primary feathers of each wing. The strength of causal factors for preening was thus increased. The results were precisely the same as for the previous experiment except that the amount and vigour of preening was enhanced. Increasing the strength of causal factors for preening therefore changed the *intensity* of preening but not its *time of occurrence*. Thus, by McFarland's definition, preening occurred by disinhibition and not by increasing its motivational status relative to feeding.

The Control of Second-in-priority Behaviour. The fact that we can show a particular change in behaviour to be due to disinhibition, does not mean that the disinhibited behaviour remains under the control of the disinhibiting system.

Figure 2.15: Different Types of Behaviour Sequence Based on Inhibition and Disinhibition. The top line, solid (behaviour A) or dashed (behaviour B) of each pair, represents the behaviour currently being expressed.

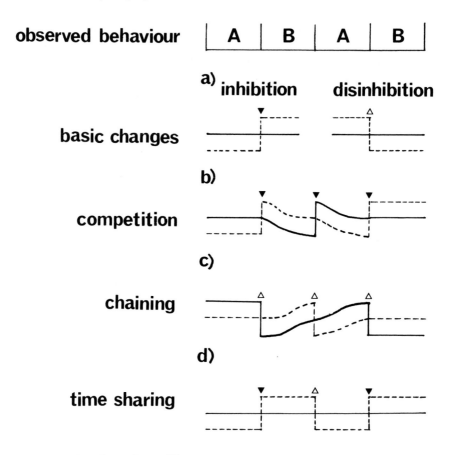

Source: Modified after McFarland[211].

When a high priority (*dominant*) system permits the occurrence of a behaviour relevant to another system, it may lose control of the BFCP and cease to be dominant. On the other hand, it may remain in control and re-establish itself after a particular period of time and a series of alternating behaviour patterns occurs. Three possible explanations of simple, alternating behaviour sequences can be put forward which depend for their operation on the two types of behavioural transition, inhibition and disinhibition (Figure 2.15a).

Competition (Figure 2.15b) implies that every transition is of the first type so that the behaviour sequence comprises a string of successive inhibitions. Miller's analysis of approach/retreat conflict is a good example. Here overt behaviour is determined by the motivational system with the strongest causal factors at the time.

An alternative to competition is the type of sequence known as *chaining* (Figure 2.15c). Here behavioural transitions result from successive disinhibitions. The current behaviour self-terminates and disinhibits a second behaviour. Each disinhibition might occur in response to either internal stimuli (e.g. satiation) or external stimuli (e.g. the disappearance of a predator). A possible example of chaining is the courtship sequence of the three-spined stickleback. Here the sequence of behaviour depends on a succession of stimuli which each fish presents to the other. Each behaviour performed by a fish is really dependent on the consequences of its previous behaviour rather than on its own causal factors.

A third and more interesting possibility (Figure 2.15d) is that a dominant behaviour which occurs by the inhibition of previous behaviour is self-terminating, thus disinhibiting another behaviour, but re-establishes itself by reinhibition after a certain period of time. In this case, the dominant behaviour not only determines the time of occurrence of the disinhibited behaviour but also its duration. McFarland calls this phenomenon *time-sharing* because, in permitting alternative behaviour, the dominant motivational system shares the BFCP with other motivational systems. In cases of time-sharing it is useful to distinguish the controlling dominant behaviour from *subdominant* behaviour. Subdominant behaviour is that which appears 'by permission' of the dominant system. Time-sharing is therefore a rather different inhibition/disinhibition process to simple competition and chaining. How important is it in determining sequences of behaviour?

Evidence for Time-sharing. A crucial feature of behaviour sequences based on time-sharing is that the opportunity and degree of motivation for performing a subdominant behaviour make no different to the pattern of performance of the dominant behaviour. The time of occurrence and duration of the dominant behaviour is under the control of its own causal factors and opportunity for performance. This would not be true, for example, where the disinhibited subdominant behaviour was self-terminating and its causal factors likely to influence its duration and thus disrupt the pattern of disinhibiting dominant behaviour.

McFarland and Lloyd[212] found that, in hungry and thirsty Barbary doves in which hunger was known to be dominant to thirst, the rate at which the birds became food-satiated in a Skinner box was not affected by the simultaneous availability of water. Similarly, when rats were deprived of food and water and then allowed to feed and drink in a Skinner box, their feeding satiation curves were similar regardless of whether water was also available. In both these cases, drinking appeared to be permitted on a time-sharing basis by the animals' feeding system. However, this kind of evidence is very circumstantial.

'Masking' Experiments. A much more rigorous way of testing for time-sharing is to use so-called 'masking' experiments. These experiments depend on the assumption that dominant behaviour controls the timing and duration of sub-dominant behaviour but that subdominant behaviour has no such influence over the performance of dominant behaviour. The procedure is very simple. If an animal is interrupted while performing dominant behaviour, the dominant behaviour should be resumed after the interruption. If it is interrupted while performing subdominant behaviour, the subdominant behaviour may or may not be resumed depending on the duration of the interruption. If the interruption is very brief, some of the subdominant behaviour time 'allocation' may remain and it can continue to be performed. If the interruption is long, it may use up the time available to the subdominant behaviour so that, when the interruption ceases, the animal reverts to dominant behaviour.

Experimentally, all we need to do is detect when an animal is changing from behaviour A (dominant) to behaviour B (subdominant) and block both A and B by a suitable interruption. If the interruption is short relative to the time available for B, the animal should perform B after interruption. If it is long, it should perform A. McFarland tested this prediction using the courtship sequence of the male stickleback.

During courtship, the male repeatedly approaches the female and performs a 'zig-zag' dance. After zig-zagging for a while, the male swims to the nest it has previously constructed, fans the nest with its fins and 'creeps through'. It then swims back to the female and resumes zig-zagging. The question we are interested in is whether the nest- and female-orientated behaviours are time-sharing. Is nest-orientated behaviour, for example, really under the control of female-orientated behaviour? McFarland set up a tank with three compartments, each with a trapdoor which could be operated by remote control. A male stickleback was allowed to build a nest in one of the end compartments of the tank and a female was placed in a jar in the other. During courtship, therefore, the male repeatedly swam between the two ends of the tank as it changed from nest- to female-orientated behaviour.

In the first part of the experiment, McFarland simply recorded the amount of time the male spent in each of the nest, middle and female compartments. In the second part, the male was trapped for a certain period (1, 5, 10, 15, 20 or 30 seconds) in the middle compartment as it swam between the two ends of the tank. Trapping occurred during both the transitions 'female to nest' and 'nest to female'. At the end of each trapping period, the trapdoors were opened and a note made of which end compartment (nest or female) the male entered. McFarland found that when there were no interruptions, the male distributed its time similarly between the nest and female compartments (Figure 2.16a, b). For example, 50–60 per cent of the time intervals spent in each compartment were shorter than 15 seconds. Figure 2.16c shows the percentage number of times the male returned to the female compartment during a 'female to nest' transition plotted against the length of the trapping period. The male was likely

Figure 2.16: The Results of a Masking Experiment for Time-shared Courting and Nest Maintenance in the Male Three-spined Stickleback (see text).

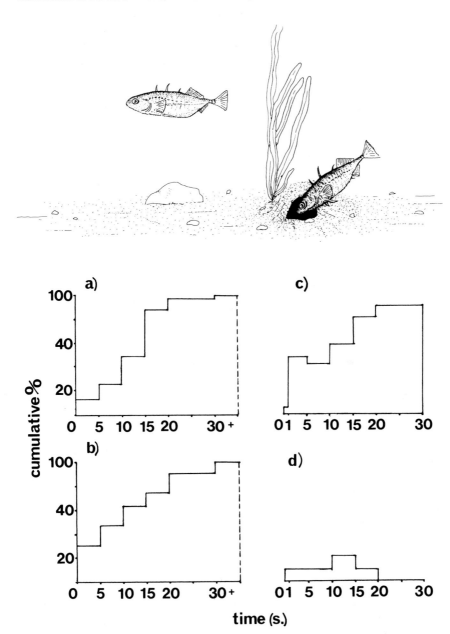

Source: Modified from McFarland[211].

to proceed to the nest compartment if the period was short (50 per cent of the time it was only 5–10 seconds) but would return to the female compartment if it was long (over 90 per cent of the time if it was longer than 20 seconds).

When the fish was trapped during a 'nest to female' transition, it entered the female compartment whatever the duration of trapping (Figure 2.16d). These results are exactly what we would expect if courtship was the dominant behaviour and nest attention was permitted on a time-sharing basis.

As a corollary, Cohen and McFarland[61] investigated the effects of increasing the strength of the causal factors for nest-orientated behaviour. They dropped snails and other debris onto the male's nest during courtship to see whether they could affect the sequence of courtship and nest maintenance. They found that dropping debris only altered the timing of nest-orientated behaviour when nest maintenance was dominant. When courtship was dominant, the male ignored increases in the causal factors for nest maintenance. Again, this is what we would expect if courtship and nest maintenance were time-sharing.

The Adaptive Value of Time-sharing. What is the 'function' of time-sharing? What advantages does time-sharing have over other ways of organising behavioural priorities? Some hint may be found by considering the two types of situation in which time-sharing has been found.

Displacement Activities. The idea of time-sharing may help to explain the performance of seemingly irrelevant displacement activities during motivational conflict (see Section 2.4.2).

Displacement activities may be a kind of 'insurance policy' against hard times. Thus displacement drinking by feeding rats and doves may allow the animal to drink during a meal and anticipate the thirst-inducing consequences of food intake. The animal could thus achieve more accurate regulation and reduce the risk of having to search for water when its water deficit was already high. Time-sharing of feeding and drinking is one way in which such regulation could be brought about.

There is also strong evidence that displacement activities may influence an animal's motivational state and help to resolve conflict. Feekes found that ground-pecking in jungle fowl decreased alarm and increased aggressive behaviour thus helping to resolve approach/avoidance conflict. Similarly Wilz suggested that the performance of displacement activities at the nest in courting male sticklebacks enhanced subsequent sexual behaviour. Built-in displacement activities on a time-sharing basis may therefore increase behavioural efficiency.

Alternating Activities. Of the three main ways in which alternating sequences of behaviour can arise, competition and chaining have no special characteristics other than depending on the processes of inhibition and disinhibition. Time-sharing, however, appears to have the additional component of temporal organisation, thus implying some kind of underlying control mechanism. Time-sharing may be a means of ensuring that multiple requirements can be satisfied in the minimum time.

In stickleback courtship, for example, McFarland suggests that the success of courtship for the male depends not only on inducing the female to lay eggs in his nest, but also on maintaining the nest in good condition. There is no point in leading a gravid female to the nest if the nest is clogged with debris. Wilz found that males used the *dorsal pricking* display to halt the female in the water while he went to inspect the nest. Furthermore, the pause in courtship also appeared to facilitate a change in the male's motivational state so that he switched from predominantly aggressive to predominantly sexual behaviours. Time-sharing may thus permit necessary maintenance activities and provide time for switches in motivational state.

Summary

(1) 'Motivation' describes an internal variable within an animal which alters the relationship between an environmental stimulus and the behavioural response it elicits.
(2) It proved useful, historically, to think in terms of a motivational drive controlling each goal-directed behaviour.
(3) Despite its economy, the drive concept has many limitations which can be overcome by substituting the concept of motivational state. Motivation can then be viewed as a behavioural analogue of physiological regulation which tends to maintain the animal's internal state at the optimum for its environment.
(4) The motivational control of behaviour can be viewed as having been shaped by natural selection so that animals optimise the mode of performance of each behaviour and the sequencing of different behaviours.
(5) It is useful to think in terms of motivational systems which control functionally related groups of behaviours. These systems interact so as to provide the animal with a flexible hierarchy of motivational priorities. In this way, urgently required behaviours can override otherwise conflicting activities. Conflict between systems may result in the performance of seemingly irrelevant behaviours.
(6) Interactions between motivational systems can be classified on the basis of behavioural inhibition and disinhibition. Two systems may time-share in that one (dominant) permits and controls the expression of another (subdominant). Time-sharing may provide a functional explanation for certain types of 'conflict behaviour' and alternating behaviour sequences.

3 FROM GENES TO BEHAVIOUR

In the previous two chapters, we have seen that behaviour is the result of complex interactions between the physiological and anatomical characteristics of an animal. Together these determine how the animal perceives and integrates environmental information and the range of behaviours with which it is able to respond. Since we know that physiological and anatomical features are coded in the animal's genes and that their phenotypic form is the result of a complex interaction between genotype and environment, it is reasonable to ask whether behaviour is similarly coded. Are animals pre-programmed to behave in the way they do, or is behaviour largely the result of experience and environmental influences? We shall leave a full discussion of the genes versus environment, or 'nature/nurture' argument until the next chapter. Let us first of all explore the evidence for a genetic basis to behaviour.

That behaviour is, at least in part, coded for genetically seems certain from various lines of investigation. Perhaps the most telling are those in which different genetic strains of a species are kept under identical environmental conditions and where individuals exhibiting a particular behavioural trait are selectively interbred to accentuate the trait.

3.1 The Relationship Between Genes and Behaviour

3.1.1 Strain Differences

Since an animal's phenotype, behavioural, physiological or otherwise, is a combined product of its genotype and the environment in which it lives, holding environmental factors constant should point to the genetically determined parts of the phenotype. To maximise genotypic differences between study animals, workers have usually resorted to using inbred *strains*. Individuals of lines which are repeatedly inbred come to be almost identical genetically and homozygous at each locus. In this way, genetic variability is minimised within strains while being maintained or enhanced between strains. In mice, inbred strains are usually produced over some 20 generations of sibling matings. There is therefore no directional selection for one particular behaviour which could exaggerate differences between strains. Such inbreeding techniques have revealed a large number of behaviours which appear to have a genetic basis. Male sexual behaviour is a good example.

Comparing inbred strains C57BL/6J and DBA/2J of mice, McGill found significant differences in 14 aspects of the male mating sequence, including the duration of mounting and intromission. C57BL, C_3H and BALB/c strains show marked differences in exploratory and 'emotional' behaviour when released into a large arena ('open field') and prefer different ambient temperatures.[261,268]

Aggression, nest-building and learning are other aspects of behaviour which differ significantly between strains of mice.

Similar studies have been carried out with birds. White and brown strains of domestic fowl differ in the amount of time they spend pacing about and incubating eggs. White birds pace more and are slower to resume incubation after pacing.[338] Behavioural differences are also apparent between strains of invertebrate species. Two strains of the mosquito *Aedes atropalus* differ in the way protein is supplied to the egg as it matures. The strain GP is homozygous for autogeny. The female mosquito does not need to feed on protein-rich food like blood for her eggs to mature. The strain TEX is homozygous for anautogeny. Here the eggs do need an external source of protein to mature. Correlated with this, GP and TEX adults show pronounced differences in mating behaviour. In particular, GP adults mate much sooner after eclosion (emergence from the pupa) than those of TEX.[124]

An intriguing difference in behaviour between naturally occurring races of honey bee was pointed out by Boch. Boch studied six races of bee: *carnica* (Carniolan), *mellifica* (German), *intermissa* (Punic), *caucasia* (Caucasian), *ligustica* (Italy) and *fasciata* (Egypt). He found that all races except *carnica* performed a so-called *sickle* foraging dance instead of the classic waggle dance (see Section 10.1 and Figure 10.1 for details). The sickle dance was so named because of the sickle-shaped path traced by the returning forager on the face of the comb. Boch also found that the transition from the round dance to the waggle or sickle dance occurred at different distances in different races. There were also differences between races in the distance of food sources indicated by the two types of dance.

Despite the many examples of clear-cut behavioural differences between strains, there is still the possibility that some subtle and undetected environmental factor is contributing towards the differences. One way to eradicate this is to *cross-foster* animals of different strains. Southwick[295] cross-fostered pups of mouse strains A/J and CFW which differed in aspects of their aggressive behaviour. A/J pups reared by CFW mothers showed a small increase in aggression over their normal level, but CFW pups reared by A/J mothers showed no difference at all. In these cases, the differences in aggression seem to be mainly genetic. Tightening control over environmental factors even further, some workers have extended cross-fostering to the level of ovarian transplants. In mice, *in utero* factors are known to have an important influence over certain physical traits like body weight. However, ovarian transplant experiments with inbred strains differing in their 'open field' behaviour, have shown that at least some aspects of behaviour are independent of the environment, even at prenatal stages.[88]

3.1.2 Selection Effects

While 'strain difference' experiments generally involve reproduction between siblings and do not deliberately set out to select for specific behaviours, *selection* experiments do just the opposite. Here individuals are chosen because

Figure 3.1: The Result of Selective Breeding for Maze-learning Ability in Rats over Seven Generations.

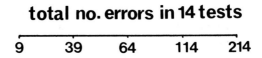

total no. errors in 14 tests

9   39   64   114   214

parents

generation 1

'bright' rats

'dull' rats

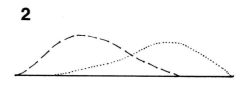

2

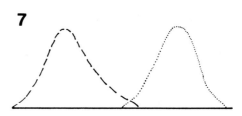

7

Source: Modified after Wilson, E.O. *et al.* (1978). *Life on Earth*, 2nd edn. Sunderland, Mass., Sinauer.

of their behavioural phenotypes. The usual procedure is to take a genetically heterogeneous population and test each individual for a particular behavioural character, say mating speed. High-scoring males are then selectively mated with high-scoring females and vice versa. Over a number of generations, therefore, two distinct behavioural lines are created. These are then compared with randomly-mated animals to see whether there is any significant divergence over generations between the high score and low score lines. One of the best known series of selection studies is that begun by Tolman[308] on maze-learning ability in rats. Here rats were selectively bred on the basis of their competence in negotiating mazes. The indices of maze learning used were the time taken to get through the maze, the number of errors made and the number of perfect runs. Although Tolman's initial results were suggestive rather than convincing, subsequent, more detailed, experiments by other workers[139,318] have implied a strong genetic basis to maze learning. Figure 3.1 summarises Tryon's results for the number of errors made by his 'bright' selected and 'dull' selected rats over seven generations.

A genetic basis to mating behaviour in *Drosophila* has been found by similar selective breeding experiments. Manning[198] selected for 'slow' and 'fast' mating speeds in *D. melanogaster* by isolating and mating the ten fastest and ten slowest pairs from a mating chamber of 50 pairs. After 25 generations, mean mating speed for pairs of the 'fast' line was 3 minutes and that for pairs of the 'slow' line, 80 minutes. In these experiments, both sexes were similarly affected by the directional selection. Thus in 'fast' lines, both males and females mated more rapidly than randomly-mated individuals (determined by mating both sexes with randomly-mated flies) and in 'slow' lines, both took significantly longer. Experiments with *D. simulans* also showed a marked divergence between 'fast' and 'slow' selected lines. Here, however, selection acted predominantly on the females. Male behaviour was not affected. Similar evidence for a genetic component in behaviour has come from a wide variety of species and behaviours including phototaxis, geotaxis and sexual responsiveness in various insects, responses during imprinting in birds and 'open field' behaviour and aggression in rodents.

3.1.3 How Many Genes?

Strain difference and selective breeding experiments indicate fairly clearly that an animal's genotype can have a lot to do with the way it behaves. The next thing we need to know is how many of the animal's genes are involved in each behaviour. In particular, can behavioural differences be explained by changes in a single gene?

Single Genes. Evidence for single gene effects on behaviour comes from a variety of sources. In *Drosophila melanogaster*, *cinnabar* and *vermilion* are mutations of single genes which affect eye colour. Both mutations produce vivid red eyes which contrast with the dull red of the wild type. During mating, *vermilion* males achieve a much lower fertilisation rate (54–61 per cent successful fertilisations) than *cinnabar* (74–80 per cent) or wild type males.[32] A similar reduction in mating success occurs in single-gene *yellow* mutants. *Yellow* is a sex-linked, recessive mutant which results in male flies possessing a yellow, instead of the normal grey, body. Bastock[18] compared the mating performance of *yellow* and wild type males. Before doing so, however, she crossed her wild and *yellow* stocks for seven generations so that the wild-type flies were genetically similar to *yellow* except for the *yellow* locus. Despite this, *yellow* males were less successful at mating with grey-bodied wild-type females than were wild-type males. The reason appears to be that the mutation causes *yellow* males to perform much shorter bouts of courtship 'licking' and wing vibration. In conjunction with visual stimuli, vibration and licking are important criteria on which females base their choice of male.[245] Males deficient or otherwise abnormal in these characteristics therefore tend to have low mating success.

Single-gene mutations have also been shown to affect mating success in male mice. Albinism in mice and certain other species (including, in some cases, man), is produced by a single autosomal recessive gene. Levine[188] compared the mating success of male *albino* mice with that of black *agouti* males. Both *albino* and *agouti* animals were drawn from inbred strains. Ten pairs of *albino* and *agouti* males were set up and each pair was placed with a single *albino* female. Matings from these mixtures produced three types of litter: all *albino*, all *agouti* and a mixture of the two. The mixed litter resulted from double inseminations. The important finding was that 76 per cent of the litters were sired solely by *albino* males, compared with 12 per cent by *agouti* males and 12 per cent by a mixture of both. Mixed litters also contained more than twice as many *albino* pups as *agouti*. Although the possibility that *albino* male sperm are favoured in an *albino* female tract cannot be dismissed, *albino* males appeared to achieve their high success through superiority in aggressive disputes. Visual preference by females for white males can be ruled out because both the females and the two strains of male used were homozygous for a gene causing retinal degeneration. All the mice were therefore blind.

Chromosome Effects. In some cases, variations in behaviour can be traced to chromosomal abnormalities. There are two basic types of abnormality, those involving unbroken chromosomes and those involving chromosome breakages and the subsequent loss or rearrangement of genetic material. The first type may involve a change in the entire set of chromosomes (*euploidy*) or the addition or subtraction of individual chromosomes (*aneuploidy*). There are four possible ways in which the second type can occur (Figure 3.2). A number of species show variation in behaviour which appears to be based on chromosomal alterations.

Natural populations of many dipteran fly species, for instance, are polymorphic for two or more inversions. (Polymorphism describes the maintenance of genetic alternatives in a population in frequencies greater than could be maintained by recurrent mutation alone.) In the South American fruit fly *Drosophila pavani*,

Figure 3.2: Changes in Chromosome Structure Which Alter the Sequence of Genes. Deletion involves the loss of all or part of a chromosome, duplication, the repetition of all or part of a chromosome, inversion, the reversal of part of a chromosome through breakage and recombination, and translocation, the swopping of material between chromosomes.

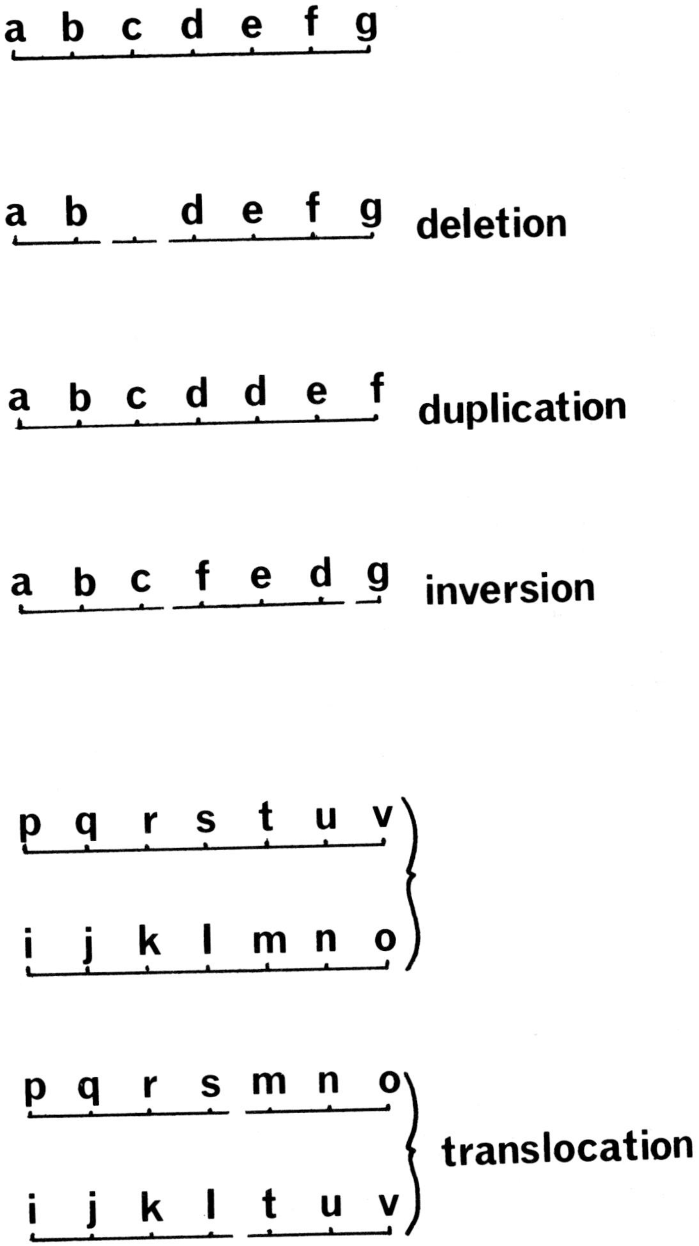

chromosomal inversions appear to affect mating behaviour. Brncic and Koref-Santibanez[36] assessed the mating ability of both heterokaryotypic (one chromosome of the affected homologous pair carries an inversion), and homokaryotypic (both chromosomes carry an inversion) male *D. pavani*. Males were mated with females of the sibling (very closely related) species *D. gaucha* and scored for (a) matings during the observation period, (b) courtings which did not lead to copulation and (c) remaining sexually inactive during the observation period. Brncic and Koref-Santibanez found there were significantly more heterokaryotypes than homokaryotypes among the males which courted and/or mated during their first few minutes with a female. Superiority in mating speed, at least up to the first mating, may be one reason inversion heterokaryotypes are maintained in *D. pavani* populations. Differences in mating speed between inversion homokaryotypes have also been found in *D. pseudoobscura* and *D. persimilis*.

Mental retardation and/or behavioural abnormalities are a common feature of chromosome aberrations in man. Down's syndrome ('Mongolism'), for instance, results from an extra copy of the small chromosome 21 (chromosomes are numbered according to their size). A fault during mitotic cell division gives rise to three instead of two pairs of the chromosome (hence the alternative term for the condition, Trisomy-21). Patients with Down's syndrome have IQs ranging from less than 20 to about 65 and are thus severely retarded. Very similar consequences of Trisomy-22 have been found in chimpanzees (*Pan troglodytes*).

Aberrations of the X and Y sex chromosomes in man may also affect behaviour. Turner's syndrome is a condition arising from loss of one of the X chromosomes in females. Females suffering from Turner's syndrome are designated XO instead of the normal XX. Behaviourally, XO females are hypertense and defective in their visual space perception. Klinefelter's syndrome, on the other hand, results from the addition of an X chromosome. Such individuals (classed as XXY) are male in appearance but sterile and show weak sexual inclinations. Klinefelter's syndrome is associated with a variety of behavioural abnormalities including seizure disorders, speech impairments, schizophrenia and deviant sexual behaviour. There is suggestive evidence that a similar XXY condition in cats affects sexual motivation.

Multiple Genes. So far we have dealt with relatively simple genetic units (single genes and whole chromosomes) controlling behaviour. Simple that is, in the sense that behavioural traits can be readily traced from the way they segregate (i.e. the way they separate during cell division and reproduction). In many cases, the genetic control of behaviour is more complicated because it involves several different genes. Here it is more difficult to trace the 'unit' of control. Inheritance of traits controlled by more than one gene is said to be *multifactorial* or *polygenic*. Polygenic control of behaviour can sometimes be demonstrated by hybridising two closely related species and observing the way behaviour

Figure 3.3: (a) Peach-faced Lovebird (*Agapornis roseicollis*), (b) Fischer's Lovebird (*A. personata fischeri*), (c) Hybrid *A. roseicollis* × *A.p. fischeri*. Graph Shows Conflicting Patterns of Carrying Nest Material Inherited by Hybrid Birds. p, strips carried in bill; q, intention movements to tuck; r, strips tucked but not carried; s, irrelevant activities; t, intention movements to carry in bill. Over the three years for which data are plotted, birds gradually learned to carry material in the bill (as in *A. fischeri*) and the number of irrelevant and inappropriate activities decreased.

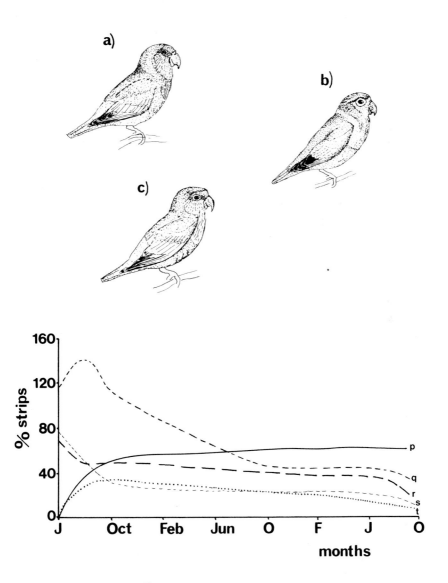

Source: Modified after Dilger[90].

patterns segregate in the offspring. Just such an approach was used by Dilger[90] in his classic study of nest-building in lovebirds (*Agapornis* spp.).

Dilger crossed two species of lovebird, the peach-faced lovebird (*A. roseicollis*) and Fischer's lovebird (*A. personata fischeri*). The females of these two species have characteristic but very different nest-building behaviour. *A. roseicollis* females painstakingly tuck strips of nesting material (paper, leaves, bark, etc.) between feathers on the rump or back. Many such strips are tucked into the feathers before the bird flies to its nesting site. If any of the strips drop out on the way to the nest, they are retrieved. *A. fischeri* females, however, carry similar materials to their nest, but this time items are transported singly in the bill. Hybrids between the two species show an interesting but highly unsuccessful compromise between the 'pure' species behaviours (Figure 3.3). Hybrid females almost always try to tuck nesting material into their feathers but fail dismally. They fail because: (a) they do not release the strips of material once they are tucked in, (b) those strips which are eventually tucked in usually fall out, (c) they try to tuck material into the wrong feathers, (d) materials are grasped inappropriately so that tucking is difficult, (e) tucking usually grades into preening activities, (f) they attempt to tuck inappropriate objects and (g) in order to get the bill near the rump, birds sometimes run backwards. Hybrids are really only successful at carrying material in the bill. However, they take three years to learn this, and even then are not as proficient as female *A. fischeri*.

It appears, therefore, that hybrids inherit behaviour patterns for carrying a large number of items from the behavioural repertoire of the *roseicollis* parent, but patterns for carrying single items from the *fischeri* parent. The fact that neither 'pure' species' repertoires is inherited by the hybrid in its entirety points strongly to polygenic control within the parent species. Hybrids are also intermediate for other behaviours. During courtship, for instance, the male lovebird approaches the female from the side, moving first towards her, then away and varying his direction of approach ('switch sidling'). Switch sidling makes up about 32 per cent of the precopulatory displays in *A. roseicollis* and about 51 per cent in *A. fischeri*. In hybrids, however, it makes up an intermediate 40 per cent. Moreover, when hybrid males are crossed with each of the parent species females, further differences appear. The offspring of F_1 hybrid males and *A. roseicollis* females spend about 33 per cent of their precopulatory time in switch sidling; those of F_1 hybrid males and *A. fischeri* females spend about 50 per cent.

One of the most detailed and quantitative investigations of the polygenic control of behaviour is Rothenbuhler's[263,264] study of so-called 'hygienic' honey bees. Some strains of honey bee are called 'hygienic' because when larvae die inside their cells, workers uncap the cells and remove them. Dead larvae of 'unhygienic' strains are simply left in their cells to decompose.

If 'unhygienic' and 'hygienic' strains are crossed, the hybrids are all 'unhygienic'. 'Unhygienic' is thus dominant. When Rothenbuhler backcrossed

Figure 3.4: Rothenbuhler's Backcross Experiment between 'Hygienic' and 'Unhygienic' Bee Strains. 'Uu' and 'Rr' represent hypothetical loci controlling uncapping and removal behaviour.

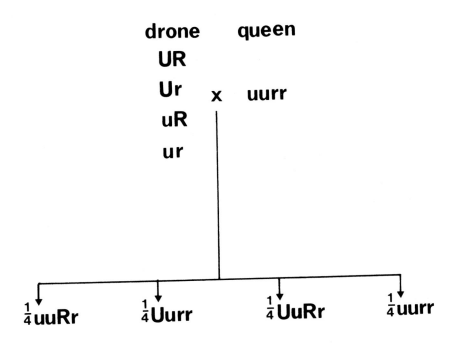

Source: From data in Rothenbuhler[263,264].

(mated with homozygous recessive individuals) his hybrids with recessive 'hygienic' bees, however, he found the following traits among the progeny:

(1) Nine of the colonies uncapped cells in which larvae had died but did not remove the corpses.
(2) Six colonies would remove dead larvae from uncapped cells, but would not actually uncap the cells themselves.
(3) Eight colonies were 'unhygienic' and thus neither uncapped cells nor removed dead larvae.
(4) Six colonies were 'hygienic' and therefore did both.

Statistically, the proportions of these four types of progeny do not differ from equality. The 'hygienic' trait can thus be explained on the basis of two genes. One codes for uncapping behaviour (the U gene), the other for larva removal (the R gene). Since worker honey bees are *diploid* organisms (that is, they possess two copies of each of their chromosomes), there are two versions (called *alleles*) of each gene. We can designate the 'unhygienic' alleles U and R because 'unhygienic' behaviour is dominant. The alleles for recessive 'hygienic' behaviour are therefore *u* and *r*. 'Hygienic' workers must have the homozygous recessive genotype *uurr* and 'unhygienic' workers the genotype *UURR*. Worker and queen hybrids of the two strains will have the constitution *UuRr* and thus be 'unhygienic'. One problem with backcrossing honey bees is that males (drones) are *haploid* (they possess only one copy of each chromosome). Rothenbuhler had to cross 29 different F_1 drones with *uurr* queens to produce the full range of backcross colonies. Among them were the four possible male genotypes *UR, ur, Ur* and *uR*. The results of the backcross in terms of the hypothetical loci are shown in Figure 3.4. The fact that the four classes of colony (1–4 above) occurred with equal frequency in the backcross progeny might suggest that single loci control each of the uncapping and removal traits. This does not appear to be so, however. 'Unhygienic' workers do in fact perform 'hygienic' activities, but at a very low frequency. It seems more likely that the *U* and *R* loci act as 'switches' which control the expression of a larger number of genes in an all-or-nothing way.

3.2 Locating the Effects of Genes

In the preceding section, we discussed ways in which a genetic element in behaviour could be demonstrated and evidence for the varying complexity of genetic control. Merely demonstrating genetic control, however, says nothing about how it is exerted. How do genes influence behaviour? Where is their site of action within the body?

3.2.1 Physical Effects and Behaviour

In some cases, changes in behaviour can be traced to specific anatomical or physiological alterations. The work of Brenner[35] and Ward[328] on chemotaxis (movement in relation to chemical stimuli) in the nematode (*Caenorhabditis elegans*) is a good example. In these experiments, the chemotactic responses of wild-type *Caenorhabditis* were compared with those of mutants (animals exposed to mutagenic cyclic nucleotides, cations or alkalis).

Mutants were worms with head or tail cuticle blisters, heads bent ventrally or dorsally, defective head muscles or short heads. What they found pointed clearly to effects of anatomical changes on behaviour. Head blister, defective head muscle and shortened head mutations reduced the efficiency of orientation compared with the wild type. Bent-headed individuals tended to track tight spirals whose direction of rotation correlated with that of the bending of the head. Genetically-produced defects in the sensory capacities of the head region thus appeared to be a key factor in changing the behaviour of experimental animals.

Chemically-induced mutations in the protozoan *Paramecium* also affect locomotory abilities. The mutant *pawn* cannot swim backwards as the normal wild type can. *Pawn* is so-called because of a similar constraint on its chesspiece namesake and results from a single-gene mutation affecting voltage-sensitive calcium cation (Ca^{2+}) conduction in the ciliated membrane. If the membrane of *pawn* individuals is disrupted using detergent, the animals can swim backwards provided enough Ca^{2+} and adenosine triphosphate is added to the medium. The mutation is thus specific to membrane conductance and does not impair the cilia themselves.

A large number of the listed mutations that affect behaviour in mice exert their effects through neurological defects (especially of the cerebellum) or defects of the inner ear. *Jerker*, *waltzer* and *quaking* mice have impaired balance and move awkwardly. *Wobbler lethal* (*WL*) mice develop slowly in various activities, like open field behaviour and are late learning to rear up on their hindlegs and to ascend an incline. *WL* results from myelin degeneration and high levels of the enzyme succinic dehydrogenase. *Looptail* mutants have enlarged brain ventricles and are defective in certain motor systems of the forebrain. Behaviourally, they show a marked reduction in face-cleaning activities compared with normal mice.

There are several problems with assessing the behavioural effects of severe neurological defects. The most obvious is the abnormality of the defects in relation to 'natural' nervous and behavioural variability. However, a few 'normal' mutations in mice have been studied for their behavioural effects.[94] Prominent among these are mutations involving changes in coat colour. The mutation *albino*, for instance, inhibits the synthesis of the enzyme tyrosinase. Tyrosinase catalyses the conversion of the amino acid tyrosine into a substance called dopa and finally into the pigment melanin. *Albino* mice differ from nonalbinos in a variety of behaviours. At least some of the differences are explicable in terms of a photophobic response resulting from reduced pigmentation in the eye. Compared with pigmented mice, albinos avoid lighted environments and show reduced activity under white light. Because of the latter, they also tend to spend longer in lighted environments once placed in them.

3.2.2 Tracing the Effects of Genes

In the examples we have just considered, it has been possible to attribute behavioural changes to obvious structural and/or biochemical alterations. As Benzer[26] points out, however, this may still be a far cry from pin-pointing the primary site of gene action, the location in the body where a gene exerts its effect. This 'focus' of gene action may be some distance from the component of the body which is outwardly affected. In man, for example, degeneration of the retina may sometimes be caused not by a defect in the eye itself, but by insufficient absorption of vitamin A in the gut. How then might we locate the 'point' at which a gene or gene complex influences behaviour? Benzer and his coworkers provided a brilliant answer.

Their technique was based on sound engineering principles. If some part of a complex operational system becomes faulty, a quick way to trace the fault is to swop components about until its source is located. This is precisely what Benzer did with mutant *Drosophila*. Instead of moving body components around surgically, however, he used an ingenious genetic technique to create *mosaic* flies. *Mosaic* in this context describes animals whose bodies comprise a number of different genotypes. Owing to peculiarities during cell division, different parts of the body (internal and external) sometimes end up with different genetic constitutions. In this way, *mosaics* provide a powerful means of pin-pointing the 'focus' of gene action within the body. The particular beauty of using *Drosophila* is that *mosaics* can be easily generated in the laboratory rather than awaited on a chance basis. The mechanism Benzer used for their generation was the loss of an unstable ring-shaped X chromosome.

The abnormal structure of the 'ring' X chromosome increases the frequency with which it is lost during cell division. Sex determination in *Drosophila* is similar to that in humans. Eggs having the constitution XX are female, those having XY, male. Now imagine female eggs which possess one normal X and one 'ring' X chromosome. Because of its unstable nature, some of the cells formed by division in a proportion of the embryos will lose the 'ring' X chromosome and come to possess only one X chromosome (Figure 3.5). They will therefore produce male tissues. If it is lost, the 'ring' X chromosome generally disappears early in development so that roughly equal numbers of XX and X nuclei are produced. After several divisions, the nuclei migrate to the surface of the egg to form the *blastula* (a monolayer of cells surrounding the yolk). Because nuclei tend to remain near their neighbours in the blastula, some areas of the blastula surface (the *blastoderm*) are female (XX), the others male (X). Furthermore, the direction of the first division in *Drosophila* is arbitrary in relation to the egg. There are therefore innumerable ways in which the blastoderm can be divided into male and female parts. Once the blastoderm is formed, its genetic composition largely determines that of the resulting embryo. The adult male/female mosaic (*gynandromorph*) can thus have a variety of arrangements of male and female components. Some possibilities are illustrated in Figure 3.6a. In general, the division into male and female parts follows the intersegmental boundaries and the longitudinal midline of the exoskeleton (Figure 3.6b). This is because each part forms independently during metamorphosis from an imaginal disc in the larva which is itself derived from a specific area of the blastoderm.

In order to trace which parts of a fly were normal (female) and which mutant (male), Benzer used 'marker' mutations. By selecting for meiotic cross-overs that resulted in a behavioural mutation (say *nonphototactic*) being combined with a physical marker (say white eyes, yellow body), 'marker' males could be mated with 'ring' X females. Among the offspring which result from such a mating will be some in which the 'ring' X chromosome bears normal genes, but the normal X chromosome, mutant genes. Those cells in which the 'ring' X

Figure 3.5: Gynandromorph Fruit Flies Can Be Produced by Mating Males with Females Bearing an Unstable 'Ring' X Chromosome Which May Be Lost during Cell Division.

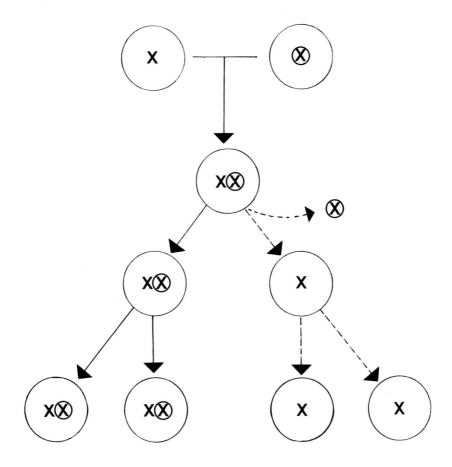

Source: Modified after Benzer[26].

chromosome is subsequently lost will therefore develop into mutant body parts (because the recessive mutant on the remaining normal X chromosome is no longer masked by its wild-type allele). If a mutant for body colour is used, identifying the normal and mutant components of the body is simply a matter of visually examining the fly.

Hotta and Benzer[146,147] used the 'ring' X gynandromorph technique to investigate different behavioural mutants of *Drosophila melanogaster*. In *nonphototactic* mutants (flies with an impaired ability to respond to light), it turned out that the 'focus' of gene action was the retina itself. In *hyperkinetic* mutants, however, the story was more complicated. When anaesthetised, *hyperkinetic* flies shake their legs in the air instead of lying still. Mosaic *hyperkinetics*

Figure 3.6: Top — Depending on the Orientation of Single (Black Dot) and Double (Open Dot) X Chromosome Nuclei at their First Division, Different Types of Gynandromorph Adult Flies are Produced. Stippled areas are mutant tissue. Bottom — Mutant and Wild-type Tissues Tend to Follow Lines of Divisions between Discrete Body Parts.

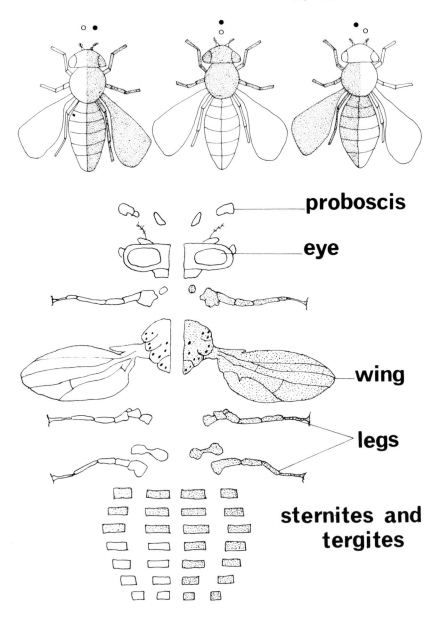

proboscis

eye

wing

legs

sternites and tergites

Source: Modified after Benzer[26].

Figure 3.7: Foci of Gene Action for Leg-shaking in Hyperkinetic Mutants. 1–3 indicate the positions of the three left legs, other numbers indicate the distance in Sturts from each leg and other anatomical landmarks to the shaking foci. pb, presutural bristle; sb, scutellar bristle; p, proboscis; fd, genetic focus for 'drop dead' mutant; a, antenna.

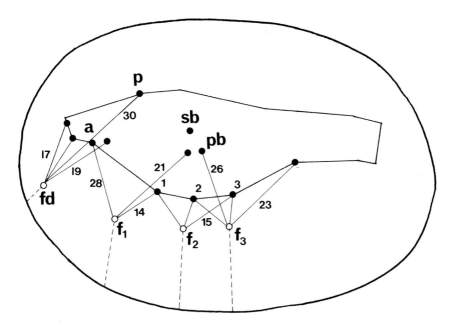

Source: Modified from Hotta and Benzer[146].

shake only some of their legs. Although there is a good correlation in mosaics between the surface genotype of a leg and whether or not it shakes, the correlation is not perfect. As Benzer points out, the markers on the outside of the fly may not necessarily reflect the genotype of underlying tissues (they develop from different parts of the embryo). Since it is likely that leg movement would be controlled from somewhere within the CNS, the tissue in which the *hyperkinetic* gene exerts its primary effect may well be genetically different from the affected legs. To find out where the 'focus' of leg-shaking lay, Hotta and Benzer developed an ingenious mapping technique.

The technique was based on an earlier method for mapping the sequence of genes along a chromosome. It depended on the simple fact that crossing-over between any given pair of genes is more likely to occur the further apart the genes are on the chromosome. It is relatively straightforward to adapt the one-dimensional mapping of chromosomes to the two-dimensional mapping of the fly blastoderm. Simply take any structures A and B in adult mosaics and record how often one is normal and the other mutant. The frequency indicates the distance between their regions of origin in the blastoderm. Taking 300 *hyperkinetic* adult mosaics, Hotta and Benzer recorded the number of times marker mutations at various surface 'landmarks' coincided with shaking in each leg. They confirmed earlier observations that the movement of each leg was independent of that of all the others and that the external genotype of a leg correlated well, but not perfectly, with its tendency to shake. Each leg therefore appeared to have a separate 'shaking focus' located some distance away. Hotta and Benzer calculated the distance from the focus to a leg and certain other bodily structures and pin-pointed the focus for each leg by a sort of triangulation process (Figure 3.7). As expected, the foci were near their affected legs but just below them in the part of the blastoderm which gives rise to the CNS. It is known that neurons in the thoracic ganglion of the CNS show abnormal activity in *hyperkinetic* mutants.

3.3 Ecology, Genes and Behaviour

In our discussion of the relationship between genetic and behavioural variation so far, we have not been concerned with functional significance. Our task has been simply to show that such a relationship exists and to give some idea of its complexity. In many cases, the mutations studied have been artificially produced (for example by exposing animals to X-rays or chemical mutagens) and would certainly be lethal in a natural environment. If studies of behavioural genetics are to help us interpret what we see in the real world, we need to be able to relate genetic variability to adaptive behaviour in the field. As the studies we have discussed indicate, measuring a genetic component in behaviour requires stringent control of environmental factors. Clearly, this becomes difficult, if not impossible, when we are dealing with natural populations. Often, a genetic basis to adaptive differences between animals has to be inferred indirectly, for example by comparing behaviour in closely related species.

3.3.1 Habitat Selection

Drosophila melanogaster and *D. simulans* are two species of fruit fly which are almost indistinguishable in their morphology. They are similar enough to be regarded as sibling species. Despite their being closely related, however, the two species show subtle differences in their behaviour and ecology. At least in part, these must be attributable to their small genetic differences. Three of the main differences occur in:

(1) *Sexual behaviour*. Males of the two species differ in their speed of court-ship. *Simulans* males take longer to start courting and show more orient-ation behaviour (see Section 3.1.3), although all the normal courtship phases are eventually performed. *Simulans* females tend to rely more on visual stimulation from male courtship rather than stimuli perceived via the antennae. The reverse is true of *melanogaster* females. The fact that *melanogaster* courtship is independent of light conditions may reflect the

broad-niched nature of the species. *Simulans* occupies a much narrower range of habitats and may therefore be more precise in its courtship requirements.

(2) *Dispersion.* The broader-niched nature of *melanogaster* compared with *simulans* is also highlighted by its greater tendency to disperse. Again species differences appear to be based on their responses to light. *Melanogaster* tends to disperse more evenly over a variety of light intensities, while *simulans* shows a more pronounced phototactic response.

(3) *Temperature preference.* *Melanogaster* appears to tolerate a wider range of temperatures than *simulans* which shows avoidance responses to extremes of temperature and humidity. Levins[189] also found that *melanogaster* acclimatised to dry heat more by developmental and physiological mechanisms than by genetic differentiation of populations. *Simulans*, on the other hand, shows less phenotypic flexibility and depends more on genetic adaptation.

Convincing evidence for a genetic component to habitat selection has also been found in rodents. The behaviour of deer mice (*Peromyscus* spp.) in the laboratory shows characteristics predictable from the habitats the mice normally occupy in the wild. Harris[133] compared the behaviours of two closely related sub-species: a 'field-dweller', *P. maniculatus bairdii* and a 'wood-dweller', *P. m. gracilis*. Both sub-species show a pronounced preference for their respective habitats and avoid habitats of the other type. Harris presented the two sub-species with a choice between artificial 'field' and 'wood' habitats in the laboratory. Not only did wild-caught mice show a clear preference for the 'habitat' most closely resembling their own but animals reared in the laboratory also showed appropriate preferences.

Bairdii and *gracilis* also exhibit preferences for different temperatures. *Bairdii* prefers temperatures around $25.8°C$, *gracilis* around $29.1°C$. These preferences make adaptive sense in that *bairdii's* prairie habitat is generally cooler than the woodland inhabited by *gracilis*. Another species, *P. leucopus*, whose natural habitat is usually $3-4°C$ warmer than the woodlands of *gracilis*, shows a preference for temperatures around $32.4°C$. Deer mice clearly tend to select habitats which are similar to those in their natal areas and their ability to do this without prior experience implies genetic control (this preference disappears after about 12-20 generations of laboratory breeding suggesting that genetic changes occur within lines kept for long periods in captivity). Further adaptive differences between the *bairdii* and *gracilis* sub-species which appear to be genetically based occur in burrowing responses, locomotor maturation and clinging ability.

3.3.2 Population Dynamics

Various aspects of behaviour, such as migration, aggression, habitat selection, mating activities, territoriality, social aggregation, etc., affect the structure of animal populations. In doing so, behaviour becomes a major force in changing

Figure 3.8: The Percentage of Different Vole Genotypes Transferring during the Population Increase Phase. C and E are alleles coding for an enzyme. Females heterozygous for C/E were most frequent among dispersers.

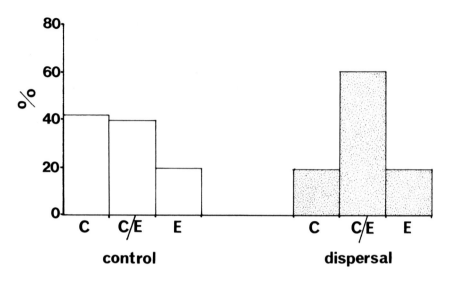

Source: Modified after Krebs *et al.*[164]

the relative frequency of genes within species' gene pools. Such changes inevitably feed back on behaviour by creating new selective pressures. A good example comes from the vole species *Microtus pennsylvanicus*.

To account for the cyclical nature of population growth and dispersal in many rodent species, Chitty[54] proposed a model based on a behavioural polymorphism. He envisaged different genotypes within a population which are either tolerant or intolerant of crowded conditions. Changes in population density set up different selective pressures on the two genotypes. As population density increases so does the level of interference and aggression between individuals. When interference reaches a certain level, crowd-intolerant animals emigrate. Subsequent field experiments with *M. pennsylvanicus* have lent considerable support to Chitty's two-genotype hypothesis.

Basically, the experiments consisted of removing voles from two areas in the field by trapping for two days every two weeks. Between trapping periods, voles were allowed to recolonise the areas from the surrounding population. As Chitty predicted, dispersal was most pronounced during periods of population increase and least pronounced during periods of decline. What is even more interesting is that the tendency to emigrate was not evenly distributed across all genotypes within the population. Using the polymorphic serum protein loci transferrin (T_f) and leucine aminopeptidase (LAP), Krebs *et al.*[164] found that (a) the frequency of a particular allele of LAP (LAP^s) dropped by about 25 per cent in males at the beginning of peak emigration, and followed suit in females about 4-6 weeks later, and (b) females heterozygous for the alleles T_f^C/T_f^E were more common among dispersing animals than those remaining resident (about 89 per cent of the loss of T_f^C/T_f^E females from the resident population was due to emigration) (Figure 3.8). There is also evidence from laboratory experiments with male *M. pennsylvanicus* and *M. ochrogaster* that a significant increase in aggressive behaviour occurs as population size peaks (the behavioural cause of dispersion in Chitty's model). Furthermore, aggression tends to be more pronounced in those male *M. pennyslvanicus* which ultimately emigrate from the resident population.

Similar correlations between behaviour and enzyme polymorphisms have been found in birds. In wintering flocks of juncos (*Junco hyemalis*), Baker and Fox[7] found that birds ranking highly in the dominance hierarchy were usually heterozygous for a certain tripeptidase enzyme. In this species, as in many others, dominance status is a good predictor of the likelihood of survival. Baker[6] also found a correlation between song dialects and genetic enzyme polymorphism in white-crowned sparrows. Populations from two counties in California had very different song types. Although they were genetically very similar, the two populations differed significantly at the locus coding for phosphoglucomutase-A.

3.4 'Selfish' Genes and the Evolution of Behaviour

Having established that genes do play an important role in shaping behaviour and that their role can be identified and measured, we come now to the implications of genetic control for the evolution of behaviour. Here, however, we have to make an important point which, at first sight, might seem obvious. As the units of heredity, it is genes which reproduce and survive in evolutionary terms. The individual bodies which house them are, in Dawkins'[79] graphic terms, their temporary, throwaway 'survival machines'. They are designed by genes simply as a means of enhancing gene survival and perpetuation (*fitness*). It is possible to view behaviour as one design feature of a 'survival machine' which genes can modify to enhance their evolutionary fitness. In this sense, genes appear to design organisms to further their own 'selfish' interests. The 'selfish' gene, however, is merely a convenient, shorthand metaphor expressing an evolutionary truism: the enhanced reproductive potential of a gene (or, more strictly, an allele) in one generation leads to its enhanced representation in the next. No sense of purpose is implied. Furthermore the metaphor does not, as it is sometimes accused of doing, assume that each behaviour is coded by a single gene. Although this may sometimes be so (see Section 3.1.3), all the metaphor assumes is that behaviour evolves by Darwinian selection between genetic 'units'. The size and complexity of the 'unit' (whether two or three loci, a large linkage group or a whole chromosome) is immaterial. What are the implications of this view of evolution for behaviour?

3.4.1 Inclusive Fitness

An individual possessing a given allele shares that allele by common descent with others in the population. The probability that any two individuals share copies of the same allele depends on their degree of relatedness. An individual has a 50 per cent chance of sharing a given allele with its parent, offspring or sibling, a 25 per cent chance with its grandparent or grandchild, a 12.5 per cent chance with its first cousin and so on. The effect an allele has on its fitness by designing an individual to behave in a certain way, depends not only on changes in that individual's reproductive potential but also on the effect of its behaviour on the potential of others sharing the allele. This extended concept of gene fitness is referred to as *inclusive fitness*.[130] When we consider the evolutionary potential of a behaviour, it is the inclusive fitness of the genetic 'unit' for that behaviour that we should use as a yardstick.

Apart from clarifying our concept of gene fitness in a general sense, the use of inclusive fitness leads to some unexpected and highly important interpretations of behaviour. One of the major stumbling blocks to Darwin's[68] original thesis of evolution by natural selection was the existence in certain species of apparently altruistic behaviour. Particular difficulties were created by the castes of sterile females among the social Hymenoptera. These females sacrifice their own reproductive potential and devote their time and energy to raising the progeny of the colony queen. Instances of apparent altruism are legion in the animal kingdom. In some bird species, like the Florida scrub jay (*Aphelocoma coerulescens*), adult birds help parental pairs raise offspring (by contributing food and defending the nest from predators) instead of setting up breeding territories of their own. Similar helpers are found in jackals (*Canis mesomelas*) and red foxes (*Vulpes vulpes*). Various species of mammalian carnivore hunt co-operatively. In hunting dogs (*Lycaon pictus*), the spoils of the hunt are shared with pups and guard adults which did not participate in the hunt. In some cases, as in the guard bees of honey bee hives, individuals may even sacrifice their lives defending the interests of others. At first sight, these behaviours, and others like them, create a knotty problem for evolutionary theory. If behaviour really is a means by which alleles increase their chances of perpetuity, coding for behaviour which sacrifices their 'survival machine's' reproductive potential does not seem the best way of achieving it.

Careful investigation of apparently altruistic relationships, however, has revealed an important common factor. In almost every case, the *donor* (the performer) and the *recipient(s)* of an altruistic act are close relatives. There is a good chance that they share the genetic 'unit' coding for altruism (nest-helping, guarding or whatever). Any benefit the recipient gets from the donor's behaviour increases its chances of passing on the gene(s) for altruism. In order to quantify the change in the altruism 'unit's' inclusive fitness, we need to take into account the net effect of the behaviour on all copies of it in the altruist's effective population (the group of individuals likely to experience the act). The sums go something like this: Net benefit in inclusive fitness of behaviour pattern = Benefit to altruist − Cost to altruist + 0.5 Benefit to sibling − 0.5 Cost to sibling (repeated for all siblings involved) + 0.5 Benefit to child − 0.5 Cost to child (repeated for all children involved) + 0.125 Benefit to first cousin − 0.125 Cost to first cousin + etc. for all individuals involved. The proportions represent *coefficients of relatedness (r)*. Selection performs a similar calculation for all the alternative behaviours open to the animal and favours the one with the highest score. Altruistic behaviour will be favoured if the joint r's of the recipients are high enough to outweigh the costs of the act to themselves and the donor.

The social behaviour of lions (*Panthera leo*) contains many apparently altruistic components. Lions hunt co-operatively, they co-operate in driving intruders out of the pride and the young are suckled communally. In a detailed study of the dynamics and genealogies of lion prides, Bertram[27] established two important points. Male lions in possession of a pride and cubs which grow up together within the pride have an average r value of about 0.22 Males and cubs are thus the approximate equivalent of half-siblings (r = 0.25). Adult females have an r value of about 0.15 and are thus a little more closely related than first cousins (r = 0.125). These r values were calculated as the mean of a range of values. Obviously, some animals are full siblings (r = 0.5), others share a father and have mothers who are sisters (r = 0.38), others may be hardly related at all and so on. Nevertheless, they indicate that, on average, animals encounter fairly close relatives within the pride and that selection is likely to lend some weight to altruistic acts between pride members.

The explanation for sterile castes in honey bee societies is more complicated. Female honey bees develop when the hive queen fertilises an egg with stored sperm. Female bees are thus related to their mother by the expected r value of 0.5. However, because of the intriguing *haplo-diploid* nature of sex determination in this species (males are haploid, females diploid), the same is not true for sisters. Since all the gametes contributed by haploid drones are identical, females and their fathers have an r value of 0.5. The gametes contributed by the queen, however, are not identical. As a diploid organism, half of them come from her mother, half from her father. The average r value for sisters is thus 0.5 (from their father) + 0.5 × 0.5 (from their mother) = 0.75. In terms of her inclusive fitness, therefore, it pays a female to help her mother rear more sisters (r = 0.75) rather than found another colony and rear daughters of her own (r = 0.5).

Because the above and other arguments hinge on the degree of relatedness between individuals, the term *kin selection* is often used to describe the mode of selection involved. While it is appropriately descriptive, the use of 'kin selection' can be misleading. At least one notable author[335] has misinterpreted it as a special case of *group selection*. One reason for confusion is the term's implicit plurality. It sounds as if selection is acting on groups of related individuals, which plainly it is not. For this reason, and because it is unnecessary, 'kin selection' should perhaps be dropped and the more accurate *gene selection* used in its place. A full discussion of the problem of 'levels' of selection can be found in Chapter 11. One further point worth mentioning is that the terms 'gene inclusive fitness' and 'individual inclusive fitness', meaning one and the same thing, are often used interchangeably. There is nothing wrong with this. When talking of individual inclusive fitness, gene inclusive fitness is understood. It is very different from confusing individual and gene *selection*, a problem we shall return to later.

3.4.2 Evolutionarily Stable Strategies

The use of cost-benefit arguments in modelling evolution has led to another important theoretical development. For each evolutionary problem facing an organism, there will usually be a range of options or *strategies* available for solving it. In evolutionary terms, we can imagine mutant alleles arising which code for each strategy. Selection will then favour the allele coding for the strategy which yields the highest net benefit. We have already encountered this kind of argument in the discussion of the optimal performance and sequencing of behaviour (Chapter 2). What will be the net result of selection between these pre-programmed strategies? The answer is neatly summed up in a term coined by Maynard Smith[203] an *evolutionarily stable strategy* (*ESS*).

An ESS is defined as a strategy which, if most members of a population adopt it, cannot be bettered by an alternative strategy. In other words, the best strategy for an individual to adopt depends on which most of the population adopt. Since each individual will be selected to maximise its own reproductive potential (and hence the inclusive fitness of its genes), the strategy that ultimately persists in the population (i.e. is evolutionarily stable) is one that cannot be bettered. If a better alternative evolves or environmental pressures change to favour a different strategy, a new strategy will become the ESS. Once evolved, the ESS is maintained by selection acting against any deviation from it.

The real value of thinking in terms of ESSs, instead of merely relatively 'good' or 'bad' strategies, is that it takes into account the evolutionary consequences of 'choices' of strategy. It enables us to compare alternatives in a rigorously quantitative way. The results we obtain often may be counter-intuitive. Strategies which at first sight seem to be highly profitable turn out to be unstable in the long term and hence not ESSs. A simple example illustrates the point.

Animals frequently contest limited resources like food, mates, oviposition sites and so on. Although the winner of a contest clearly benefits, animals often use *ritualised*, non-damaging (*conventional*) tactics to decide the outcome, even if they possess dangerous weapons (like horns, large canines, etc.). At first sight it would seem more sensible to fight in an aggressive, potentially damaging — even lethal — way. An animal could then make sure of winning and perhaps permanently rid itself of a competitor into the bargain. We should therefore expect a population of conventional contestants to be quickly invaded by any mutant escalating contestants which arise. Let's see if this is what would happen.

In order to develop an evolutionary model of conventional versus escalated strategies, we must first decide on the relative costs and benefits of the two strategies when they are played against each other. The conventional, ritualised strategy we shall call *dove*, and the escalated, aggressive strategy, *hawk*. *Dove* uses only conventional tactics and retreats from a contest at the first sign of escalation. *Hawk* escalates immediately it enters a contest. When *dove* meets *dove* there is no clearly predictable winner. Either contestant could win. On average, therefore, each gets half the value of the resource (V) devalued by the time costs (T) of ritualised display. The *pay-off* in 'fitness units' to each *dove* is thus ½V − T. When *dove* meets *hawk*, however, *dove* always loses but does not waste any time displaying. The pay-off to *dove* is thus zero, and the pay-off to *hawk*, V (the total value of the resource). When *hawk* meets *hawk*, the outcome is again unpredictable. As in contests between two *doves*, each contestant gets on average ½V, but this time devalued not by display costs, but by the costs of injury (W). Since both contestants escalate, it is likely that one or both will be seriously injured. The pay-off to each *hawk* is thus ½(V − W). These pay-offs are represented in the matrix in Table 3.1.

Table 3.1: Pay-off Matrix for Contests between 'Hawk' and 'Dove' Strategists (see text). Numbers represent hypothetical fitness units.

Cost of injury (W)	= −20
Value of winning (V)	= +10
Time cost of long contest (T)	= −3

		Against	
		Hawk	Dove
Pay-off to	Hawk	− 5 (½(V − W))	+10(V)
	Dove	0	+2(½V − T)

The ESS is where the proportions of 'hawk' and 'dove' strategists are such that the average pay-off to a 'hawk' = the average pay-off to a 'dove', i.e. when

$$P \, ½(V − W) + (1 − p)V = p(0) + (1 − p)(½V − T)$$

Source: Modified after Dawkins and Krebs[85] and Maynard Smith, J. (1978). The evolution of behavior. *Sci. Am., 239*: 136–45.

Now, if we assume (quite reasonably) that the cost of injury is greater than the value of the resource (W > V), then clearly *hawk* does worse than *dove* in a population of hawks (½(V − W) < 0). As we might expect, however, *hawk* does much better than *dove* in a population of doves (V > ½V − T). The net benefit accruing to each strategy is thus *frequency-dependent*. Each when common is open to invasion by the other and so is not an ESS. However, we can find the ESS for the matrix in Table 3.1. It is that mixture of hawks and doves in which the two strategies achieve equal pay-offs. The equation in Table 3.1 expresses the ESS mathematically. The equal pay-off mixture is an ESS because if either strategy increased in frequency, the other would immediately be at a selective advantage.

In this case the ESS consists not of a *pure* strategy (*hawk* or *dove*) but *mixed* strategy (both *hawk* and *dove*) at a certain relative frequency. Moreover, this mixed ESS can be realised in two ways. Individuals may be either *hawks* or *doves* and therefore make up their stable proportions of the population (say 5/9 of individuals playing *hawk* and 4/9 *dove*), or they may play both strategies but for a stable proportion of the time (all individuals play *hawk* 5/9 of the time and *dove* 4/9 of the time). Of course, a pure ESS might exist under different conditions. For instance, if the cost of injury does not exceed the value of the resource $(V > W)$, *hawk* may be a pure ESS. This appears to be the case among elephant seals (*Mirounga angustirostris*) where bulls indulge in long and sometimes highly damaging fights but achieve an enormous pay-off in terms of the number of cows they fertilise if they win.

The ESS may also be a *conditional* strategy. This is equivalent to saying 'if faced with situation x, play strategy a; if faced with y, play b'. We can complicate the *hawk/dove* example above by including a simple conditional strategy in the analysis. This strategy is called *bourgeois*. *Bourgeois* follows the simple rule 'fight conventionally if usurping a resource, but escalate if holding it'. Through its conditional strategy, *bourgeois* risks only half the injury cost of a *hawk*. Assuming the same fitness units as before, Table 3.2 shows that *bourgeois* is an ESS.

Table 3.2: Pay-off Matrix for Contests between 'Hawk', 'Dove' and 'Bourgeois' Strategists. Under these conditions, bourgeois is an ESS (see text).

		Against		
		Hawk	Dove	Bourgeois
Pay-off to	Hawk	$-5(\frac{1}{2}(V-W))$	$+10(V)$	$-5(\frac{1}{2}(V-W))$
	Dove	0	$+2(\frac{1}{2}V-T)$	$+2(\frac{1}{2}V-T)$
	Bourgeois	$-2.5(\dfrac{\frac{1}{2}(V-W)}{2}+0)$	$+6(\frac{1}{2}V+\dfrac{(\frac{1}{2}V-T)}{2})$	$+5(\frac{1}{2}V+0)$

Source: Modified after Maynard Smith, J. (1978). The evolution of behavior. *Sci. Am.,* *239*: 136–45.

Calculating the ESS for the above examples is relatively straightforward because of the simplified strategies and 'fitness units' we used. In reality, reducing all the relevant costs and benefits to the same currency is not easy. The ESS may also be complex, perhaps conditional on a large number of different factors, or even cycling between two or more unstable equilibria. Nevertheless, the principle is always the same, however much it varies in complexity. The *hawk/dove/bourgeois* model provides a useful introduction to the concept and calculation of ESSs.

3.4.3 The Extended Phenotype

Our view of an animal as merely the gene's mechanism for ensuring its own survival leads to another interesting idea. For obvious reasons, an individual's phenotype is usually regarded as being restricted to the characteristics of its physical body. The phenotype is the entity we see, touch, hear and smell. However, a major determinant of an animal's survival and reproductive potential is the way it influences the biotic and abiotic environment around it (for example, by building a shelter or attracting a mate). Selection will therefore favour alleles which influence the environment so as to maximise their chances of propagation. In short, we can extend our concept of the phenotype to include the adaptive influence of genes outside their own 'survival machines'. Dawkins[80,82] discusses some examples of such *extended phenotypes*.

In the freshwater gastropod *Limnaea peregra*, the shell may coil in either a right- or a left-handed helix. The direction of coiling is inherited in a simple Mondelian fashion. There is nothing special about the control of coiling except that it is mediated not by alleles present in the individual but by alleles present in its mother. The expression of coiling is constantly delayed by one generation. Maternal alleles thus extend their phenotypic expression into the next generation.

A different kind of example is the protective 'house' that caddis fly (Insecta: Trichoptera) larvae build around themselves. These 'houses' are made up of sticks, stones and other debris which the insect gathers from the bottom of the stream or pond. Clearly the 'house' is the product of the insect's behaviour and the behaviour, in turn, a product of its genes. The 'house' can thus be viewed as an additional link in the embryological chain of events leading from genes through protein synthesis to the adult phenotype. Although not organic, it is an outer casing created by the caddis fly's genes upon which the survival of those genes depends. A similar argument can be made for the elaborate bowers built by male bowerbirds (Aves: Ptilonorhynchidae). Although the bird does not live inside the bower, its reproductive potential depends very much on the bower's mate-attracting qualities.

A more subtle form of extended phenotype occurs in the *manipulation* of host behaviour by parasites. The fluke *Leucochloridium* spends part of its life cycle in a snail. While in the snail, the fluke migrates into a tentacle and forms a brightly coloured, pulsating cyst. It also affects the snail's nervous system in such a way as to reverse its response to light. Instead of being negatively photo-tactic as usual, the snail becomes positively phototactic and moves towards bright sources of light. In its natural environment this means it tends to move up blades of grass at the top of which its now gaudy tentacles are visible to predatory birds, the final host in the fluke's life cycle. The snail's behaviour is thus not controlled by its own genes but is part of the fluke's extended pheno-type. Many other such examples of host manipulation are known among parasites.

What should be clear from the above examples is that the concept of the extended phenotype can be applied to a wide range of behaviours. Behaviour is often a means of manipulating objects or other individuals in the environment to suit the survival requirements of the performer. In particular, we shall see in Chapter 10 how the vast and complex series of behaviours known collectively as communication can be regarded as a means of manipulation.

Summary

(1) Many different lines of investigation have indicated that behaviour may be coded genetically. Chief among these have been inbred strain and selection techniques.

(2) In some cases, behavioural differences can be traced to changes in a single gene. More commonly, control is complex, involving many loci or even whole chromosomes.

(3) Some genes exert their influence over behaviour directly, e.g. by causing some defect in an organ required for behaviour. Others affect behaviour less directly, perhaps by complex embryological pathways.

(4) It is possible, in some cases, to infer genetic control for certain adaptive behaviour patterns in the field. In other cases, behavioural differences are suggestively correlated with genetic changes in other characteristics.

(5) Behaviour can be viewed as one means by which genes shape animals to enhance their survivorship in the gene pool. Key concepts in this 'selfish' gene approach are inclusive fitness and evolutionarily stable strategies.

4 EXPERIENCE AND LEARNING

In Chapter 3, we mentioned that the behavioural phenotype of an animal was the joint product of its genes and life-time development. While we have illustrated the role of genes in shaping behaviour, we have not yet considered the effects of experience and development, nor the way these interact with genes to produce behaviour. In this chapter we shall examine these effects.

The most widely held view is that an animal's behaviour is formed during its ontogeny as the integrated product of both internal (genetic) and external (environmental) influences. Because they are dependent on environmental conditions for their ultimate form of expression, genes can produce a variety of end-products, according to prevailing constraints. The ontogeny of behaviour is thus seen as a complex and changeable interaction between organism and environment. The term ontogeny covers all aspects of developmental change in behaviour from the time before birth or hatching through to adult life. This chapter deals with the two major components of behavioural ontogeny: changes in behaviour due to physical maturation effects and changes due to experience (*learning*).

4.1 Maturation and Behaviour

Functional movements which we call behaviour are the product of intricately co-ordinated sensory and motor systems. When animals are born, these systems may not be fully developed. This is especially true of *altricial* species (e.g. passerine birds) where the young are born at a relatively early stage in development (as opposed to *precocial* species – e.g. bovids, charadriiform birds – where they are born physically advanced). Behavioural co-ordination and aptitude can develop in two distinct ways. Progressive improvement may be the result of the animal practising behaviour for which the appropriate muscle systems and nerve connections are already established, or it may occur with the maturation of these systems and connections. In the following sections we shall discuss the role of both processes in the development of different aspects of behaviour.

4.1.1 Development and Experience

Development without Experience. Many complex movement patterns characteristic of adult animals are performed for the first time in their entirety. Practice plays little part in their development in the young. The intricate patterns of movement used by the large white butterfly (*Pieris rapi*) during flight are fully developed by the time it emerges from the pupa. The small improvements which are shown during the first day or two of adult life are probably due to the hardening of the wing cuticle rather than to changes in behaviour. Young axolotls (*Amblyostoma* spp.) show recognisable swimming motions before they even leave the egg. Here the progressive development of co-ordination in swimming appears to depend on growth and differentiation within the nervous system. The progression: 'reflex flexure' – 'coiling' – 'S-bending', for instance, occurs synchronously with the establishment of certain neural connections.

Such examples are not limited to invertebrates and lower vertebrates. Working with dromedaries (*Camelus dromedarius*), Gauthier-Pilters found that an enormous range of behaviours appeared 'spontaneously' without prior practice, including chewing (10 minutes after birth), rolling (after 74 minutes), head-jerking (after 120), yawning (after 160), tail-beating (after 198), shaking (after 304), head-shaking (after 100), sucking (after 100), kicking (after 156), urinating (after 185) and the three phases of rising (after 10, 87 and 95).

We are already familiar with the fact that a behaviour may develop without regard to its functional consequences. The hybrid lovebirds in Dilger's experiments (see Section 3.1.3), for example, performed acts characteristic of their two parent species but which were functionally useless in hybrids because of their incorrect sequencing or mode of performance. Other cross-breeding experiments between mallard and pintail (*Anas acuta*) have shown that, when both parent species possess similar courtship displays, these are unchanged in the hybrid offspring, but when the parent species show slightly different displays, hybrids are intermediate. When displays are present in only one parent, however, hybrids perform them in an all-or-nothing manner or exhibit aberrant versions of them. No progressive alteration occurs in hybrid displays even though they may be inappropriate for their original purpose.

Even more extreme examples are provided by various surgical techniques. Weiss[329] interchanged the left and right forelimb buds of salamander embryos when the anterior-posterior axes of the limbs had been determined. The transplanted limbs developed normally except they faced backwards instead of forwards. Moreover when neural connections developed, the limbs moved in a manner appropriate to their normal orientation. That is, they worked to push the animal backwards while the hind limbs and the rest of the animal worked to move it forward. Even after a year of experience with the transplanted legs, the animal had not corrected their reversed action.

In some cases, it has been possible to trace the relationship between the development of movement patterns and the stage of embryological development in detail. In developing chicks, for instance, the ontogeny of movement can be viewed through 'windows' cut in the egg shell. Spontaneous movement by the embryo starts at around 3.5 to 4 days after the onset of incubation, but at this stage it comprises only the sporadic and slight bending of the head region. The duration and frequency of movement increase over the next 8 to 9 days reaching a peak at around the thirteenth day. Now the embryo is capable of a greater range of movement involving not only the head but also the trunk, limbs, toes, eyes and bill. However, movements are still disorientated and jerky. After about the seventeenth day they become more controlled, eventually resulting in the orientation of the body ready for hatching on day 20.

Development Involving Experience. While many behaviour patterns appear in their fully functional form without any previous practice by the animal, many only become functional after a process of trial and improvement. A good example is the typical food-holding behaviour of many small passerine birds. Adult tits (Paridae) hold down food items with one or both feet while they peck. While young birds perform foot-holding on their first encounter with an appropriate object, their initial attempts are clumsy. Moreover, the rate at which holding improves depends on how much opportunity birds get to practice with suitable items.[143] Similarly, squirrels (*Sciurus vulgaris*) require experience in order to open a hazel nut efficiently.[95] Young squirrels possess all the necessary gnawing and prying movements but cannot deploy them effectively. Experienced animals gnaw a vertical furrow in the broad side of the nut, then drive their incisors into the aperture and prise the nut open. Naive animals gnaw haphazardly, sometimes creating several furrows, until the nut happens to break. By trial and error, however, the squirrel learns that nuts break open more easily when the furrow is gnawed parallel to the grain. Similar improvement with experience occurs in the dehusking of seeds and hard-shelled fruits by finches and rodents.

Some of the most detailed investigations of experience in behavioural ontogeny have involved studies of bird song.[201,303] If young male chaffinches are isolated from conspecifics so that they hear no vocalisations by other birds, the song they eventually develop is far simpler than that of normal adult males (Figure 4.1a). If they are instead kept in pairs, their song is intermediate in complexity between those of the isolates and the normal adults (Figure 4.1b, c). Birds provided with an adult male 'tutor', however, end up singing normal adult song.

Hearing *models* is therefore important in the ontogeny of the young chaffinch's song but is it all that is important? Song development proceeds through three well-defined stages. The first is known as *subsong* and is little more than a soft, featureless babble. This is sung at the end of the bird's first summer. In the following spring, subsong matures into the second stage, *plastic song*. Plastic song is similar to the final stage (*full song*) in having most of the right syllables, but it lacks organisation into phrases. Once full song emerges, it remains unchanged throughout the bird's life. This developmental progression suggests that the bird needs to practise its song in order to mimic the model it has heard.

To see how important such feedback is, Nottebohm[227] deafened birds at different stages of song development. Deafening had very little effect on adult birds whose songs had already matured (Figure 4.2a). If birds were deafened while singing plastic song (Figure 4.2b), however, the normality of their song depended on how long they had been singing plastic song before being deafened. If they had been singing for some time, full song was almost normal; if only for

Figure 4.1: Sonographs of Chaffinch Song. (a) Male chaffinch kept in auditory isolation from hatching. (b) Bird reared with a peer. (c) Bird provided with an adult male chaffinch tutor.

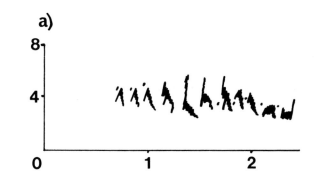

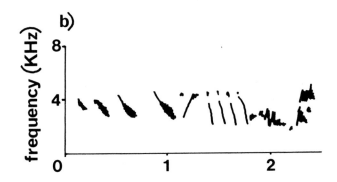

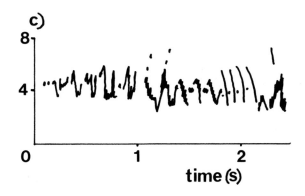

Source: Modified after Thorpe[303].

Figure 4.2: Sonographs of Songs of Deafened Chaffinches. (a) Male deafened as an adult. (b) Bird deafened during plastic song. (c) Bird deafened during subsong.

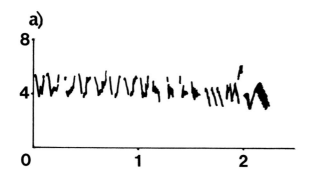

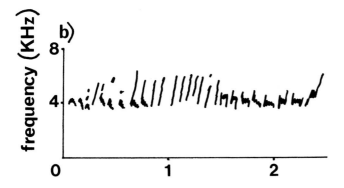

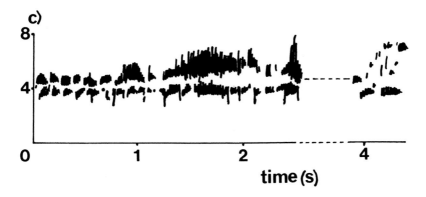

frequency (KHz)

time (s)

Source: Modified after Nottebohm[227].

a short time, the song was less complex than usual. Birds deafened before or during subsong ended up with a long screech (Figure 4.2c). The type of song chaffinches finally sing therefore seems to be a joint product of the songs they hear when young and the attempts they make to match them.

Finally, although it may be obvious that experience underlies the development of a particular behaviour, it may not always be clear how experience effects the changes we see. For instance, cockroaches clean their antennae with the front pair of legs. If the front legs are amputated, the cockroach eventually (after 8 to 10 days) learns to use the middle pair. However, it is not the ability to clean with the middle legs that requires practice (if a cockroach is forced onto its back after amputation of the front legs, it can clean with the middle legs immediately), but the ability to balance on three legs while doing it.

4.1.2 Maturation and Motivation

The performance of a behaviour by a young animal may occur in a completely different context to its performance by an adult. The motivation underlying a behaviour varies with the developmental stage of the animal. A young great tit pecking at a small conspicuous object in its environment is unlikely to be hungry. Hungry fledglings solicit food from a parent rather than securing it for themselves. Cichlid fish (*Etroplus maculatus*) approach conspecifics for different reasons when young than when adult. Approach and 'glancing' behaviour by young fish are expressions of feeding (the fish skim mucus from the side of a parent) and approaches are only made to larger individuals. Overtly similar behaviour by adult fish is part of pair bond maintenance and is directed towards individuals of similar size.[326]

Motivational development can be complex, as illustrated by the development of aggression in the jungle fowl (*Gallus gallus*). Week-old chicks, hopping about, happen, by chance, to collide occasionally with other birds. By the time they are two weeks old, hopping has begun to be deliberately orientated towards other birds. By three weeks, a variety of aggressive behaviours have become included in the chicks' repertoire including 'frontal threat', 'leaping', 'aggressive pecking' and 'kicking'. At this stage, aggression appears to be motivationally linked with locomotion and feeding (active and/or hungry chicks fight more). After about 21 days, fighting starts to become more complex involving several *ambivalent* (see Section 2.4.2) approach/retreat postures. After four weeks it is also more specific in context, being more likely when birds are in close proximity, after an alarm and after feeding and/or preening. By 50 to 80 days, fighting is motivationally divorced from locomotion and feeding with which it is associated in chicks.

Fear and avoidance responses show clear patterns of development in both birds and mammals. Before and at hatching, domestic chicks show little response, apart from reflex blinking, to the kinds of stimuli which will produce fear or avoidance when they are adult. However, responsiveness to auditory, visual and painful alarm stimuli increases rapidly after hatching, reaching a plateau after only a few days. In mammals, fear responses to visual stimuli take much longer to develop, ranging from two and three weeks in dogs and monkeys to six months in humans. However, distress reactions to pain and low temperature can be elicited at birth.

Feeding behaviour in neonatal mammals appears to be unrelated to the degree of food deficit. Puppies fed through artificial nipples with large apertures, which thus obtain their required amount of milk very quickly, suck more at objects not associated with feeding than those fed through nipples with small apertures. Similar results have been obtained with kittens in which the amount of time spent sucking is independent of the rate of satiation over the first three weeks of life. Human babies actually suck more when satiated or aroused in some way than when hungry.

From our discussion in the preceding two sections, it is clear that behaviour, in the sense of an adult animal's response to some stimulus, is the result of progressive modification at a variety of developmental levels. Stimulus-response relationships may change with the maturation of physiological systems and the potential for new modes of expression. For example, neural maturation may lead to enhanced perceptual abilities. In salamanders, the orientation of responses to visual stimuli depends on precise neural connections between the retina and the brain. Disturbance of these connections (e.g. by rotating the eye through 180°) may lead to a permanent re-orientation of movement (e.g. locomotory movements may be towards the opposite side of the visual field to an approach-eliciting stimulus). Physical maturation is one way in which stimulus-response relationships can alter through development. Hormonal effects is another (see Section 2.2). Hormones are particularly important in changing the relationship between motivational state and behaviour. However, important as these maturation effects are, it is through various types of learning that behaviour shows the greatest developmental changes.

4.2 Learning

A parallel is often drawn between the processes of learning and the evolution of behaviour by natural selection. Both are seen as means by which animals improve their behaviour and become more effective at solving the environmental problems facing them. Learning shapes behaviour *within* generations, while selection operates *between* generations. We have already mentioned several examples of behavioural improvement through experience. In this section, we shall examine the various ways in which experience is thought to effect such changes and the constraints that may be levied on an animal's ability to learn. The section also provides a conceptual framework for interpreting the many consequences of learning discussed or implied in later chapters.

4.2.1 Types of Learning

The development of learning theory and the means of measuring learning has been the result of many different workers using almost as many different approaches. The pioneers of learning theory include some of the greatest names in the history of the study of behaviour – Pavlov, Thorndike, Watson, Hull, Tolman, Skinner, to mention a few. Not surprisingly, this plethora of great minds has resulted in complicated disagreement over the fundamental issues in learning. Readers wishing to examine these historical aspects in more detail are referred to Bolles[30]. However, despite the polemics, it is possible to distinguish between several different types of learning or, more accurately, different ways in which learning has been shown to occur. The classification we shall use was first proposed by Thorpe[304] and is only one of many. Like others, it has its critics. (As a brief illustration, one of its categories, classical conditioning, was first proposed by Pavlov, discounted by Thorndike, modified by Watson and Guthrie, regarded as unimportant by Tolman and Hull and accepted by Skinner.[30])

Thorpe defines learning quite simply as 'that process which manifests itself by adaptive changes in individual behaviour as a result of experience'. As Manning[199] points out, this definition acknowledges two important points; (1) that learning is adaptive, and (2) that we can not usually measure it directly. What we measure are the consequences of learning. Thorpe divides learning into six categories:

(1) Habituation.
(2) Classical conditioning (also called conditioned reflex (CR) Type I, respondent or Pavlovian conditioning).
(3) Operant conditioning (also called CR Type II, instrumental conditioning or 'trial and error' learning).
(4) Latent learning.
(5) Insight learning.
(6) Imprinting.

Habituation. Habituation is, perhaps, the simplest type of learning in that it involves the loss of old responses rather than the development of new ones. In fact Razran has defined habituation as 'learning what not to do'. Animals are constantly bombarded by a host of different stimuli emanating from the environment. The time and energy costs of responding to every one would clearly be prohibitive. Furthermore, only a small proportion of the stimuli require a response. In Chapter 2, we saw how the animal's perceptual mechanisms filter out some of the noise. Habituation is a way of filtering out some more. More specifically, it is a way of eliminating responses to stimuli which are sometimes important but which, in a particular case, are irrelevant. Rustling leaves, for instance, are worth reacting to (e.g. by hiding) because they sometimes indicate the approach of a predator. However, repeated rustling without the

Figure 4.3: (a) Habituation in Territorial Male Sticklebacks. a–d as in text. (b) Habituation in Rats to Tones Presented at Different Intervals. Closed circles, rats trained to expect 16-second intervals; open circles, rats trained to expect 2-second intervals. Left: prehabituation phase in which animals were presented with 2, 4, 8 and 16 second interstimulus intervals at random. Centre: habituation training phase where animals received either 2 or 16 second ISIs. Right: posthabituation phase which was identical to first phase.

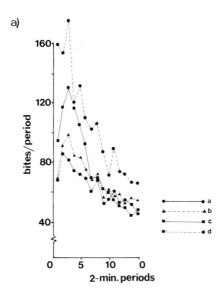

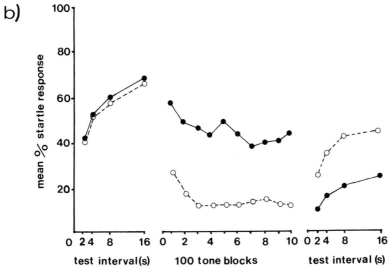

Source: (a) Modified after Peeke and Veno[242]. (b) Modified after Davis[77].

appearance of a predator is likely to be caused by the wind. In this case, the animal ceases to respond (i.e. habituates) to the rustling.

An important feature of habituation is its stimulus specificity. A decrease in responsiveness occurs only with reference to the *habituating stimulus* or one very like it. This stimulus specificity is shown clearly by habituation in territorial sticklebacks. Peeke and Veno[242] observed decreases in the level of aggression between neighbouring territory owners. Fish were aggressive to one another when they first established territories but quickly ceased to respond. The adaptive sense of this habituation is clear; there is no point wasting time and energy chasing fish which are neighbours rather than intruders. Equally clearly, territory owners cannot afford to generalise this habituation to all conspecifics. It should be selectively restricted to neighbours.

To test this, Peeke and Veno introduced 'intruder' fish in glass tubes to various locations within a territory owner's tank. Owners were exposed to the 'intruder' for 30 minutes, given a 15-minute 'rest' and then tested with one of four stimuli: (a) the same fish in the same location, (b) the same fish in a different location, (c) a different fish in the same location or (d) a different fish in a different location. The results were clear-cut (Figure 4.3a). Owners showed least aggression towards stimuli (a) and (b) though (b) elicited more than (a). More aggression was shown towards stimulus (c) and most towards stimulus (d). Decreasing familiarity in terms of both individuals and their spatial location therefore decreased the degree of habituation although, interestingly, not the rate at which it occurred (the lines in Figure 4.3a have approximately the same slope).

Another important factor influencing habituation is the length of the inter-stimulus interval (ISI). The longer the ISI, the less habituation we expect. Davis[77] investigated the effects of increasing the ISI in rats. He trained animals to expect different ISIs for a 50 ms alarm tone of 120 db, then compared their tendencies to habituate under identical conditions. He found that long ISIs were less effective in producing habituation but had longer-lasting effects overall (Figure 4.3b).

Habituation is the selective reduction in responsiveness to a stimulus as a result of what the animal learns about it. It is similar in some ways to sensory adaptation (for instance, we cease to notice our clothes soon after putting them on because tactile receptors in the skin adapt and cease firing). In many cases, however, as in the classic retraction response of *Nereis* to mechanical stimulation with a probe, habituation is clearly not a function of sensory adaptation. Habituated worms still perceive the probe because they try to seize it in their jaws.[55] Habituation-like processes occur in all taxonomic groups of animals from the Protozoa upwards. Indeed, as Manning[199] points out, it is almost a universal property of living matter.

Classical Conditioning. Classical conditioning is regarded as the simplest form of an heterogeneous set of phenomena known collectively as *associative learning*.

Figure 4.4: Pavlov's Classic Conditioning Experiment Linking a Conditional Stimulus (Bell) with the Salivation Response in Dogs.

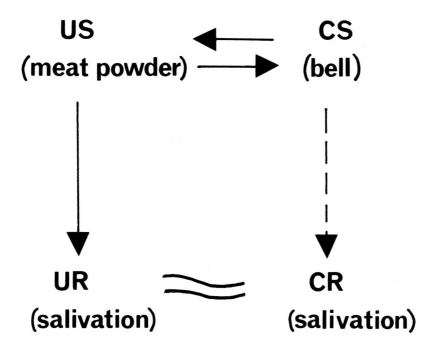

Source: Modified after Fantino and Logan[104].

It was first proposed by the physiologist and pioneer learning theorist Pavlov. A stimulus which does not initially elicit a response comes to do so by association with a stimulus which does. The eliciting property of the stimulus is *conditional* upon its association with an established stimulus-response relationship. Since the stimulus in the established relationship does not require special conditions to elicit a response, it is known as the *unconditional stimulus* (US). The response it elicits is the *unconditional response* (UR). Because the otherwise ineffective stimulus only elicits a response through its association with the US, it is called the *conditional stimulus* (CS). The response it elicits is the *conditional response* (CR). Although the CR is functionally similar to the UR, there may be subtle qualitative and quantitative differences between the two types of response. Pavlov's original classic experiment is summarised in Figure 4.4.

The temporal and/or spatial assocation of the CS and the US is known as *reinforcement*. The close proximity of the US eventually leads the animal to 'anticipate' it on perceiving the CS and hence perform the CR. The important relationship in classical conditioning is that between the two stimuli. Reinforcement occurs even if the animal does not perform the CR (in early trials the CS is almost always ineffective in eliciting the CR). This contrasts with operant

Figure 4.5: Types of Classical Conditioning Schedule. US, unconditioned stimulus; CS, conditioned stimulus.

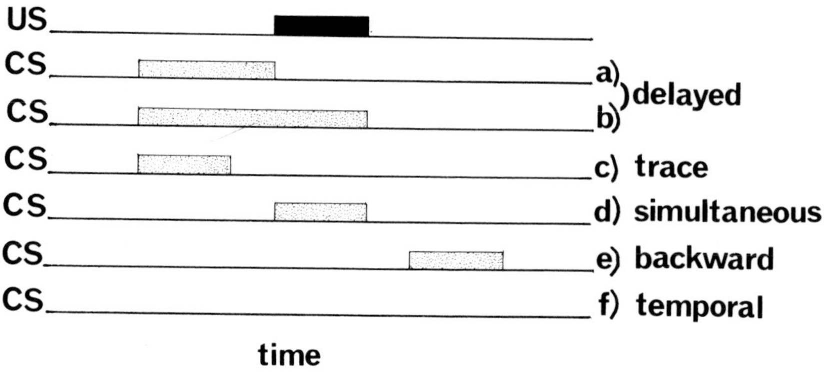

Source: Modified after Fantino and Logan[104].

conditioning where the animal experiences the reinforcing stimulus only if it performs the CR.

From the above discussion, we might conclude that the CS has to precede the US if stimulus association is to occur. In fact, this is not so, although the relative positioning of the two stimuli does affect the efficacy of conditioning. Figure 4.5 shows a range of six possible temporal relationships between the CS and US. So far, we have only considered situations a and b in the figure (so-called *delayed conditioning*). Compared with situations c to f, delayed conditioning is by far the most effective method. *Backward conditioning* can be made more effective by presenting the US and CS contiguously, but *simultaneous conditioning* is remarkably ineffective. Of particular interest, especially from the point of view of biological clocks (see Chapter 1), is situation f. *Temporal conditioning* relies on the presentation of the US at predictable time intervals so that time itself acts as a CS.

Classical conditioning shows several other characteristic features. In particular, *repetition* of the CS–US association improves the CR. Pavlov found that his dogs' salivary CR improved with successive reinforced trials until it was comparable with the UR. Although repetition only enhances the CR up to a maximum, further repitition or *overtraining* increases the CR's resistance to *extinction*. Extinction refers to the degree of persistance of the CR after reinforcement ceases. When rewards are no longer given, the CR decays and disappears. For some reason, however, it appears that classically conditioned responses are more easily extinguished than those produced by operant conditioning. A CR is not completely lost with extinction. The later provision of a reward will resurrect it, as will combining the CS with a novel stimulus (e.g. by combining the bell with a flashing light in Pavlov's experiment). Two further properties of classical conditioning are *generalisation* and *discrimination*. Animals conditioned to a given CS (say a tone of 1,000 cycles) may also perform to different, but qualitatively similar, stimuli (e.g. tones of a different number of cycles). The CR is thus generalised to stimuli other than the CS. Conversely, animals can be extremely selective in the stimuli to which they respond. Dogs, for instance, can be trained to respond to just one tone, especially if responses to other tones are punished slightly.

Originally, classical conditioning was assumed to reflect the learning of a response to the CS. As Bolles points out, however, it may not involve the learning of a response at all. Animals show convincing signs of being able to *anticipate* the US. Classical conditioning may therefore reflect not the learning of a CR but the animal learning to expect the US. Experiments by Rescorla and coworkers[253] have lent considerable weight to the idea that classical conditioning occurs not when the CS is *paired* with the US, but when it *predicts* the US. This smacks of cognitive processes (see later). Whether or not we accept this view, conditioning cannot be tacitly assumed to produce a CS–CR relationship.

Operant Conditioning. A different form of conditioning occurs during so-called trial-and-error learning. Instead of starting out with a US-linked UR which becomes associated through experience with a novel stimulus, an animal may become conditioned through initially chance reinforcement to respond to a previously ignored stimulus. For example, a hungry animal wandering about in search of food is likely to perform a number of different behaviours. If one of these happens to procure food and is associated with finding food sufficiently often, the animal learns to perform the behaviour regularly in that particular situation.

The central concept of operant conditioning is reinforcement. A reinforcer is defined as *any event which increases the probability that the behaviour it follows will recur in the future*. There are two kinds of reinforcer: *positive* and *negative*. Positive reinforcers make the actions they follow more likely by their occurrence, while negative reinforcers make the actions they follow more likely by their termination or non-occurrence. An electric shock which induces a dog to jump a hurdle and thereby escape the shock, acts as a negative reinforcer because it increases the likelihood of the response that terminates it. Logically, it can be difficult to distinguish between positive and negative reinforcers. For instance, eating by a hungry animal can be viewed as positive reinforcement (getting a food reward) or negative (escaping from deprivation or hunger). Similarly, the negatively reinforced jumping of the dog could be explained as positive reinforcement as the dog escapes to safety.

A negative reinforcer, however, must not be confused with a *punisher*. When we are dealing with reinforcement, whether positive or negative, we are dealing with factors that *enhance* the behaviour they follow. Punishment *weakens* the behaviour it follows. In our dog example, electric shock would be a punisher if it decreased the likelihood that the dog would remain in the cage until the floor was electrified. Punishment thus constitutes the presentation of a noxious stimulus. A rather more contorted notion is that of a *negative punisher*. Here, a response is made less likely by the *omission* of a positive reinforcer. A dog can be trained not to jump at people by withholding a food reward when it jumps. Only by not jumping does the animal receive the reward.

Avoidance behaviour of the type implied by the dog jumping to avoid shock, poses an intriguing problem for reinforcement theory. The main difficulty is that nothing can be explicitly designated as the reinforcer for avoidance. The avoided stimulus clearly cannot act as a reinforcer, neither can we really view its non-occurrence as reinforcement. As Bolles points out, many things logically do not occur. The non-occurrence of the stimulus in question (say shock) can no more be claimed as a reinforcer than the non-occurrence of any other stimulus (e.g. a loud noise).

An interesting feature of avoidance learning is the difference in the readiness with which a given animal will acquire an avoidance response to different stimuli. Thus rats readily learn to avoid running in a wheel when this results in shock but hardly ever learn not to press a bar. A possible explanation is suggested by Bolles[29]. His argument hinges on the fact that some natural hazards like predators present little opportunity for an animal to learn. If it makes a mistake, it is dead. Intense selective pressure of this sort is likely to result in genetically-coded rather than learned avoidance responses. Like other aspects of behaviour, avoidance may be highly stereotyped so that we can talk of *species-specific defence reactions* (*SSDRs*). In the rat, therefore, the SSDR may be predominantly fleeing or 'freezing'. When expecting a noxious stimulus the animal can only flee or 'freeze'. Hence the rat's ability to learn with the running wheel but not with the bar-pressing. Activities akin to bar-pressing do not come within the rat's repertoire of avoidance reactions. Although an over-simple explanation for avoidance characteristics, Bolles' interpretation highlights an important problem in learning: animals do not learn different tasks with equal facility. We shall return to this point in Section 4.2.2. The adaptive value of avoidance responses is obvious. Scrub-jays and rats learn to avoid toxic or otherwise noxious food once they have suffered the ill-effects of eating them, small passerines bunch together or retreat into cover when they perceive a hawk-like silhouette and so on. Avoidance behaviour can also have far-reaching evolutionary effects. Predators which learn to avoid distasteful prey, for instance, create the selective pressure for Batesian and Mullerian mimicry complexes. Alarm displays and chemical defence mechanisms may similarly evolve through predator deterrence. Avoidance behaviour carries a high survival premium for both predator and prey.

Early views of operant conditioning saw the consequences of responses as automatically reinforcing any preceding stimulus-response association. However, the bulk of more recent evidence shows that this is too simplistic. Mowrer[218], for example, sees the enhancement of a response as entirely attributable to increased levels of motivation as a result of reinforcement. In Mowrer's view, food does not progressively enhance an animal's approach response to a feeding site, but increases its motivation to approach. The operantly-conditioned stimulus-response association can thus be reinterpreted as classically-conditioned motivation. It is now generally agreed that many instances of learning involve both types of conditioning. Operant conditioning determines qualitatively the response performed, while classical conditioning determines the level of motivation with which it is performed. For good detailed discussions of classical and operant conditioning see Bolles[30] and Fantino and Logan[104].

Latent Learning. Thorpe's fourth learning category can also be regarded as associative learning, but of a special sort. Thorpe defines it as 'the association of indifferent stimuli or situations without patent reward'. Learning thus occurs in the absence of any obvious reward and the occurrence of learning may not be immediately apparent. Learning remains unexpressed or *latent* within the animal. Latent learning is nicely illustrated by Tolman and Honzik's[309] experiment with rats.

Figure 4.6: Latent Learning by Rats in a Maze. Open circles, animals trained without food; squares, animals trained with food present; closed circles, animals trained without food until day 11, then provided with food in the goal.

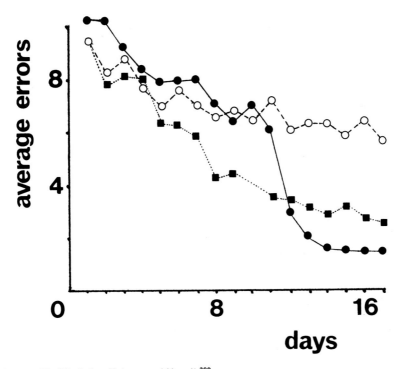

Source: Modified after Tolman and Honzik[309].

Three groups of rats were tested in a maze. Animals in a conventionally rewarded group received food when they successfully completed the maze. *No reward* rats were allowed to wander in the maze at will but were never presented with food in the goal box. A third, *delayed reward*, group received no food reward for the first ten days of the experiment but was given food on successful completion of the maze on the eleventh and subsequent days. Figure 4.6 shows the progress of maze learning in the three groups.

Not surprisingly, the *no reward* group showed little evidence of learning (the number of errors remained high). The *conventionally rewarded* groups showed a typical learning curve. *Delayed reward* animals, however, learned at a rate comparable with the *no reward* group until day 10, but then showed a precipitous decrease in error-making. This sudden decrease when food was presented was far more dramatic than would be expected if animals only started to learn about the maze at that point. The data imply that the rats developed some kind of 'map' of the maze during the ten days in which they were wandering about without reward. This latent knowledge, however, was only put into practice when a reward was attainable.

The adaptive implications of latent learning are obvious. A detailed geographical knowledge of the home area could be crucial in escaping from predators. Some animals which make long foraging trips show preliminary 'reconnaissance' movements before departing. Particularly when colonising a new area, many insects perform pre-foraging 'orientation flights' during which they establish the position of the home area relative to the sun and other reference points.

Insight Learning. Insight learning is widely regarded as the most advanced form of learning. Responses produced by insight are those resulting from a rapid appreciation of relationships in which animals solve problems too quickly to have gone through a trial-and-error process. The animal seems to arrive at a solution by *reasoning* (defined as 'the ability to combine spontaneously two or more separate or isolated experiences to form a new experience, which is effective for obtaining a desired end[199]). There is still considerable argument, however, about whether insight really is different from a trial-and-error process, or whether the animal simply substitutes a mental elimination exercise for a physical one.

A common experimental procedure used to test for reasoning involves an animal's ability to make detours in a maze.[197,280] Shepard[280], found that once rats had learned a maze, they quickly capitalised on newly created short-cuts (created by removing partitions, so that what was previously a blind alley became a quicker route to the goal box). Tool use, e.g. chimpanzees reaching bananas high up in their cage by piling boxes on top of each other or joining lengths of stick together,[162] provides another possible instance of reasoning.

Thorpe also includes within his definition of insight learning, various forms of *imitation*. Imitation covers an heterogeneous range of behaviours including *social facilitation* (the performance of a behaviour already within an individual's repertoire as a result of its performance by another individual), *local enhancement* (the increased tendency to respond to a particular part of the environment as a result of responses to it by another individual, e.g. the spread of milk-bottle opening through tit populations) and *true imitation (the copying of otherwise improbable utterances or acts,*[303] e.g. vocal mimicry in birds, food washing in primates). In Chapter 10, we shall discuss the ability of chimpanzees to learn human sign languages, an ability which, as we shall see, strongly suggests insight learning.

Imprinting and 'Critical Periods'. The final type of learning in Thorpe's classification, *imprinting*, has a number of unique features which set it apart from the others. These include (a) its long-lasting effects, (b) its irreversibility and (c) its occurrence during a 'critical period' in the animal's development.

Basically, imprinting results in the narrowing down of the range of stimuli to which certain social responses are directed. More specifically, Salzen[266] defines it as a process 'in which innate social responses that can be elicited by and directed to a wide variety of objects, come to be elicited by and directed to only the classes of objects experienced in a limited neonatal period'. At least in some species, there seems to be no limit to the variety of objects on which individuals will imprint. For instance, birds have been imprinted on objects ranging from humans and canvas hides to balloons and match boxes. However, birds frequently show 'naive' preferences for objects with certain characteristics (e.g. colour, shape, size, movement).

Attachments formed during this limited neonatal period are referred to as *filial imprinting* because the responses established are those of an offspring to a parent object. However, far from being limited in their effect to juvenile stages of development, imprinted attachments influence behaviour well into adult life. In particular, they may affect the animal's choice of a sexual partner (so-called *sexual imprinting*). Schutz[273] reared mallard ducklings with foster parents of other species. When the mature birds were released onto a lake containing various duck and goose species, many attempted to mate with members of their foster parents' species. However, the tendency was more pronounced in drakes. A possible explanation is that females of many duck species are drably coloured and difficult to distinguish. Drakes therefore need a hard and fast rule for identifying the correct species, a need amply catered for by imprinting. Females, on the other hand, choose between brightly coloured drakes whose appearance differs markedly between species. Discrimination is much easier and need not rely on learning. In species where both sexes are drab, as in doves, both males and females acquire their sexual preferences through early imprinting. The irreversibility of these imprinted attachments is reflected in the inability of 'counter-experience' to change them. Even when birds reared by a different species have been mated for several years with conspecifics, they prefer to mate with their foster species when given a choice.

The most intriguing and characteristic feature of imprinting, however, is its

Figure 4.7: Duration of the Critical Period in Mallard Ducklings as Measured by Post-training Responsiveness to Objects at Different Post-hatching Ages.

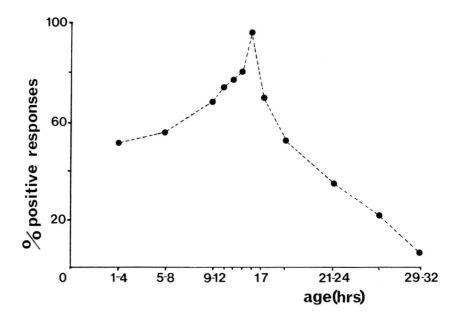

Source: Modified after Hess, E.H. (1958). Imprinting in animals. *Sci. Am., 198*: 81–9.

occurrence within a proscribed *critical period*. The critical period refers to a phase in the animal's development during which it is particularly sensitive to certain stimuli. It is also variously called the optimal, vulnerable, crucial or susceptible moment, point or stage. Figure 4.7 shows Hess's[142] classic data for the relationship between post-hatching age of exposure to certain stimuli and the degree of imprinting revealed by subsequent tests in mallard ducklings. There has been much argument about the nature of critical periods. While most workers agree that imprinting does not occur equally well at all stages of an animal's development, many suggest that changes in sensitivity are more gradual and flexible than those shown in Figure 4.7. In some cases, for example, the timing of sensitivity can be changed by altering rearing conditions. A useful model of critical periods has been proposed by Bateson[19].

Bateson uses the analogy of carriages in a train (Figure 4.8). The train represents a developing animal travelling one way from 'Conception' to a point where it vanishes from the tracks. Each compartment and its occupants represent a behavioural system which is sensitive to the outside environment at a certain stage of the animal's development. The occupants are temporarily screened from outside stimuli by opaque windows. Figure 4.8a shows the extreme situation in which all the windows are closed for the first part of the journey but are flung open at a particular moment to expose the occupants. Later, all the

Figure 4.8: Bateson's Train Carriage Analogy of Critical Periods (see text).

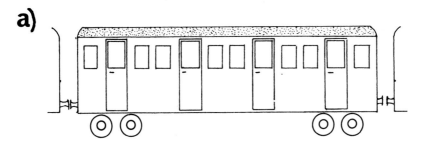

a)

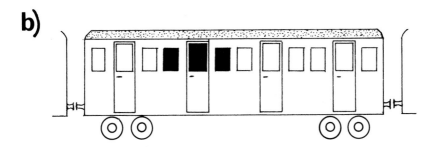

b)

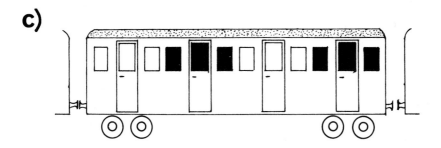

c)

Source: Modified after Bateson[19].

windows are closed again. The occupants of all the compartments are thus exposed to the environment for the same period. Alternatively (Figure 4.8b), the windows of different compartments may be opened and closed at different times. In this case, different behavioural systems become sensitive at different stages of development. A third possibility (Figure 4.8c) is that the windows of a compartment may be opened but never closed. Now any end to the critical period results from changes to the occupants (behavioural systems) themselves rather than a cut-off mechanism. A variety of different mechanisms can thus produce overtly similar results. What we need to know now is how the onset and offset of the critical period is controlled.

Intuitively, we might expect two factors to be important: (a) the developmental stage of the animal and (b) its postnatal post-hatching experience. Gottlieb[120] found that the extent to which mallard ducklings would narrow their preferences for moving objects was better measured in terms of developmental age than age after hatching (which Hess had used). The ensuing argument was resolved by Landsberg[180] who, by keeping the time between training and testing, rather than post-hatching age, constant, showed that both post-hatching and developmental age had significant effects.

When it comes to the offset of the critical period, imprinting is more or less a self-terminating process. By limiting preferences to those stimuli which are 'familiar', it effectively prevents new experiences from modifying them. However, some stimuli are more effective at restricting preferences than others. Birds, for instance, respond more to conspecifics than to the cages they are kept in. Likewise, birds reared with a 'near-optimal' stimulus, such as a sibling, form preferences more quickly than birds reared in isolation. An interesting paradox occurs when a young bird's experience is experimentally attenuated. On the one hand, attenuation can increase the age at which birds respond to novel objects; on the other, short bursts of various social experiences can increase the tendency to avoid novel objects. What the paradox suggests is that stimulation can have both general and specific effects.[19] It initially raises responsiveness to all objects and then restricts attachments to those with which the animal has become familiar.

An important question begged by Bateson's train analogy concerns the extent to which the sensitivity and exposure of compartments are jointly controlled. For example, are filial and sexual imprinting governed by the same behavioural system? The evidence is conflicting. Some birds certainly direct their sexual attentions to objects on which they have been previously imprinted. Others prefer different objects. When presented with a range of models at different ages, cockerel chicks showed sexual preferences for the models to which they showed least filial attachment.[322] Mallard drakes have been shown to prefer sexual models which were not encountered until well after their presumed filial critical period,[272] although this effect was confounded with several other variables.

There have been many suggestions as to the function of imprinting. It has been seen as a species recognition mechanism, a 'fail-safe' back-up to other behavioural systems involved in species recognition and a means of appropriately modifying stimulus preferences as a species' characteristics change during evolution. However, all these have the shortcoming that they lump filial and sexual processes. Bateson[19] prefers to consider imprinting as a two-fold kin recognition mechanism. First, imprinting facilitates the formation of a parent-offspring bond. Parents who selectively raise their own offspring tend to have greater reproductive success than those which are indiscriminate and thus promote the survivorship of genes for parental care. Filial imprinting is thus a by-product of selection for discriminating parents. Secondly, sexual imprinting allows an individual to learn the characteristics of close kin and subsequently to choose a mate which is slightly different. In this way, the animal can achieve a balance between inbreeding (with its benefits of maintaining advantageous gene complexes intact but risks of resulting in lethal recessives) and outbreeding (with its advantages of increased offspring variability in changeable environments). Recently, Bateson and coworkers have shown that some bird species prefer to mate with individuals having slightly different physical characteristics to their own parents. Japanese quail (*Coturnix coturnix japonica*) prefer mates with slightly different plumage colour[20] while Bewick's swans (*Cygnus columbianus*) prefer mates with slightly different facial markings.[21]

4.2.2 Limits to Learning

For a long time, learning theorists, particularly Skinner, believed that a universal law underlay learning. It was believed that principles used in learning by one animal in one situation could be generalised to all types and conditions of learning. Despite an enormous amount of research, this universal law has proved elusive. During our discussion in Section 4.2.1, there were suggestions that learning may not occur as easily in some situations as in others. We must therefore question the universal or equipotential nature of learning. In this section, we shall discuss two aspects of learning which illustrate its dependence on a variety of factors, not least the type of animal used, the nature of the animal's experience and the task being asked of it.

Constraints on Learning. Attention turned to the possibility that learning might be geared to species-specific ways of life, an idea already touched on in the discussion of avoidance learning. Broadly what investigations have shown is that (a) inflexible species-specific responses sometimes exist which prevent the learning of certain tasks, (b) some responses are more easily altered than others and (c) there are marked differences between species in the tasks that are learned and the ways in which they are learned. Such variation clearly cannot be embodied within a single generalised principle.

The first solid evidence for the existence of *constraints on learning* came from the work of Garcia. Garcia and Koelling[111] tested the efficacy of two punishers (nausea and shock) in suppressing drinking in rats. However, the punishers were paired with two artificial characteristics of drinking water. Water was either 'bright'-'noisy' (a loud noise or flashing light was produced when the rat drank) or 'tasty' (the water was flavoured with saccharin). Although both punishers conditioned avoidance, their effects depended on the characteristics of the water. Nausea was only effective with 'tasty' water and shock with 'bright'-'noisy' water. The rats only associated punishers and water cues which were qualitatively related (gustatory cue with digestion-related punisher and mechanical cue with mechanical punisher). A similar effect was found in chaffinches by Stevenson-Hinde[298]. She could train chaffinches to perch at a particular point in her apparatus to produce playback of chaffinch song. She could also train them to peck a key for food. It was extremely difficult, however, to train them to perch for food or peck for song. The functional 'relevance' of a behaviour therefore played an important part in the birds' ability to link it with a particular reward. The 'Garcia effect', however, is not a simple all-or-nothing process. Rats and many other species appear to have a hierarchy of cue characteristics which allow different degrees of learning.

Constraints on learning are also apparent in insects. A classic instance was reported by Schneirla[270]. Schneirla tested the maze-learning ability of ants (Hymenoptera) by having the ants' nest as the starting point and food in a goal box as the finish. During trips *from* the nest to the food, ants showed little sign of learning the maze. If the maze was reversed so that the ants carried food *back* to the nest, they learned very quickly. The ants' maze-learning ability thus depended on their motivational state.

Similar limits to learning have been shown in birds. Kramer[163] had great difficulty training captive starlings to change the orientation of their movements according to the position of the sun. However he trained them, the birds always went to the east of the aviary (where they were accustomed to finding food). It transpires of course that starlings have an internal clock with which they compensate for the sun's apparent movement and thus migrate in a constant direction (see Chapter 5). The interesting point is that this precludes them from learning a much simpler task using the same cue. Another intriguing example comes from the ontogeny of bird song. Some birds can be made to learn a wide variety of song types, some very different to those of their own species. Others are remarkably specific in what they will learn. If swamp-sparrows (*Melospiza georgiana*) are played artificial songs which are similar in construction to their own species' song but consist partly of marsh-sparrow syllables and partly of syllables from song-sparrow (*M. melodia*) songs, they will selectively learn only the swamp-sparrow syllables.

Limitations on learning such as these can lead to problems in operant situations. Animals fail to learn a task when their performance with other, often more difficult, tasks would lead us to expect otherwise. Breland and Breland[34] discuss a number of such 'misbehaviours' in various species. What we now know is that this failure is not a sign of poor learning ability, but evidence that learning is tailored to the animal's needs. The term *constraints on learning*, however, is perhaps not wholly appropriate because it suggests that learning is limited by its design features. As Bolles[30] points out, this tailoring can equally well result in *facilitations* of learning. We should thus think in terms of learning design and structure rather than constraints.

Attention. The potential to learn may be restricted in another way. Animals do not *attend* equally to all the stimuli reaching them. They preferentially attend to those which are usually reinforced. Pigeons or doves prefer to peck the lighted key of a pair in a Skinner Box, cats attend more to movement rather than colour and so on. How does attention to a stimulus affect its role in learning?

One important effect of attention is that it may bias the capacity for different stimuli to become associated with a response. The control of a response by a stimulus may depend not only on the relevance of the stimulus and its past association with the response but on the relevance and past association of any other stimuli present. If stimulus A has been associated with reinforcement in the past, it may prevent the development of response control by stimulus B, when A and B are presented together. This is known as *blocking*. Sometimes the effectiveness of a stimulus may simply be reduced when it is presented in association with others, with the degree of reduction dependent on the relative strengths of the other stimuli. Miles and Jenkins[213] found that a tone stimulus could be more effective than lights of different intensity when the light intensities were very similar. When the lights were easily distinguishable, light was better than tone. In the first case the tone is said to *overshadow* the light; in the second, the light overshadows the tone. Pavlov found that, as a rule, the more intense of any pair of stimuli tended to overshadow the other. Also stimuli with which the animal has had prior experience overshadow those which are novel. As Rescorla and Wagner[254] suggest, this may be because a given reinforcer can effect only a limited degree of conditioning. Once its potential is used up by one stimulus, it is not available for another.

Attention can also be conditional on certain correlations between stimuli. Pigeons presented with circles or triangles on red or green backgrounds can be trained to attend to shape when a blue lamp is shining, but to background colour when the lamp is yellow. For obvious reasons this is known as *conditional stimulus control*. Sometimes, an animal can be trained to discriminate between two stimuli by using an already effective stimulus and *transferring stimulus control* to one of the new stimuli. Frequently, this can be done by establishing attention to the old stimulus and then 'fading' in the new so that attention is uninterruptedly transferred. *Feature-value effects* (the attention-facilitating effect of some distinct feature of a stimulus, e.g. a spot on a Skinner Box key) also influence stimulus-conditioning efficiency.

Attention, by being directed towards stimuli associated with reinforcement, is likely to emphasise stimuli which are important biologically. Attention to different stimuli may be modified by experience. Through changes in reinforcement, previously irrelevant stimuli may become selectively attended to, while others cease to correlate with reinforcement and lose their responses. However, as we have seen, previous experience can also inhibit attentional shifts to new stimuli through the 'inertia' of past associations.

4.3 Choice

Animals are continually faced with choices between alternative courses of action. At any given time, an animal could make one of a number of responses. We have seen how internal motivational 'choices' are made between alternative behaviours; in this section we shall briefly examine the way different types of reinforcement influence choices between external stimuli once an animal has decided which behaviour to perform. Making a choice often depends on experience of alternative reinforcement schedules over a period of time, rather than on an instantaneous decision. If so, the relative rate and/or duration of reinforcement are likely to be important.

4.3.1 Matching

Herrnstein[140] and Catania[48] showed that pigeons in a Skinner Box in which pecks to different keys resulted in different food reward schedules (*concurrent variable-interval schedule*), delivered most pecking responses to the key yielding the highest rate or duration of food presentation. In each study, the pigeons' pecking response is said to *match* the pattern of reinforcement (more pecks are delivered to the key giving the highest reward). Response matching has led to the formulation of quantitative laws of reinforcement/response relationships (see Fantino and Logan[104] for an overview) which, while not perfect, successfully predict absolute as well as relative response rates to both single and concurrent reinforcement schedules. The so-called 'matching law' has important functional implications. As Staddon[296] points out, matching is a 'rule of thumb' which tends to maximise an animal's overall reward rate during feeding. It thus conforms to the predictions of optimal foraging models which we shall discuss in detail in Chapter 6. However, a better fit to an optimal foraging model might be produced by 'over-matching' (spending all the time at the key yielding the highest reward rate). As we shall see in Chapter 6, this is exactly what some animals do.

An important alternative to the response rate matching argument, however, is proposed by Shimp[282]. While Shimp acknowledges that animals allocate their time and effort according to the pattern of reinforcement, he suggests that they select the alternative with the highest instantaneous probability of reinforcement at the time. In Shimp's view, animals are not matching their response rates with the rate of reinforcement, but are choosing the stimulus which is likely to be reinforced first. It has not yet been possible to make a satisfactory distinction between these two possibilities.

4.3.2 Choice and Reinforcement Schedule Control

While response matching with concurrent schedules of reinforcement may seem a rather obvious phenomenon, there is a problem. If very different reinforcement schedules or different reinforcers are used, the subsequent measure of choice may confound matching with schedule-controlled responses *per se*.

A simple example illustrates the point: a predator will spend more time hunting for and consuming food during the day than it does sleeping. Despite the obvious difference in time investment in the two behaviours, however, we cannot infer that the predator prefers hunting to sleeping. Clearly, the requirements of hunting behaviour mean that it must be performed more often and for longer periods during the day than sleeping. Similarly, reinforcement schedules may demand specific rates and durations of response.

At a broad level, this is certainly true. Response rates on concurrent variable ratio schedules (where the relative rather than absolute relationship between reinforcement schedules is important) are usually much higher than those on fixed interval schedules. Comparing rates of response as a measure of choice here would be highly inappropriate. At a more subtle level, reinforcement schedules which differ only slightly might even confound choice measurement. Imagine an animal faced with two reinforcers. Both reinforcers are highly desirable but one is just a little more desirable than the other. Imagine also that the two reinforcers can be obtained by pressing keys that operate on equal variable interval schedules. If the schedule is short (say ten seconds), the animal might be expected to persist in pressing the key yielding the most preferred reinforcer. Even though, by trying both keys, the animal would have been reinforced after five seconds, it might be worth its while waiting a few seconds more to be sure of getting the best reward. The animal might therefore show a clear preference for one key. If the variable interval schedule is ten minutes, however, the value of waiting an hour for a slightly preferred reward when the animal could obtain one of the rewards in five minutes by pressing both keys, is questionable. Here, we might expect the animal's effort to be distributed more evenly between the keys.

Caraco and coworkers[43] investigated the functional implications of this in feeding juncos (*Junco phaeonotus*). They trained birds to feed at two feeding stations, one with a certain (2 seed) reward, the other with an unpredictable (0 or 4 seeds) reward. Birds previously starved for a short period chose the 'certain' station because this provided just enough seed for their daily requirement. Those starved for a long period, however, chose the 'risky' station because only there could they obtain sufficient seed to offset their deficit.

4.4 Play

A prominent feature of behavioural ontogeny in many species, particularly mammals, is the performance by young animals of seemingly purposeless activities collectively termed *play*. Kittens frequently 'attack' small, movable objects, batting them about with their paws. Young primates chase each other and engage in sham fights. Puppies chase their tails. Play is one of the most familiar aspects of development, but one of the hardest to explain.

Even defining play satisfactorily is difficult. To begin with, the term covers a wide and heterogeneous array of behaviours, including diverse forms of social interaction, manipulation of inanimate objects and solitary movement patterns. As good a working definition as any is that proposed by Bekoff and Byers[24]. They see play as all motor activity performed postnatally which appears to be purposeless and in which motor patterns from other contexts may be used, often in modified forms and/or altered temporal sequencing. This definition suggests that play affects the structuring of behaviour. A *structural* view of play has been stressed by Loizos[142]. She pointed out that play often involves the rearrangement of adult behaviour sequences. Behaviours within sequences may be exaggerated, repeated and/or incompletely performed. Alternatively, whole sequences may become reordered or fragmented. Behaviours usually performed in quite different contexts may even become involved.

However, Bekoff and Byers argue that play should not be considered solely with respect to adult behaviour, but as an adaptation in its own right. They see play as forming a continuum with prenatal motor activity. Selection favours the early 'practice' of motor patterns because it enhances neuromuscular development. Prenatal movements may even be the precursor of play.

There is little doubt that practice enhances the development of specific behaviours (see Section 4.1). What is not clear is whether playful practice enhances development. Some argue that playfulness is unnecessary. If you need to practice, practice earnestly. Playing merely wastes extra time and energy. Another point against the practice/perfection argument is that play sequences are often extremely similar to their adult form (and thus unlikely to be improved upon by practice) or so dissimilar that their function cannot be one of practice. Bekoff[23] qualifies the argument by saying that play experience enhances 'smoothness' of performance rather than the ultimate goal-achieving properties of a behaviour (e.g. finding food or copulating). For an excellent, critical discussion of these issues see Symons[300].

It has also been suggested that play generates new behaviour patterns, or enhances motivational arousal by providing novel stimuli. By playing, young animals may not so much learn established adult behaviour patterns as 'create' new and uniquely individual ones. This is based on the observation that species which are good learners are generally the most flexible behaviourally. It can thus be argued that a function of learning (and hence play) is to allow the acquisition of novel behaviour. By allowing free combination of different activities, play might come up with a new behaviour pattern which is useful to the animal, a process reminiscent of genetic mutation, recombination and selection. As Symons[300] points out, however, such arguments attempt to deduce function from structuralist comparisons between play and non-play behaviour, rather than being based on studies of the development of new skills. Even if an adaptive behaviour does arise from play, it may simply be an occasional by-product rather than indicative of an underlying function.

Finally, there is a welter of hypotheses, backed by rather qualitative evidence, that play enhances socialisation. Socialisation hypotheses suggest the young animals learn to interact profitably with conspecifics through play. The main arguments are that play helps the development of communication, the reduction of aggression and/or the development of social 'bonds' (see Symons[300] for a fuller discussion). Most of these arguments, however, skirt around the important question of whether such experience results in a selective advantage to play and thus provides a driving force for its evolution.[24] Such an evolutionary tack has been taken by Fagen[102,103]. Fagen modelled the likely costs and benefits of play and pointed to various instances in which changes in the frequency or intensity of play correlates with changes in relevant environmental pressures. Thus young lambs play less when the ewe's milk supply is no longer adequate. At this point, lambs have less energetic reserve available to spend on play. Fagen sees such cost/benefit-linked schedules of play as directly increasing the reproductive potential of the player.

4.5 The Nature/Nurture Argument

We come now to one of the most persistent and intractable arguments in the study of behaviour. To what extent is an animal's behaviour a product of the environment in which it lives? After reading this and the preceding chapter, one might be tempted to say that the answer is obvious. Behaviour clearly involves elements of both. If the answer was that simple, the 'nature/nurture' debate would have died long ago. Nevertheless, as Dewsbury[89] points out, many workers have dismissed it as a 'pseudoquestion' on precisely these grounds. Why then does the argument continue?

At one level, it persists because, by demonstrating a genetic contribution to behaviour in animals, it is quite reasonable to suggest that genes might influence human behaviour. Unfortunately, this perfectly scientific suggestion has precipitated a deluge of sterile criticism which is based largely on the political views of its exponents rather than any sound scientific argument (see Caplan[40]). Nevertheless, as a result, the relationship between genes and behaviour enjoys a wide public interest.

At quite another level, the argument persists because behaviour is not simply a blend of 'nature' and 'nurture'. It is clear that some behaviours develop with little or no environmental influence (see Section 4.1). Their frequency and intensity may be modifiable but they always occur in a qualitatively species-specific form.[89] As such, they constitute what ethologists used to call an *instinct*. 'Instinct' is a useful term which dropped out of use as ethology and comparative psychology compromised over the genes-versus-environment issue. If we take the view, as does Dewsbury, that genes and environment do not represent alternatives in ontogeny but extremes of a continuum, 'instinct' terminology once again becomes appropriate. This does not mean that experience is irrelevant in shaping an 'instinctive' behaviour. The fact that the behaviour is complete on its first performance and is specific to certain eliciting stimuli, however, means it must be regarded as internally programmed. The bill-pecking response of herring gull chicks is a good example.

In some classic experiments, Tinbergen and Perdeck[305] found that chicks preferentially pecked at a red spot on the parent's bill to elicit regurgitation. Fewer pecks were directed towards a red spot placed on the forehead of a model bird. The specificity of this response at an early age was a good reason for calling it instinctive. Hailman[127], however, used a wider range of parent models and found that Tinbergen and Perdeck's results could be explained as chicks responding to the most conspicuous (most rapidly moving) stimulus. When forehead spots were made as conspicuous as bill spots (by moving them closer to the chick), young chicks responded equally to both. If older chicks were tested, however, they preferred bill spots whatever their relative conspicuousness. Hailman's experiments showed that the pecking response and its elicitation by a moving red spot is instinctive in the sense discussed above, but that the narrowing of the response to red spots on the bill requires experience. The involvement of learning, therefore, does not preclude a behaviour from being instinctive. Of course, there are behaviours which are apparently purely instinctive or purely the result of experience. The habitat preferences of deer mice (see Chapter 3) are good examples of the former and the development of tool use and food-washing in primates examples of the latter. In between are behaviours like bird song the development of which depends on various combinations of genetic and experiential factors. The important thing is to recognise and quantify the relative contributions of genes and environment rather than treating them as irreconcilable alternatives.

Summary

(1) Some behaviour patterns may develop in their entirety without practice or other experience. They may also develop as a result of maturation effects on motivation.

(2) Other behaviour patterns depend on experience (learning) for their development. In many cases experience may simply 'polish' behaviour rather than result in qualitative changes.

(3) Different types of learning can be recognised which depend on different stimulus-response relationships. These learning categories are not hard-and-fast, however. In some cases the distinctions between them break down. They are probably better seen as different ways in which learning can be shown to occur.

(4) Animals do not learn different tasks with equal facility. Constraints may be levied on learning ability, for instance by the species-specific design of behaviour and attentional processes.

(5) Learning about relative reinforcement schedules can influence the choice an animal makes between alternative behaviours. The effect of a reinforcer on choice, however, is complex and may depend on the specific requirements for exploiting particular reinforcers.

(6) Play is an apparently purposeless activity, characteristic of behavioural ontogeny in certain species. Several functions have been suggested for it, none entirely satisfactory. However, like other behaviours, play can be investigated using cost/benefit analysis.

(7) The continuing 'nature'/'nurture' debate can perhaps be resolved by viewing instincts and behaviour patterns resulting from experience as extremes of a continuum rather than opposing alternatives.

5 FINDING A PLACE TO LIVE

So far, we have examined the way behaviour is instigated, mediated and developed *within* the animal. In the chapters which follow, we shall consider the shaping of behaviour from a *functional* point of view. How does behaviour provide a solution to the animal's problem of survival? We have already seen how motivation is geared to the animal's survival requirements, but now we need to see how it copes with various 'ecological' problems — finding a living area, food, mates and so on. Once an animal has 'decided' on a particular goal-directed behaviour, its problems have just begun.

One of the most pressing needs an animal may face is finding a suitable habitat in which to live. For various reasons, animals may be forced to leave the area in which they have been living and to explore the environment for a new one. Young animals may have to leave their natal area to avoid competition with their parents. An area which is good for feeding may not be good for breeding. Animals may be obliged to move between a number of habitats according to their changing motivational priorities. In a general way, species distributions also tend to be associated with particular kinds of habitat. Species have unique ecologies which distinguish them from one another. How are these associations between species, behaviour and habitat, brought about?

5.1 Habitat Selection

Before we can answer that, we need to clarify what is meant by 'habitat'. 'Habitat' can mean different things according to the context in which it is used. In this chapter, we use it in the broad sense defined by Partridge[240]. A 'habitat' is simply *the conglomerate of physical and biotic factors (e.g. shade, humidity, prey items, potential nesting sites) which together make up the sort of place in which an animal lives*. Clearly the quality of its habitat, in terms of resource availability, protection from predators etc., is likely to affect the animal's chance of survival and hence its reproductive potential. It is unlikely, therefore, that evolution will leave habitat selection to chance. Of course, in some cases, as in plants and zooplankton, the degree of voluntary mobility may be small. The distribution of these organisms is determined largely by abiotic factors like air and water currents. Most independently mobile animals, however, might be expected to exercise some *choice* over the habitat in which they live.

5.1.1 Habitat Choice

What is the evidence for habitat choice? A number of field investigations have attempted to relate habitat preferences and geographical distributions in various species. Douglass[91], for instance, found that two species of vole (*Microtus montanus* and *M. pennsylvanicus*), maintained in enclosures of natural vegetation, spent most of their time in different types of vegetation. Moreover, the types chosen by each species were similar to those in which they occur naturally. Similar results have been obtained in deer mice (see also Section 3.3.3).

Lindauer[191] carried out a detailed study of habitat selection in honey bees. In late spring, just before the new queens emerge, the old queen leaves the hive with about half of its total population. The swarm flies to a vantage point near the hive where it aggregates into a seething cluster. From here, 'scout' bees fly out to inspect various potential sites for a new hive and return to the swarm. On their return, they 'dance' on the outside of the swarm. Rather like the dance performed after foraging flights (see Section 1.3.1 and Chapter 10), the 'scouts' dance movements convey information about the position and suitability of potential sites. Other bees (recruits) then fly out to inspect whichever locations have been indicated. The relative merits of different sites appear to be assessed by the way the scouts and recruits dance (e.g. by the vigour of movement). The dances deemed as indicating the best site attract more and more recruits until, eventually, the whole swarm 'agrees' on a location for the new hive. In this way, the bees can investigate a range of sites and choose the best. Making a careful choice of a site for the hive is clearly important. The choice process costs a lot of time and energy, particularly for the egg-laden queen whose powers of flight are poor.

Habitat choice has also been demonstrated in the laboratory. Blue tits (*Parus caeruleus*) and coal tits (*P. ater*), provided with a mixture of vegetational types in an aviary, show clear-cut species-specific preferences.[113] Blue tits prefer broad-leaved vegetation and coal tits prefer coniferous. As with the deer mice and voles, these preferences correlate with the types of vegetation the tits normally inhabit.

Similar preferences have been found in invertebrates. The polychaete *Spirorbis borealis* settles on different species of seaweed according to how sheltered (in terms of protection from wave action) an area it is occupying. Larvae taken from populations in sheltered areas prefer to settle on *Ascophyllum nodosum*, while those from exposed areas prefer *Fucus serratus*[196] (which provides firmer attachment). Of course laboratory experiments are only an approximation of the real world, but the fact that, in some cases, preferences demonstrated in the laboratory do not reflect a species natural pattern of distribution, does not necessarily mean the results are artificial. As Gale[110] points out, while local factors such as predation and competition may modify habitat choice, the basic choice criteria remain the same.

5.1.2 Choice Criteria

How do animals choose their habitat? Where choice is involved it is reasonable to suspect that it will be designed to maximise the animal's inclusive fitness. Habitats should be chosen which best accommodate the animal's biological requirements. Is there any evidence that this is so?

In his study of honey bees, Lindauer was able to show that the site for a new hive was chosen on the basis of a variety of factors. Through a series of choice experiments, he discovered three which appeared to be of primary importance. These were:

(1) The degree of protection from climatic changes.
(2) Size (sites just big enough to accommodate the whole swarm were preferred).
(3) Distance from the swarm's original hive (the farthest of otherwise equivalent sites was generally preferred).

The latter criterion presumably minimises competition between hives for limited nectar and pollen supplies.

Other species also seem to choose habitats for protective benefits. Various hummingbird (Aves: Apodiformes) species choose nest sites which are sheltered by overhead branches. Such an arrangement appears to reduce heat loss by radiation and convection. The pale and melanic forms of the peppered moth (*Biston betularia*) choose to rest on substrates similar in colour and tone to themselves, thereby decreasing the risk of predation by visually-hunting birds.

Habitat choice may also reflect the particular skills of an organism. Partridge[239] found that the choice of foliage in which to forage by blue and coal tits, correlated with the species' respective feeding skills. Blue tits, living mainly in deciduous trees, feed on insects which require pulling, tearing and hacking abilities to extract them from the bark. Conifer-dwelling coal tits, however, take prey for which visual acuity and probing are more appropriate. When tested on various artificial feeding sites, the two species did best on sites demanding the skills assumed appropriate in their natural habitat. Of course, it is difficult to distinguish cause from effect in this case. Tits may choose a habitat because it suits their already developed skills or they may choose a habitat for some other reason, and then modify their feeding behaviour accordingly.

Even though an animal can choose its habitat and thereby enhance its inclusive fitness, there is no guarantee it will be able to obtain its most preferred habitat. Competition from other animals may constrain it to settle for second, or even third, best. Also, like any other behaviour, searching for a preferred habitat has costs as well as benefits. Searching too long before making a choice may mean all the suitable sites are taken by other individuals, while remaining too long in an inappropriate habitat may increase the risk of predation and reduce feeding efficiency. Waiting for the right habitat to come along therefore needs to be traded off against the penalties of delay. Exactly how the two sides of the equation are balanced depends on the situation considered, but animals do appear to respond as if to the relative costs and benefits of choosiness. The planktonic larvae of *Spirorbis borealis*, for example, become less fussy about sites on which to settle as time spent searching increases.[161]

Figure 5.1: The Effect of Population Density on Habitat Suitability and thus Distribution. Two habitats are shown, 'a' having a higher suitability than 'b'. Horizontal lines represent two population sizes A and B and intersect the suitability lines at points of equal suitability. Assuming animals distribute themselves so that each experiences the same habitat suitability, the density of animals in the two habitats at the two population sizes shown are indicated on the habitat distribution lines below the x-axis. d_a, density in habitat a; d_b, density in habitat b.

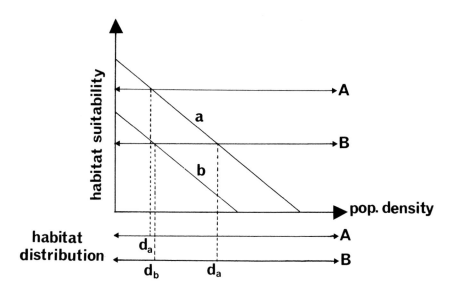

Source: Modified after Fretwell, S.D. and Lucas, H.L. (1970). On territorial behaviour and other factors influencing habitat distribution in birds. *Acta Biotheoretica, 19*: 16-36.

Although we can envisage optimal and sub-optimal habitats for an animal, the distinction may not be easy to make in the real world. One reason is that good habitats are likely to attract a lot of residents. The reduction in net benefit incurred by feeding in a less suitable type of vegetation may be smaller than that incurred by feeding in an overcrowded preferred type. The criterion for an optimal habitat becomes complicated by differences in population density across habitat types. The combined effects of habitat suitability and competition on habitat choice are modelled in Figure 5.1. Experience in a sub-optimal habitat may also modify later choice. In great tits (*Parus major*), individuals inhabiting sub-optimal breeding habitats (hedgerows) will readily invade vacancies in the optimal habitat (woodlands) if they are juvenile, but not if they are adult.[165] Preference for sub-optimal habitats apparently increases with experience of reproductive success in those habitats. Also learning the skills appropriate for surviving in an originally sub-optimal habitat may eventually raise the utility of that habitat. The optimal habitat type may therefore change as a function of experience.

The fact that the relative utility of habitats can change is important. While experience may enhance habitat utility, resource depletion, competition, predation, abiotic changes and other factors may decrease it. Animals may therefore need to change habitats at intervals through their lifetime. Changing habitats, however, involves two distinct decisions: (1) when to leave the current habitat and (2) what sort of habitat to move to. In the next section, we shall discuss how animals make these decisions and the ways in which movement or *migration* between habitats is effected.

5.2 Migration

'Migration' conjures up images of a particular kind of animal movement, usually taking place over a long distance and on some sort of seasonal basis. However, this popular interpretation is only one of many used by workers to distinguish real migration from more 'trivial' non-migratory movement. In many cases, the definition of migration has depended on the taxonomic group and its range of mobility.

The problem lies in defining what animals move between when they migrate. If we assume that migration entails movement from one habitat to another (which is the essence of at least one dictionary definition), we run into the old problem of defining a habitat. For instance, a bird moving from the older, lower branches of a tree to the upper younger branches could be said to change habitat. On the other hand, the same bird flying from a bunch of berries on one tree to a similar bunch on another does not appear to change habitat even though the distance moved in the second case may be greater than that moved in the first. As Baker[8] points out, the distinction betweeen inter- and intra-habitat movement is clouded by the open-endedness of the definition of habitat. For animals whose home range is a particular tree, we might regard movement within the tree as 'trivial', but movement between trees migratory. For those whose home range is a whole wood, movement between trees becomes trivial and only that between woods migratory.

Baker gets round the problem by viewing any movement from one place to another as migration. Migration thus becomes *the act of moving from one spatial unit to another*.[8] Baker suggests that the same basic principles apply to movement at all levels of localisation. Migration is thus a property of all animals. The interesting question is what makes one pattern of movement appropriate for species A, but another appropriate for species B? In Section 5.1 we saw that habitat choice has survival value so that animals are most likely to move between habitats only if doing so improves their inclusive fitness. This assumption forms the basis of a detailed evolutionary model of migration recently developed by Baker[8]. Animals are seen as assessing the utility of their present habitat (h_1) relative to that of another potential habitat (h_2). They migrate only when the utility of h_1 drops below that of h_2 multiplied by a *migration factor* (m)

Figure 5.2: Types of Migration

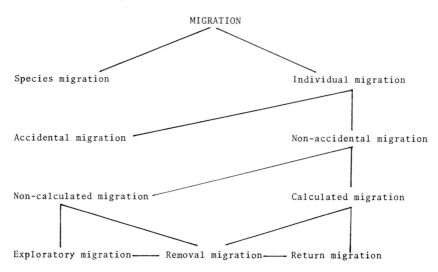

Source: Modified after Baker[8].

(i.e. when $h_1 < h_2m$). The factor m is an expression of the cost:benefit ratio of the migratory act itself. This is a similar kind of decision rule to the incentive X deficit rule discussed in Chapter 2.

Since we are now talking about migration as the result of an adaptive decision, it stands to reason that it is *non-accidental*. The distinction between this and 'accidental' migration (involuntary displacement) is the first in a 'hierarchy' of migration categories proposed by Baker (Figure 5.2). Within the non-accidental category, Baker includes a range of different types of migration.

5.2.1 Calculated Migration

Calculated migration is migration to a particular destination the utility of which is known to the animal before it migrates. The animal may become familiar with its potential destination by various means; for instance through direct perception, prior acquaintance or social communication. Selection is likely to favour individuals which make the most accurate comparisons between h_1 and h_2m. Both those which migrate too soon or to an inappropriate habitat, and those which delay in a poor habitat when they could do better by migrating will be penalised.

In some cases, animals may be able to gather the relevant information while still in h_1. Among social primate species, the composition of the social group may be an important determinant of individual reproductive success. Departure from the optimal composition decreases the inclusive fitness of the group's component individuals. This appears to be the case in gorillas (*Gorilla gorilla*). Very often, groups of gorillas encounter one another on the mutual border of

Figure 5.3: An Instance of Calculated Removal Migration in a Black-tailed Prairie Dog Town. Solid boundaries, established territories; dashed boundaries, new territories; arrows, migrating individuals.

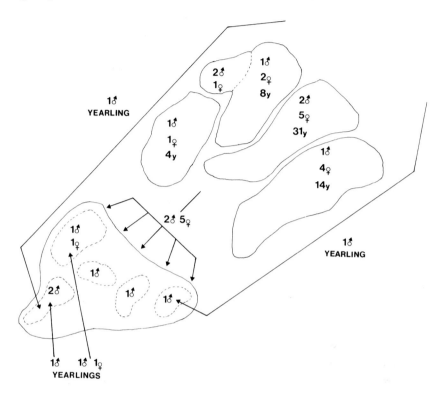

Source: Modified after Baker[8] and King[160].

their home areas. During these meetings, animals have an opportunity to compare the compositions of the two groups. Although individuals of both sexes migrate between groups, females tend to migrate only when groups are actually in contact and their relative utilities can be compared. Using data collected by Schaller[267], Baker examined the circumstances under which females transferred (showed *removal migration*) between particular groups. His analysis showed that females were most likely to transfer when the alternative group had a higher ratio of silver-backed males (SBMs) to females without infant young. The adaptive value of a high SBM:female ratio apparently lies in the protection afforded by males to females and young.

A more complex form of calculated migration is shown by the black-tailed prairie dog (*Cynomys ludovicianus*). Prairie dogs live in 'towns' which occupy

anything from 3 to 30 hectares. The towns are in turn subdivided by topograph-
ical features like trees and ridges into 'wards' and the wards into territories of
about 2,000–3,000 m^2. Each territory is occupied by a 'unit' of dogs comprising
1–2 adult males, 2–3 adult females and up to 40 young. Animals in the 'unit'
feed on the grass, broad-leaved plants and seeds within the territory. Migration
from individual territories takes place mainly in June and July, when the young
of the year become independent. Curiously, it is the adults rather than the
young which emigrate. Figure 5.3 shows a particular instance of migration[160]
which illustrates the general pattern of events.

In early spring a variety of individuals (of all age classes, both sexes and a
number of different territories) may visit an uninhabited feeding area outside
their territory but return to their territory to roost or escape predators. Later
in the spring, some animals begin to dig burrows in the new area. By late summer,
most of the animals using the area are adults from one territory which gradually
spend more time in the area, until they finally leave their old territory altogether.
Through July and August, the males divide the new area into incipient territories.
As adult females move through the area, they choose a male and begin new
territorial 'units'. Invasion of an already occupied area follows a similar pattern
except that exploratory trips are usually followed by aggressive disputes. Assessment of h_1 and h_2m in the pairie dogs' case appears to be based on the relative
availability of food and escape burrows and the distance between the two areas.
When h_2 is occupied, the chances of success in territorial disputes are also
important.

5.2.2 Non-calculated Migration

An animal's reproductive potential in a new habitat is clearly likely to be greater
if it has had a chance to assess the relative merits of the habitat beforehand. On
average therefore, calculated migration will result in a greater increase in inclusive
fitness than *non-calculated migration*. In some cases, however, exploration may
not reveal any better habitats. Indeed, exploration may not even be possible.
This may be so in, for instance, populations which are surrounded by large
expanses of unsuitable or dangerous habitat. Here, non-calculated migration
may be obligatory if the utility of the present habitat deteriorates.

Just such a problem faces the North Pacific sea otter (*Enhydra lutris*).[159]
Otters inhabit the coastal waters of the North Pacific. Their range extends
roughly from 1 to 16 km from the shore and is restricted to water less than 60 m
deep. When an area is colonised, the usual pattern of events involves a gradual
rise in population density to about 16 animals per km^2. A crash then follows
with a subsequent increase in density to about 8 animals/km^2. The population
eventually settles at around 4–6/km^2. Migration appears to be a response to
decreased feeding efficiency within the present habitat and is mostly calculated.
Exploratory migrations made when population density is low and which increase
in number as density rises, provide experience of the relative suitability of
nearby habitats. Already occupied habitats are usually assessed as the most

suitable. As the population density in the present habitat nears 16 animals/km^2, calculated removal migration may occur to one of the previously visited alternatives. Sometimes, however, suitable habitats may be separated by stretches of very deep (over 60 m) water which otters do not normally cross. Under these circumstances, non-calculated migration may occur, with animals traversing 100 km or more of unsuitable water to reach areas of unknown utility.

Another example of non-calculated migration occurs in the small tortoiseshell butterfly (*Aglais urticae*).[8] Here, butterflies periodically migrate across areas of unsuitable habitat which are outside their range of sensory perception. Several migrations may be undertaken in a day. Because each migration is in the same direction relative to the sun's azimuth, butterflies are constantly crossing unexplored country. In this way, serial non-calculated migration may take the insects over several thousand kilometres during their lifetimes. These numerous migrations appear to take place for several reasons. Sometimes they are due to changes in time-budgeting priorities, e.g. when feeding or oviposition becomes more important than roosting. At other times they are apparently responses to declining habitat utility for a given activity (e.g. when food availability decreases during feeding).

5.2.3 Seasonal Return Migration

Instead of moving to a new habitat and remaining there, animals may return to the original habitat sometime later. The best known *return migrations* are those performed on a seasonal basis. Here, animals may feed in one type of habitat and breed in another, making the return journey between the two each year. In some cases, the distance between the two habitats may be enormous. In the arctic tern (*Sterna macrura*), it is half the circumference of the world. Nevertheless, seasonal return migration can be explained using the same evolutionary arguments as before. The familiar long distance migrations of African ungulates are a good example.[8]

In the Serengeti region, long seasonal return migrations are performed by zebra (*Equus* spp.), wildebeest (*Connochaetes taurinus*) and Thomson's gazelle (*Gazella thomsoni*). These species spend the dry season (July to December) feeding in open thorn woodland where, owing to the low rainfall, they are limited in their movements to the vicinity of water holes. In December, when the first rains come, the herds move into the central plains of the Serengeti and then migrate in a predominantly anticlockwise direction following the rains and new vegetation.

Although the direction of movement is purely opportunistic (animals simply move towards areas where rain can be seen or heard falling), the sequence in which the different species move round is determined by species-specific food requirements. In the Serengeti, the sequence is generally zebra first, wildebeest second, gazelle third. Stomach analyses have shown that zebra take a high proportion of grass stem material. By trampling the grass and removing the tough stems, zebra open up the herb layer so that wildebeest, which feed mainly on grass leaf, can then move in. The joint action of the zebra and wilde-beest exposes the broad-leaved plant layer preferred by the gazelles. By following the rains, the zebra continually move from a depleted habitat to one with fresh, abundant vegetation. In their turn, the wildebeest and gazelle capitalise on the selective feeding of the previous species and also move from poor to good habitats. The survival value of seasonal return migration is illustrated by the mortality figures for migrant and non-migrant species. The most important mortality factor in the migrant zebra, wildebeest and gazelle populations is predation. Few animals die of starvation. In non-migrants like impala (*Aepyceros melampus*) and wart hog (*Phacochoerus aethiopicus*) the opposite is true. Starvation often outweighs predation as a cause of death.

A more complex form of seasonal return migration is shown by the red-billed quelea (*Quelea quelea*),[327] an African ploceine which feeds mainly on seeds and insect larvae. During the dry season, birds take seeds which are lying dormant on the ground and, as the season progresses, large numbers build up at the better feeding areas. When the rains begin, however, the seeds germinate and are no longer available as food. The flocks then perform the 'early rains' migration back along the path of the rains to a region where they had begun a sufficient time previously for the plants to have produced a new crop of green seeds. The birds then return to their dry season habitat by keeping pace with the green belt as it moves along the path of the rains. This return phase is also a breeding migration. Like the ungulate species in the previous example, the quelea con-stantly move from a poor feeding area to a richer one and the relatively small migrations between successive feeding sites eventually add up to long-distance round trip.

5.3 Orientation and Navigation

Choosing when to move between habitats and which habitats to settle in is clearly important to an animal. Equally important is a means of orientating and navigating through the environment so that the aims of migration can be achieved. The animal needs to direct its behaviour with respect to relevant environmental stimuli. As we shall see, the stimuli used are diverse and sometimes extremely subtle.

5.3.1 Taxes and Kineses

Orientation on a local scale (over a few centimetres or so) and to easily defined stimuli, like a light source or humidity gradient, can be seen in many inverte-brates. It falls into two main types: (1) movements in which animals attain their preferred habitat without orientating with respect to the source of the stimulus (*kineses*) and (2) movements which are made at some fixed angle to the stimulus source (*taxes*). These two categories contain a number of different types of response.

Figure 5.4: Klinokinesis by Lice on Different Materials. Lice turn violently when they move from stockinet onto silk and less violently but more often when they move onto blotting paper.

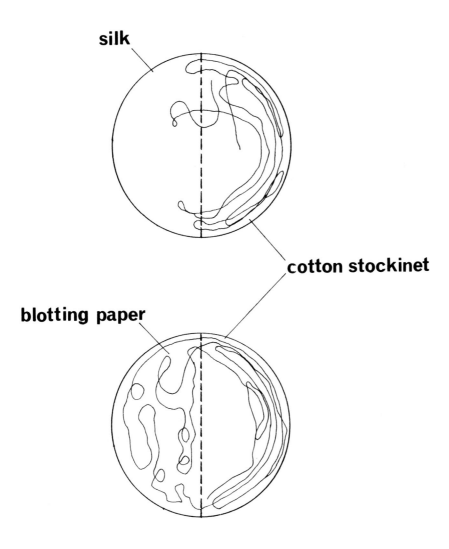

Source: Modified after Carthy, J.D. (1971). *The Study of Behaviour*. London, Edward Arnold.

Orthokinesis describes the tendency for animals to gravitate to their preferred habitat by moving more slowly there than elsewhere. When woodlice (*Oniscus porcellio*) are allowed to move about on gauze, one half of which is maintained at high humidity and the other half at low humidity, they scuttle rapidly over the dry half but slow down on the damp half. After a while a cluster of animals builds up on the damp part of the gauze. Similar distributions can be brought about by increased rates of randomly directed turning (*klinokinesis*) in unfavourable habitats (Figure 5.4). By turning rapidly as soon as it leaves its preferred habitat, an animal increases the likelihood of re-entering it.

While ortho- and klinokinesis are not orientated with respect to the stimulus source, taxes are. *Klinotaxis* is shown by blowfly larvae moving away from a directional light source. As the larva moves, it swings its head from side to side alternately exposing left and right photoreceptors to the light behind. When the left receptor is stimulated, the animal responds by bending to the right and vice versa, thus moving away from the light.

Tropotaxis is the result of symmetrically-placed receptors whose relative stimulation may be compared centrally. If one of the receptors is masked, the animal moves in a tight circle keeping the good receptor towards (positive stimulation) or away from (negative stimulation) the stimulus source. By contrast, *telotaxis* involves orientation to a directional stimulus when only one of a symmetrical set of sense organs is operating. Honey bees with one eye masked will walk up a beam of light. Orientation is achieved by bending the body until the good eye is maximally stimulated. By reducing the amount of turning as stimulation of the eye increases, the bee maintains a steady course up the beam. Telotaxis covers a heterogeneous range of orientation behaviours including the familiar dorsal and ventral light responses of many marine organisms. For further discussion, and a wealth of examples, the reader is referred to Fraenckel and Gunn[108].

5.3.2 Navigation Mechanisms

The ways in which animals navigate between two given areas, (either by *homing* back to the home area, or migrating to a new one), sometimes thousands of kilometres apart, has long intrigued biologists. The term 'navigation' implies two things: first a degree of familiarity with the destination or goal, and secondly, a lack of familiarity with the geography of the area between the destination and the present position. Navigation is thus a means of determining the direction of a familiar goal across an unfamiliar area. As such it is a form of *goal orientation*[9] to be contrasted with simple *orientation* which is concerned solely with taking up a particular direction rather than seeking a specified destination. 'Navigation' and 'orientation' are sometimes confused and used incorrectly in an interchangeable way in the literature. While they are quite distinct processes, however, orientation to some reference cue usually plays a vital role in the initial stages of navigation. For this reason, the term 'orientation' necessarily occurs in our discussion of animal navigation. Sun and star orientation are good examples.

While we have known for some time that celestial cues like the sun and stars are used as navigational aids, it is only relatively recently that the manner in which they are used has been clarified. It has also become apparent that animals can use other, sometimes subtle, environmental cues and hitherto unsuspected sensory modalities. Although long distance navigation is known in several taxanomic groups, including insects, fish, reptiles, and birds,[269] it is with birds that most progress has been made.[157] Consequently, the following discussion refers largely to bird navigation.

Celestial Cues. One of the earliest, and now most familiar, discoveries about animal navigation was that the sun and stars provide important cues for orientation. At one time it was suggested that birds displaced from their home area might compare the sun's arc (the apparent path traced by the sun across the sky) in the new area with that expected at home. However, it now seems clear that the sun is used as a simple compass and that only its azimuth (direction from the observer) provides information for orientation. This was shown by Kramer's[163] classic experiments in which the locomotory orientation of caged starlings could be altered predictably by using mirrors to 'move' the sun (Figure 5.5).

A problem with using the sun as a reference point is, of course, its apparent movement during the day. To orientate accurately, birds need to compensate for the sun's changing position. They need some kind of internal clock (see Chapter 1) with which to pace its movement. Schmidt-Koenig[268] found that homing pigeons (*Columba livia*) possess just such a clock. When pigeons had their internal clocks shifted by six hours (a quarter of a day) relative to true sun time, they chose bearings roughly 90° (a quarter of a circle) from control birds when released away from home (Figure 5.6). A point which will become important later, however, is that clock-shifting had no effect on overcast days when the pigeons could not see the sun. This implies that, in the absence of the sun, birds switch to a back-up orientation mechanism to compensate.

The ability of birds to use the sun compass has some interesting ontogenetic characteristics. An extensive series of experiments by Wiltschko, Keeton and coworkers has shown that pigeons, at least, synchronise their internal clocks on the basis of their early experience with the sun. Experience facilitates an accurate coupling of the clock with the sun's apparently changing position. By means of clock-shifting experiments, young pigeons could be taught to read a southerly sun as an early morning sun, a westerly sun as a midday sun and so on. Birds therefore appeared to have no inherent ability to interpret the sun's azimuth at a particular time as indicating a particular direction. Instead the compass is established by experience. Interestingly, even before they have learned to use the sun compass, young pigeons are able to navigate back to a home loft accurately.[337] There does therefore appear to be an inborn navigation mechanism, but one which is substituted at an early age by perhaps a more reliable and convenient compass rule of thumb.

Figure 5.5: Orientation of Caged Starlings in Relation to a Deflected Light Source. Dashed arrows, direction of light; solid arrows, mean orientation direction of birds.

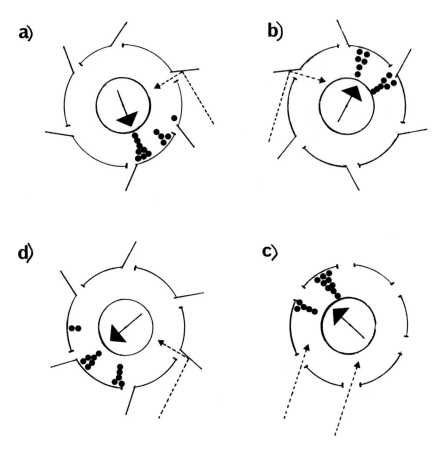

Source: Modified after Kramer[163].

Like the sun, the stars also seem to provide only compass information, although it would be possible to use them for bico-ordinate navigation.[99] Stellar orientation has been shown in a number of species. Bellrose[25] released mallard at night with small lights attached to their feet. Birds released under clear skies flew almost directly north. Those released in overcast conditions flew in all directions. Unlike the sun compass, however, stellar navigation does not require time compensation. Using artificial constellations in a planetarium, Emlen[98] found that indigo buntings (*Passerina cyanea*) used learned star patterns rather than the azimuth of individual stars to gauge the axis of apparent rotation of the night sky. There is a critical period (see Section 4.2.1) just prior to the first autumnal migration, during which young buntings must learn the star patterns

Figure 5.6: Vanishing Bearings (Direction of Last Sighting from the Point of Release) for Pigeons Clock-shifted Six Hours Fast (Closed Circles) Compared with Control Birds (Open Circles). Data for releases far (more than 1.5 km) from the home loft, close (less than 1.5 km) to the loft and on overcast days. Open and closed arrows show the mean bearings for clock-shifted and control birds respectively. Clock-shifted birds were 90° out in their orientation except on overcast days. Their ability to orientate on overcast days indicates that a back-up mechanism is being used.

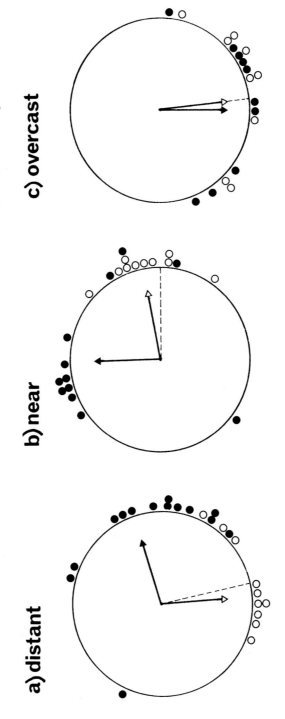

a) distant

b) near

c) overcast

Source: Modified from Schmidt-Koenig, K. (1960). Internal clocks and homing. *Cold Spring Harbor Symp. Quant. Biol.*, 25: 389–93.

in order to be able to navigate later. However, while birds initially respond to the rotation of stellar constellations, they use it only to learn the star patterns which indicate the axis of rotation. Thereafter, they rely on the learned patterns. The axis of rotation is merely a reference point against which star patterns are calibrated.

There are two long-recognised problems with using celestial compasses. First, they are not always available (for instance, under cloudy conditions) and, secondly, a compass alone cannot tell an animal where it is or the direction to a particular location. Since animals plainly do navigate between given points and under conditions of poor visibility (even pigeons wearing frosted lenses can still home), other environmental cues must provide a backup 'map' system. Surprisingly, obvious reference points like landmarks seem not to be used, so what other cues might provide the essential map?

Magnetic Cues. As long ago as 1882, Vignier suggested that the Earth's magnetic field could provide a navigational grid. Birds might be able to detect and measure the three essential components of the field: *intensity*, *inclination* (the angle made by the field with the horizontal) and *declination* (the angle between magnetic and true north). Since isoclines of intensity and inclination tend to cross at oblique angles, the magnetic field provides a poor navigational grid unless declination is also incorporated. Have any species evolved to capitalise on the potential magnetic map?

Wiltschko[336] tested the magnetic sensitivity of captive robins (*Erithacus rubecula*) by placing Helmholtz coils around their (visually cueless) cage during periods of spring and autumn Zugunruhe. By altering the magnetic field impinging on the birds, he could change the directions they regarded as north and south in a predictable manner (Figure 5.7). Wiltschko found that birds used the alignment of the vector of the magnetic field rather than its polarity. Thus, they orientated north in spring whether the vector pointed north and down or south and up, and south in autumn when it pointed in the opposite directions. They were unable to orientate at all with a horizontal vector. The birds' magnetic navigation system is thus very different in principle from our own magnetic compass.

Keeton[156] showed that homing pigeons were also sensitive to magnetism. Bar magnets, attached to the backs of experienced birds, caused disorientation on overcast days but not on sunny days. In pigeons at least, the magnetic field seems to provide a compass which is used mainly when the sun compass is not available, although there is evidence that it still operates under clear conditions. Magnetic orientation in pigeons may be facilitated by particles of magnetite which have recently been discovered inside the head. The sensitivity of birds to magnetism seems to be quite high. The orientation of gulls and pigeons can be affected by natural levels of magnetic disturbance. Some insects are astonishingly sensitive. Honey bees, for instance, can detect magnetic changes of less than 10^{-3} gauss.

Figure 5.7: Changes in the Orientation of Caged Robins in Response to Manipulated Magnetic Fields. Open arrows, expected orientation; closed arrows, mean observed orientation; MN, ME, MS, MW, manipulated magnetic north, east, south and west. Birds orientated towards apparent magnetic north at the time of their normal spring migration and towards apparent magnetic south at the time of their autumn migration.

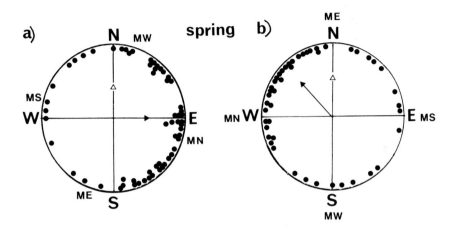

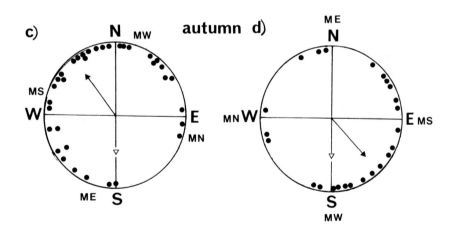

Source: Modified after Wiltschko[336].

Since the pioneering work on animal magnetic sensitivity, the wealth of examples from diverse taxonomic groups suggests that an immense variety of, if not all, animals possess a magnetic sense. Particularly intriguing is the recent work of Baker[9] comparing magnetic sensitivity in man with that of other animals, especially rodents and homing pigeons. Baker found that disturbance of the magnetic field about the head (using electromagnetic helmets) during a long outward journey significantly reduced the ability of people to point to a particular destination (in this case the place they had been driven from wearing blindfolds). Moreover, Baker found interesting differences between the sexes in their navigational abilities. Women tended to be better at navigating 'passively' over short distances; that is they appeared to have a more consistent 'sense of direction' during short tortuous journeys than men. Men, however, were more adept at translating their journey-based estimates of direction into a proposed direction of travel to reach a specific destination. These results are consistent with the assumption that, during evolution, males have tended to carry out more long-distance exploration and thus have been subjected to more intense selective pressure for an accurate navigation mechanism. The suggestion that magnetic sensitivity was the basis for the human sense of direction has received dramatic support in the discovery of magnetically sensitive tissue in the bones of the sphenoid sinus behind the eyes. This appears to be homologous with similar tissue found in the facial bones of rodents which have been suggested to possess magnetic sensitivity.

Gravity Cues. When animals show sensitivity to magnetism, they often respond to gravitational cues as well. Larkin and Keeton[182] tested the sensitivity of pigeons to gravity by looking for the influence of the monthly gravitational cycles caused by the changing relative positions of the earth, sun and moon. They found a significant correlation between the bearings taken by birds on release and the day of the lunar month. These findings are not conclusive, however, because other factors correlating with the lunar month may have influenced orientation.

Nevertheless, gravity could be an extremely useful navigational aid. It varies temporally and geographically and the geographical variation is both regular (in a north-south gradient, owing to the earth not being a perfect sphere) and irregular (because the density of the earth's crust varies in different places). Gravity could therefore provide a north-south axis and an additional 'topography'.

Barometric Pressure Cues. Sensitivity to changes in barometric pressure would clearly be of advantage to organisms which spend much of their time in the air. Its greatest use would be in providing a form of altimeter. Perhaps not surprisingly therefore, pigeons have recently been shown to respond to pressure changes. Kreithen and Keeton[178] trained pigeons to expect a shock immediately after a pressure change and then measured changes in the birds' heart-beat when the pressure was subsequently altered. The sudden acceleration they found showed that pigeons are sensitive to pressure changes equivalent to a 10 m, or even smaller, change in altitude. Sensitivity of this sort could help to compensate for the reduction in gravity with increased altitude.

There is also anecdotal evidence that birds register changes in barometric pressure. It is well known that birds show changes in behaviour as a weather front approaches. Birds about to set off on migration seem able to gauge wind conditions at high altitude, taking off only when they are favourable. Similarly, certain eastern American migrants fly to the east of a high pressure area in autumn but to the west, ahead of a cold front, in spring.

Other Cues. A variety of other cues might also be used as navigational aids. It is known that birds can detect infrasonic frequencies (at least down as far as 0.05 Hz). *Infrasound* can travel many hundreds, even thousands of kilometres, thus allowing birds to orientate to, say, distant mountains (via the wind blowing through them) or shorelines (from the sound of breaking waves). The sensitivity of pigeons to infrasound is great enough for them to detect Doppler shifts as they fly towards and then away from the sound source. In this way they could achieve the directional information necessary to orientate to the source. Pigeons, like honey bees, can also detect *polarised* and *ultra-violet light*. They are thus in a position to use their sun compass on overcast days, provided some blue sky is visible.

A more contentious possibility is the use of *odours*. While it is well known that some animals, e.g. slugs (Mollusca: Gastropoda) and salmon, can navigate by olfaction, the suggestion that birds are able to do so is surprising. Birds are usually regarded as microsmatic (possessing a poor sense of smell). Nevertheless Papi and coworkers have suggested that pigeons can home by scent. Pigeons might learn the directions of different odours arriving at their home loft; odour A, for instance, may come from the south, B from the east and so on. If a pigeon is displaced some distance south, odour A will be stronger. The bird therefore determines its position to be south of home and uses one of its 'compass' systems to fly north.

Papi tested his hypothesis by deflecting the air currents entering his birds' home loft. Pigeons which had experienced deflected currents, chose bearings to the right or left of control birds when released away from home depending on whether air deflection had been in a clockwise or anti-clockwise direction. Unfortunately, it has been difficult to duplicate Papi's results elsewhere and it may be that his deflector lofts incidentally altered some other factor influencing orientation. Nevertheless, other work does lend some support to the olfactory hypothesis. Wallraff and Foá, for instance, transported pigeons in airtight containers which contained special air filters. They then compared the homing accuracy of birds from containers with charcoal (very effective) filters which removed most olfactory cues from the air during transportation, with that of birds transported with paper (poor) filters or with no filters at all. Birds were treated with xylocaine to remove olfactory sensitivity before release and so could only use olfactory information gathered during the outward journey. On release charcoal filter birds showed significantly less accurate vanishing bearings than paper filter or no filter birds.

While most progress has been made with birds, it is clear that other groups also possess a complex array of navigation mechanisms. Fish, for instance, can navigate using carbon dioxide concentration gradients, temperature changes, sounds from other fish, geoelectric information, olfaction, magnetism, polarised light and the sun. The problem now is to find out how all these different cues interact to facilitate navigation, and what priority of use exists among them. It seems clear that different cues interact in the development of navigational abilities in some species. Young pigeons, for instance, need both the sun and magnetic cues to navigate. Also, different cues seem to be used for short and long distance navigation, at least in pigeons and man[9] and experience greatly modifies the cues which are used. Some cues like magnetism appear to be mainly back-up systems for use when a simpler cue, like the sun, is not available, but the functional priority and flexibility within complex navigational systems is far from understood.

Summary

(1) The natural distribution of animals may to a large extent reflect adaptive habitat choice.

(2) A variety of criteria may influence habitat choice, including the degree of protection the habitat affords and the behavioural skills of the animal.

(3) Migration between habitats is likely to reflect an adaptive decision-making process which takes into account relative habitat utility and the costs of migration. Although there are many types of migration, each tends to result in animals moving from lower to higher utility habitats, even when the utility of the destination cannot be assessed prior to migration.

(4) For migration to be effective, animals require means of orientating within and navigating through the environment. An enormous diversity of environmental stimuli have been shown to act as orientational and navigational aids. Some of these have revealed sensory modalities within animals which had not been recognised until recently.

6 FINDING FOOD

Most animals are predators of one sort or another, even if only on plant material or immobile seeds. Any animal which has to harvest food from the environment faces broadly similar problems. Having decided to feed, it must then decide where to feed, what sort of food items to take, when to move to a new feeding area and so on.[172] This applies as much to herbivores and detritus feeders as it does to animals which hunt mobile prey.

Just as with other survival problems, the decision an animal makes in each case is likely to be shaped by natural selection. Animals choosing the most efficient or optimal solutions will have the highest inclusive fitness. Decisions about the day-to-day organisation of foraging compound to determine the animal's long-term chances of surviving to reproduce (see Chapter 2).

Nevertheless, it is not always easy to see why an animal should forage optimally. While it is clear that a small endothermic vertebrate like a shrew or a tit will need to be an effective predator (in winter, tits must find an item of food roughly every three seconds to survive through the day[114]), it is not so clear why, say, an herbivorous insect should feed efficiently. Surely, here is an animal surrounded by a plethora of food. As long as it feeds *adequately* it will survive, so why bother to optimise? There is no obvious competition forcing it to feed optimally. The flaw in the 'adequate will do' argument is that it focuses on the wrong kind of competition. While its component individuals will certainly survive to reproduce, a population of adequate feeders is open to invasion by an optimally-feeding mutant. By definition, the most efficient feeder will have more time and energy to spend on other activities, like avoiding predators and finding a mate. Its expected reproductive potential will therefore be higher than that of the adequate feeder and the trait will spread through the population. The important level of competition is the evolutionary competition between alternative feeding strategies. The 'winner' is the strategy which is *evolutionarily stable* (see Section 3.4.2) and hence resistant to invasion by alternatives. Having said why we expect animals to forage efficiently, how do they make the various decisions necessary during a foraging bout?

6.1 Foraging in Patchy Environments

Natural food supplies are seldom distributed in a purely random fashion through the environment. Resource requirements, patterns of social interaction and so on within species mean that prey are often clumped in time and/or space.[301] When prey are clumped, we can talk of food supplies being *patchily* distributed.[195] Patchiness can occur in different ways. Patches may be discrete entities like

175

swarms of insects or inflorescences or they may be local variations in the density of continuously-distributed items like earthworms in a pasture. Whatever form patchiness takes, the problems it poses for the predator are broadly similar.

6.1.1 Choosing Where to Forage

The first decision a hungry predator must make is where to start feeding. As it moves through its environment, the animal encounters a range of food patches, some of which have a higher utility (net value in terms of food availability, risk of predation and so on) than others. Clearly, if it is to feed efficiently, the predator should select the patch with the highest utility. However, differences between patches may not be readily apparent when the patches are first encountered. In order to select the best, therefore, the animal may first have to sample a range of patches. This poses a tricky problem. If the animal spends too long exploring different patches before exploiting what it considers the most profitable, its net rate of food intake for the foraging bout will be reduced. Conversely, if it doesn't explore for long enough, it may settle for a patch which is not the best. Is there any evidence that animals compare the utility of food patches and opt for the best?

Smith and Sweatman[292] examined the way great tits sampled a series of grids in an aviary. Six grids in which mealworms (*Tenebrio molitor*) were hidden represented six food patches. Patches contained different densities of food which could be swopped about by the experimenters. Tits soon learned to concentrate their feeding effort on the grid which had the highest density, but continued to spend some time searching the others. The adaptive value of this seemingly sub-optimal behaviour became apparent when the food density on the best grid was suddenly reduced. Now, tits switched their attention primarily to the grid which had previously contained the second highest density (Figure 6.1). The implications are that the birds allocated their time as a trade-off between feeding in the best patch and gaining information about the rest of the environment which could be put to use later if necessary.

Krebs and coworkers[175] examined the 'exploration versus exploitation' dilemma in a different way. They trained captive great tits to hop up and down on two perches at opposite ends of a large aviary. These caused a computer-controlled disc containing mealworm prey to rotate. The two perches resulted in a different rate of prey presentation per hop and thus simulated two different quality patches. Krebs and coworkers measured how long birds sampled the two patches before deciding which to exploit. If they foraged optimally, birds should have hit on the mixture of exploration and exploitation which maximised their reward rate for the foraging bout. As it turned out, the number of hops the birds took to reach a decision agreed very closely with the predicted optimal number based on the average length of the birds' foraging bouts (Figure 6.2a). That birds were also very good at deciding which of the two patches was the best is shown in Figure 6.2b. The problem facing the tits was analogous to one of choice on a variable ratio schedule (see Section 4.4.2) and results similar to those

Figure 6.1: Great Tits Learn to Forage in the Densest Food Patch on a Grid (a), but Switch to the Second Densest when Food is Reduced on the Previously Best Patch (b).

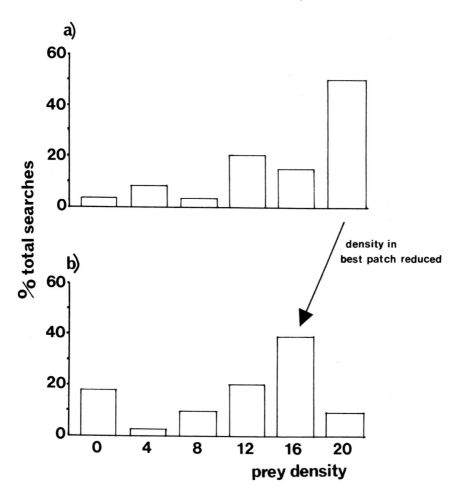

Source: Modified after Smith and Sweatman[292].

in Figure 6.2b have also been obtained from pigeons pecking at Skinner box keys programmed on variable ratio schedules.[141]

6.1.2 Exploiting Food Patches

In the experiments described in Section 6.1.1, food depletion was either minimal[292] or non-existent[175]. In the wild, a predator's feeding activity usually results in some sort of prey depletion. Clearly, as the number of prey in a patch is reduced, the utility of the patch relative to others drops. Eventually, there comes

Figure 6.2: (a) (i) Computer-predicted Number of Hops for a Tit to Decide on a Patch to Exploit (see text). a for 250-hop, b for 150-hop, c for 50-hop tests. (ii) Frequency Distribution of Hop Numbers per Test. (b) Great Tits Switch to Perch A when the Relative Reward Rate There is Greater than 50 Per Cent (see text).

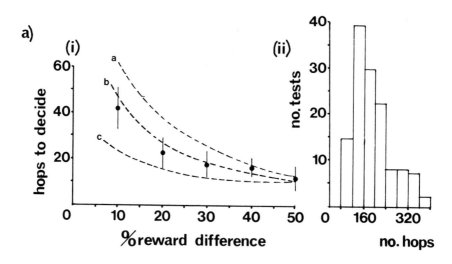

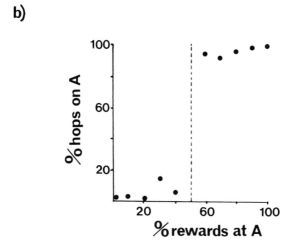

Source: (a) Modified after Krebs *et al.*[175] (b) Modified after Krebs[171].

a point where the predator could do better by moving to a new, undepleted, patch. How does it decide when that point has come?

The 'Marginal Value' Theorem. Several factors need to be taken into account. To begin with, the predator requires an estimate of the average utility of food patches in the environment. This provides a yardstick against which to measure the utility of the patch currently being exploited. The amount of time and energy the predator must spend travelling between patches is also important. As we saw in Chapter 2, the cost of changing behaviour (here meaning the change from one place to another) affects the decision an animal makes about when to change. Thirdly, the predator needs some index of the declining utility of the current patch which can be related to the expected utility of other patches in the environment. Charnov[51] combined these factors in a simple predictive model (Figure 6.3a).

The curve in Figure 6.3a represents the predator's cumulative food intake in a patch as a function of time spent in that patch. At first, the intake curve rises steeply as the predator encounters a high density of undepleted food supply. As items are removed, food density drops and with it the predator's intake rate. If it is to maximise its rate of food intake over the foraging bout, the predator should move to a new patch when its intake rate drops to a level equal to the average for the environment (i.e. when it reaches the 'marginal value'[51] for the environment). The average intake rate for the environment is represented by the tangent to the intake curve in Figure 6.3a. The optimal predator should therefore stay in a patch just long enough to maximise the slope of the tangent.

Of course, this does not necessarily mean the predator will stay for the same time in each patch. Patches within an environment are likely to vary in their characteristics, and the predator will experience different rates of food intake as it moves between them. If it retains the same marginal value for each patch (because they are all in the same environment), the optimal time to leave a patch will vary (Figure 6.3b).

Giving-up Times. One way of testing the 'marginal value' theorem is to measure the time between the predator's last capture in a patch and the moment it leaves the patch. The reciprocal of this *giving-up time* (*GUT*) indicates the threshold capture rate (marginal value) at which the predator decides to move to a new patch.[177] A predator should leave later in poor (low food availability) environments because it can deplete the current patch further before it is worth incurring the cost of travelling to a new but low quality patch.

To test this idea, Cowie[62] provided great tits with pieces of mealworm hidden in sawdust-filled plastic cups (patches). The cups were arranged on the 'branches' of five artificial dowelling trees set in a large aviary. Cowie tested his birds in two types of environment, one with a short travel time between patches, one with a long travel time. Travel time was manipulated by fitting tight or loose-fitting cardboard lids onto the cups. If birds were maximising their net rate of

Figure 6.3: (a) Charnov's Marginal Value Theorem Predicting the Optimal Stay Time in a Patch. A predator should stay until its capture rate in a patch falls to the average for all patches in the enviroment (t_{opt}). (b) In Environments Where Patches Vary in Quality, Predators Should Use the Same Marginal Capture Rate in Deciding When to Leave a Patch and Thus Show Different Stay Times in Different Patches.

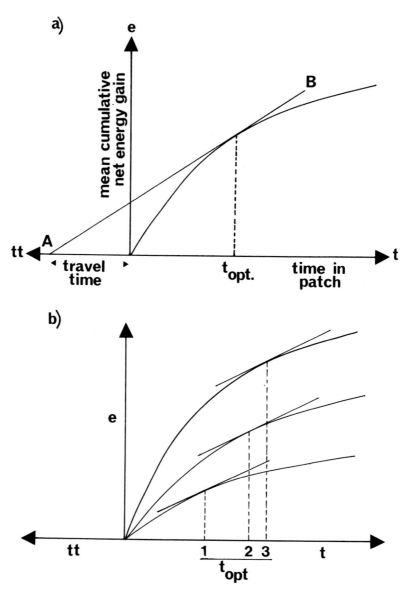

Source: Modified after Krebs[171].

Figure 6.4: (a) If Travel Time (tt) is Increased, the Predator Should Stay Longer in a Patch. (b) Data for Time Spent per Patch by Great Tits When 'Travel Time' Was Varied by Making it Easy or Hard to Obtain Food. Dashed curve, expected relationship if birds were taking only the time cost into account; solid curve, expected relationship if birds were also taking energy expenditure into account.

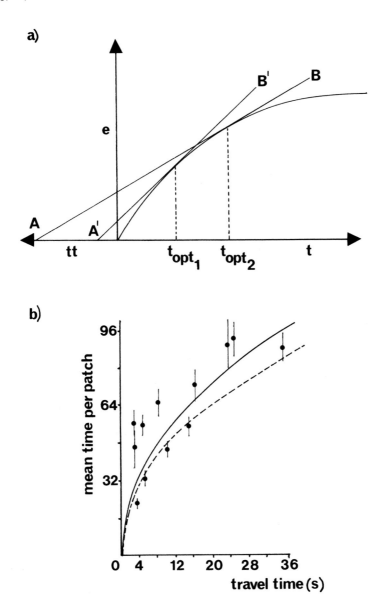

Source: (a) Modified from Krebs[171]. (b) Modified after Cowie[62].

food intake during a foraging bout, they should have remained longer in long-travel-time patches (Figure 6.4a). Figure 6.4b shows that this was the case. However, Cowie made an important distinction between two possibilities. He compared his data with the stay times predicted if birds were only taking into account the *time* spent travelling between patches (dotted line in Figure 6.4b) and with the times expected if they also took the *energy* costs of travelling into account (solid line). The better fit of the time plus energy curve implied the birds were responding to the actual energetic costs rather than 'estimating' them by the amount of time spent. As we stressed before, there is no suggestion that birds are intelligently counting the calories as they behave. They are simply designed by natural selection to respond to environmental stimuli in such a way that they minimise important cost functions (see Chapter 2). Other species also forage in accordance with the marginal value rule.

Davies[70] studied fly-catchers (*Muscicapa striata*) flying out from a vantage point to attack passing swarms of insects. The fly-catcher's problem is deciding whether or not to return to the same perch (functionally equivalent to a patch) after each sally. In this case, patch depletion occurs through the behaviour of the prey (when the swarm moves off) rather than the feeding success of the predator. The success of a sally decreased sharply with increasing distance of the swarm from the bird's perch. The bird must therefore make a trade-off between leaving a perch too soon, when a swarm is still nearby, and waiting too long for another swarm to arrive if the present one has moved away. The birds' GUT for leaving a perch should reflect this trade-off. The mean duration of GUT measured by Davies in the field was around 25 to 30 seconds. When the expected GUT was computed on the assumption that birds were maximising their net rate of food intake (using empirical measures of prey handling time and travel time between suitable perches), the agreement with the observed mean GUT was very close. A detailed example of an insect species foraging according to the marginal value rule is given in the discussion of coevolution in Chapter 11 (see also Krebs[171] for a review).

Memory Windows. An interesting, and as yet unresolved, problem is that of how the predator estimates the average utility of patches in its environment in order to compare it with the feeding performance in individual patches. As we have seen, animals are able to register important differences between environments and modify their GUTs accordingly. The most likely way in which they keep track of environmental changes is by using a sliding 'memory window' of the patches visited in the past. We might expect shorter memory windows when the environment is changeable, because only the last few patches will provide a reliable yardstick for the current patch. One piece of evidence for this comes from laboratory mice. Mice trained to expect regular and predictable food distributions on a grid take longer to respond appropriately to changes in distribution than those trained on randomly distributed items (T.J. Newing and C.J. Barnard, unpublished data). Sometimes the memory window may be very short. Barnard[12] found that GUTs for house sparrows (*Passer domesticus*) feeding on barley seed were best predicted by the average interpeck interval for the preceding two to three pecks prior to birds departing from the food source.

6.2 Choosing What to Eat

So far, we have discussed how predators respond to the spatial properties of their food supply. In doing so, we have said nothing of the choices facing an animal while it is actually feeding in a given place. Prey supplies are seldom homogeneous. Predators are likely to be faced with a choice between several different types of prey. Which type should they choose?

6.2.1 Optimal Diet Selection

As with food patch exploitation, we might expect predators to choose food items so as to maximise their rate of food intake during a foraging bout. They should therefore be sensitive to the various costs and benefits of taking different types of food. Benefit may be measured in many ways: the energy value of an item, its nutrient quality, trace element level and so on. The cost of taking different items falls into two main categories: *handling costs* and *searching costs*. Handling costs take into account the time and energy spent and the risk (of predation, aggression, etc.) incurred between the moment an item is picked up until it is swallowed. Removing undigestible integuments, killing struggling prey, pulling resisting prey out of the ground and so on are all ways in which the handling costs of prey may be increased. Searching costs refer to the time, energy and risk involved in seeking out prey. The utility of a given prey item can thus be expressed as the value (V) of the item (in terms of energy, nutrients, etc.) divided by the sum of the handling (h) and searching (s) costs, i.e. as $V/(h + s)$.

From this simple relationship, we can make predictions about the range of prey types a predator should include in its diet (Figure 6.5a, b). The graphs in Figure 6.5 model the effects of varying searching costs on the number of prey types which should be included in the diet. In rich environments, where searching costs are low, the predator does best by taking only the two highest quality types. In poor environments, the increased searching costs mean it must broaden its diet to include more of the poorer types. Three general predictions therefore emerge from the model:

(1) Predators should prefer more profitable prey.
(2) They should feed more selectively when profitable prey are abundant.
(3) They should ignore unprofitable prey, regardless of how common they are, when profitable prey are abundant.

Krebs and coworkers[174] tested these predictions using great tits. Captive birds

Figure 6.5: Optimal Diet Selection in Rich (a) and Poor (b) Environments. More items should be included in the predator's diet in a poor environment (the peak of V/h + s occurs further to the right than in a rich environment).

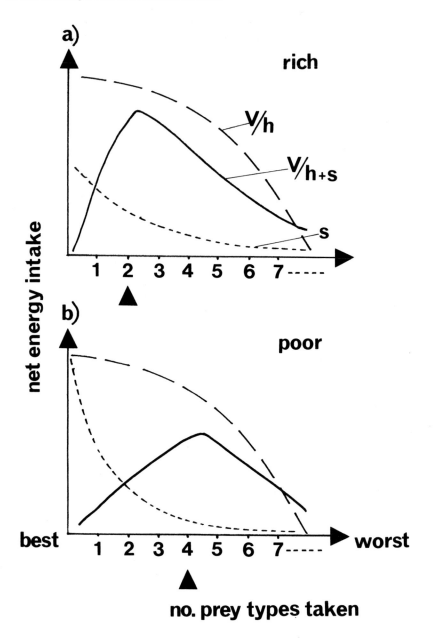

Source: Modified after Krebs[171].

Figure 6.6: Ratios of Profitable Large (Open Columns) to Unprofitable Small (Shaded Columns) Prey Presented to Great Tits. 1, Low encounter rate; 2, high encounter rate with large prey; 3, high encounter rates with both large and small prey; 4, high encounter rate with large prey but even higher encounter rate with small prey. (a) Ratios of large to small prey encountered by bird. (b) The predicted proportion of large and small prey in the diet on the basis of an optimal diet selection model. (c) Ratios of prey sizes taken.

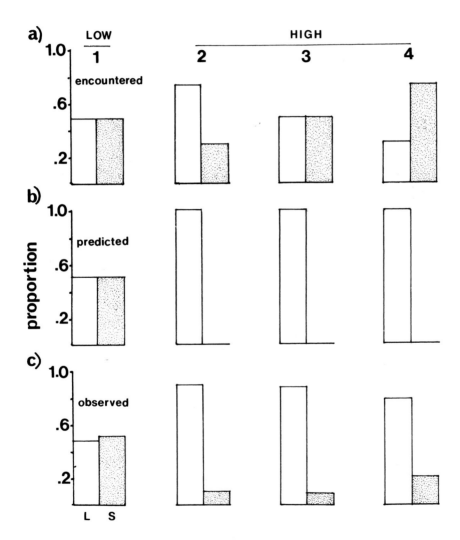

were presented with large (eight-segment) and small (four-segment) pieces of mealworm on a conveyor belt. Small pieces had strips of card attached to them to increase their handling time and reduce their profitability relative to large pieces. The tits' encounter rate with the two types (functionally equivalent to their abundance) was controlled via the moving belt. The results of their experiment are shown in Figure 6.6. When the birds' encounter rate with both prey types was low, they were unselective, taking large and small prey with equal probability. When their encounter rate with large prey was increased, so that they could do better by ignoring small prey, the birds preferentially selected large prey. If the high encounter rate with large prey was kept constant but the encounter rate with small prey was increased to *twice* that with large, the birds remained selective. However, agreement with the predictions of the optimal diet selection model was not perfect. Birds still took some small prey when they should have taken only large. Two possibilities are (a) the birds needed to sample both types to be sure of taking the most profitable and (b) they may have needed time to *recognise* small prey.[171] The second possibility will be discussed further below.

Similar results were obtained by Werner and Hall for bluegill sunfish (*Lepomis macrochirus*) feeding on *Daphnia*. As Krebs[171] points out, however, their results would have been the same if fish were simply selecting the *largest* rather than the most profitable prey. Nearby small *Daphnia* might appear the same size as more distant large *Daphnia*. At high prey densities, it is likely that a large *Daphnia* will happen to be near enough to appear the largest available to the fish.

Although selection on the basis of prey size will sometimes result in unprofitable prey being taken, size may be a good indicator of profitability. The increase in calorific value with the size of an item may outweigh the searching and handling costs involved in eating it. If so, size alone may provide a quick and easy means of prey assessment, especially for predators taking a wide and varying range of prey types.

Barnard and Brown[14] have distinguished between selection for prey size and selection for profitability in common shrews (*Sorex araneus*). Shrews were given a choice of two- and four-segment mealworm pieces distributed randomly in wells on a grid. Because of handling characteristics however, four-segment pieces were less profitable than two-segment pieces (although over a wide range of segment numbers profitability increased with size). If shrews selected prey on the basis of profitability they should have followed predictions 1–3 above of the optimal diet selection model, selecting small prey when common and ignoring large prey. If they selected by size, they should have followed the same predictions but preferred large prey. In fact, shrews showed clear selection for large prey. The mixture of large and small in the diet was determined by the shrews' encounter rate with large prey and was independent of encounter rate with small prey once their encounter rate with large reached a certain point.

Field studies have also suggested that predators tend to select the most profitable prey. Redshank (*Tringa totanus*), feeding on mudflats, tend to take large (profitable) worms in proportion to their density in the mud. The proportion of small (unprofitable) worms taken, however, is more closely related to the density of large worms than to their own density.[118] Lapwings (*Vanellus vanellus*) feeding on earthworms (*Lumbricidae*) in agricultural pasture, also show selection.[17] Birds tend to bias their intake towards worms which are around a lapwing bill-length in size. These turn out to be the most profitable in terms of food energy gained per unit searching and handling time. Selection results in lapwings taking a more profitable mixture of worm sizes than if they simply took worms as they occurred in the ground. The picture is not as clear-cut as that in the redshanks, however, because the degree of selection shown depends on lapwing flock size and the presence or absence of kleptoparasitic gulls (see Chapter 9).

Selection for Specific Nutrients. In the particular cases we have described, the selection criterion has been the net energy value of the prey. This is unlikely to be so in all cases. Sometimes specific nutrients or other qualities of the prey may be more important. If so, it may be very difficult to decide on the correct cost function for foraging behaviour. A possible example of selection for a specific nutrient is found in redshank. When both nereid worms and the crustacean *Corophium volutator* are available in the mudflat, birds select *Corophium*, even though they are energetically less profitable than any size class of worm.[119] One possibility is that *Corophium* is rich in some nutrient, like calcium, which is required by the birds.

Selection for specific nutrients is well known in breeding animals. During periods of egg-laying, spotted flycatchers take a number of calcium-rich prey (e.g. woodlice (Isopoda), millipedes (Diplopoda) and snail shells (Gastropoda)) which do not occur in the diet at other times.[70] Similarly female weaverbirds take a higher than normal proportion of calcium-rich grit on the day of eggshell deposition. Pollen-feeding heliconiid butterflies have been found to show selection for nitrogen.[115] Nitrogen from cucurbit pollen is important for egg production by females. Normally it is only the females that feed at times of peak pollen production, nitrogen intake being less important for males. An interesting exception occurs on the day following a mating when males would benefit from replacing lost spermatophore material (nitrogen compounds in the spermatophore contribute directly to egg production). On these days males forage for pollen more vigorously and successfully.

6.2.2 Prey Recognition Time and Diet Selection

A component of searching cost which we have not yet considered and which is likely to affect prey selection, is the time and effort needed to recognise prey. If prey types take a long time to distinguish from one another or from irrelevant features of the environment, searching costs are increased. Functionally, this has a similar affect to reducing the density of prey and increasing the predator's travelling time between prey. We should therefore expect the predator to

Figure 6.7: The Percentage of Unprofitable Small Prey Taken by Great Tits When There Was a High (Open Histogram) and Low (Shaded Histogram) Recognition Cost for Taking Apparently Large Prey. Line indicates per cent small prey expected on a chance basis.

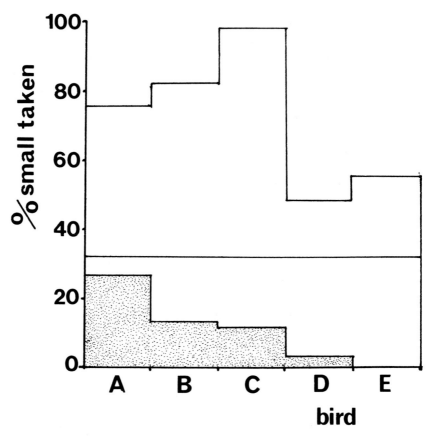

Source: Plotted from data in Erichsen *et al.*[101]

include less profitable prey in its diet when prey are difficult to recognise.

Erichsen and coworkers[101] examined the effects of increased prey recognition time on diet selection by great tits. Using the same conveyor belt apparatus as Krebs *et al.*[174], but this time giving the birds a choice of two (unprofitable) and four (profitable) segment mealworm pieces, they compared prey selection when profitable prey were difficult to distinguish from inedible 'twigs'. Small mealworm pieces were placed inside lengths of transparent drinking straw and were easily distinguishable from large pieces concealed in opaque lengths of straw. Also concealed in opaque lengths of straw, however, were pieces of brown string ('twigs'). Birds therefore had to spend time peeling open the straws to see what was inside and risked wasting time and effort on an inedible object.

Birds were given two different treatments. In Treatment A, 'twigs' were four times as common as large mealworm pieces. In Treatment B, the ratio was reversed. The ratio of 'twigs' to worms in Treatment A was high enough to change the optimal policy of the birds from selection for large prey to selection for small (predicted from an optimal diet equation[174]). In Treatment B, the optimal policy was still selection for large prey. As Figure 6.7 shows, the percentage of small prey taken was significantly higher than would be expected by chance in Treatment A and significantly lower in all cases except one in Treatment B.

Houston and coworkers[149] performed a similar experiment, this time making it difficult to distinguish small from large prey. When discrimination time between the two prey types was zero, birds selected large prey as expected from their relative profitability. When the two types were difficult to distinguish, birds became unselective depending on their individual prey recognition abilities and on their encounter rate with large prey.

6.2.3 Competition and Diet Selection

Animals seldom have sole access to a food supply. Even territorial individuals are subject to intrusion by neighbours or itinerants. Competition for food from other animals is likely to alter a predator's criteria for prey selection in several ways. In particular, competitors will change the availability of some, if not all, the prey types in the animal's environment. Thus both the *absolute* and *relative* abundance of each prey type may be reduced and prey type mixtures become less predictable. Under these circumstances, a predator's past experience of its food supply tells it little about the nature of the supply in the future. Since future prey availability becomes uncertain, we might expect competition to reduce the selectivity of the predator. Barnard and Brown[14] tested this prediction with shrews.

Shrews were presented with two- and four-segment mealworm pieces on a grid. At the same time a 'competitor' individual was introduced which had not previously been trained to forage on the grid. This meant the test animal was aware of the 'competitor's' presence, but the 'competitor' did not actually deplete the food supply. In this way, Barnard and Brown could measure the test animal's *expectation* of competition and its subsequent response, rather than simply its response to a more rapidly depleting food supply. Mealworm pieces were presented so that test animals' encounter rates with large pieces were high enough for them to selectively take large prey (see Section 6.2.1). The degree of selection for large prey observed, however, was significantly lower than that expected from the animals' encounter rates with large prey (Figure 6.8). Shrews responded to apparent competition by reducing their feeding selectivity.

Figure 6.8: Common Shrews Took Fewer Large Prey When a Competitor Was Present (Shaded Histogram) than Expected on the Basis of their Encounter Rate with Large Prey (Open Histogram).

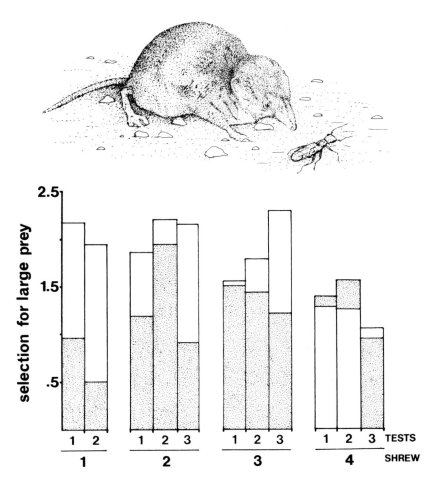

Source: Plotted from data in Barnard and Brown[14].

6.3 Central Place Foraging

The predators we have dealt with so far have been those which consume their prey more or less where they find them. Many predators, however, have to return to some fixed point to consume prey. This is true, for instance, of predators gathering food for nestling young or those diving to secure benthic prey which they can only eat on the water surface or on land. The key difference between such *central place foragers*[232] (CPFs) and other predators is that a foraging bout for a CPF consists of a round trip. The round trip has three components, (1) an outward journey, (2) a period of foraging and (3) a return journey. The predator spends time and energy in all three, but acquires food only in the second. Furthermore, the return trip is likely to be more expensive than the outward trip because of the food load being carried and because the predator might consume food on the trip and increase its body weight. We might expect these extended travelling costs to influence (a) the way predators select and exploit food patches, (b) the way they select their diet and (c) the load size with which they return to the fixed point ('home base'). A series of predictive models along these lines has been developed by Orians and Pearson[232].

6.3.1 Central Place Foraging and Patch Selection

Bearing in mind the additional burden of travelling to and from feeding areas (*commuting costs*), we might expect CPFs to feed as near to their home base as possible. However, this would provide only a temporary reduction in total foraging costs. Nearby patches would quickly become depleted and the animal would have to search further afield. This problem leads to one of the more unexpected predictions of central place foraging theory.

Morrison points out that, because of high commuting costs, it pays a CPF to minimise the risk of revisiting previously-depleted feeding areas. While revisiting may not be too serious for a non-CPF (for whom localised searching around the site of a find may be profitable), commuting costs mean a CPF must streamline its searching behaviour as much as possible. By means of a mathematical model, Morrison showed that the optimal search strategy involved a reduction in the amount of turning during locomotion. This has two effects. First, it reduces the risk of revisiting during a foraging bout; secondly it means the CPF is more likely to discover and exploit distant patches. Orians and Pearson also predicted that depletion of nearby patches would lead to CPFs exploiting patches distant from the home base. If nearby patches are depleted before the CPF gathers its optimal load, the efficiency of subsequent searching and/or the return trip is reduced. CPFs should thus start foraging at a distant patch and work back towards the home base. By the time the optimal load is gathered, the return trip will have been considerably shortened.

Counter-intuitive as these predictions are, they are borne out by the behaviour of the CPSs studied.   Morrison found that the fruit bat *Artibeus jamaicensis* fed on fig trees that were some three times further away than the nearest ripe fig to their roost. More striking still, the African bat *Epomophorous gambianus* flies off to feed on distant ripe fruit while ignoring apparently similar ripe fruit hanging in its own roost tree.

6.3.2 Central Place Foraging and Prey Selection

Just as the round trip costs of central place foraging affect the way CPFs exploit food patches, they also affect their choice of prey. Unlike in non-CPFs, the criteria governing prey selection by CPFs revolve around the optimal food load for the return trip. This is likely to involve a trade-off between the optimum load in terms of travelling energetics and the physical limitations (e.g. mouth or crop size, gripping ability) of the predator's feeding apparatus. In many cases, as in raptors and carnivorous mammals, CPFs may be constrained to take only one prey per round trip. If so, we might expect these single-item CPFs to be more selective than their multi-item counterparts. Suggestive evidence comes from red-winged blackbirds (*Agelaius phoeniceus*).

Blackbirds nesting near marsh vegetation in Costa Rica appear to be mainly single-item CPFs. The main prey items are large grasshoppers (Orthoptera), small grasshoppers occurring far less frequently in the diet than expected from their relative density in the habitat. In contrast, blackbirds nesting in eastern Washington bring around twenty items to the nest. At least in terms of energy content, these prey are highly variable, ranging from dragonflies and cicadas at 800 calories per item to small flies of around 10-15 calories. Furthermore, the number of small prey in the diet does not decrease with increasing capture rates of large prey, as might be expected in a non-CPF. The probable reason for the difference in loading strategy in the two populations is the difference in pursuit characteristics of their prey. The emerging aquatic insects taken in Washington require negligible pursuit because their flight is weak or non-existent. The prey taken in Costa Rica, on the other hand, are either concealed, mixed up with other material on the surface of the water or active. They thus require investment in probing and pursuit behaviour.

Another prediction of central place foraging theory is that load size should increase with distance of the feeding site from the home base. Clearly, higher commuting costs must be offset by increased feeding returns. Figure 6.9 shows the mean number of prey returned to the nest by Brewer's blackbirds (*Euphagus cyanocephalus*) breeding in Washington. Birds gathered more prey per trip the further they foraged from the nest. Carlson and Moreno[44] found the same relationship for wheatears (*Oenanthe oenanthe*). Their birds also spent longer collecting food when they fed in distance patches. Similarly Davies[70] found that spotted flycatchers tended to return to their nestlings with larger prey after a long-distance sally. At a more general level, adult auks (guillemots (*Uria aalge*), puffins (*Fratercula arctica*), etc.) show a positive correlation between foraging distance from the nest and the number of prey brought back.[59] Additional pressure on the size of the prey loads is levied by the attacks of kleptoparastic gulls and skuas. The extreme design characteristics of auk bills can be interpreted as adaptations to increase the efficiency of central place foraging in response to these various pressures.

Finally, CPFs bringing food back to the home base for consumption by other individuals, must clearly feed themselves as well. They could do this in two ways. They could either devote some trips to feeding only themselves, or they could feed at the beginning of a foraging period and leave the feeding apparatus free for the remainder. The second possibility seems more likely because there is nothing to be gained by returning to the home base empty-handed. In line

Figure 6.9: Nesting Brewer's Blackbirds Tend to Bring More Prey Back to the Nest (Central Place) When They Have Travelled a Long Way to Find It. Open and closed circles represent data for different years; open triangles, estimates of prey number based on crop size.

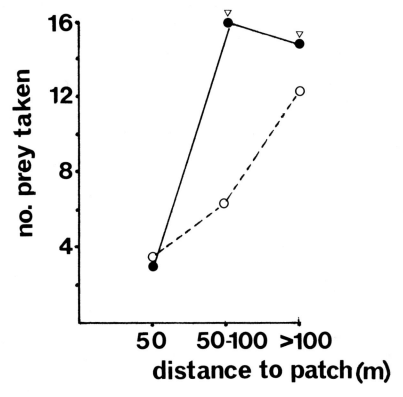

Source: Plotted from data in Orians and Pearson[232].

with this expectation, yellowheaded (*Xanthocephalus xanthocephalus*) and Brewer's blackbirds tend to feed themselves first on over 50 per cent of round trips in which they forage for nestlings.[232]

6.4 Foraging Mechanisms

In the last three sections, we have seen how predators organise their foraging behaviour so as to maximise their intake of important dietary constituents like food energy and specific nutrients. However, apart from the discussion of GUTs, we have said little about the behavioural mechanisms involved. In this section, we shall discuss some aspects of behaviour which help a predator optimise its feeding behaviour.

6.4.1 Area Restricted Searching

One way in which predators may be able to increase the efficiency with which they exploit patchily-distributed food supplies, is by altering their pattern of movement after finding prey. If they increase their rate or degree of turning and/or decrease their rate of movement, they will tend to increase the amount of time they spend in areas of high food density. Such changes in the pattern of movement are known collectively as *area restricted searching*. They represent a simple mechanism by which predators can maintain themselves in good quality patches.

Murdie and Hassell[223] found that houseflies (*Musca domestica*) searching for tiny droplets of sugar solution in an arena, showed marked changes in searching behaviour after encountering a droplet. The changes were of two sorts. First, there was a change in the angle of turning (klinokinesis). Immediately after flies found food, there was a dramatic increase in the angle of turning which gradually decayed back to pre-feeding levels if no further droplets were encountered. The second change was in the speed of movement. After finding food, the rate of movement dropped sharply and then slowly increased. The net result was that flies tended to concentrate their movement near spots where they found sugar.

Similar results have been obtained with other species. House sparrows feeding on mealworms distributed on a grid, showed decreased rates of hopping and increased tortuousness (measured as the ratio of actual distance travelled between two points and the straight line distance between those points) in their foraging paths after finding a worm[10] (see Table 6.1).

Table 6.1: Meander Ratios (Straight Line Distance between Two Points Divided by the Actual Distance Travelled by the Animal between Those Points) for House Sparrows before and after Finding a Mealworm on Uniform, Random and Clumped Prey Distributions. Higher numbers mean more search path tortuousness and area restricted searching. Sparrows showed most area restricted searching after finding items on clumped distributions.

	Before	After	
Uniform	1.39 ± 0.148	2.31 ± 0.315	t = 2.64, n = 24, p < 0.05
Random	2.49 ± 0.794	3.18 ± 0.842	t = 0.60, n = 32, n.s.
Clumped	1.44 ± 0.149	4.33 ± 1.04	t = 2.75, n = 26, p < 0.05

Source: Barnard[10].

Adaptiveness of Area Restricted Searching. A mechanism like area restricted searching is only useful as long as the predator is dealing with clumped prey. There is little point in a predator area-restricting its search with an overdispersed prey supply. We should therefore expect predators to vary the degree of area restricted searching according to the distribution of prey.

Smith[290] tested this idea by observing the behaviour of foraging thrushes (*Turdus* spp.). Like the previous examples, thrushes changed their searching behaviour after finding prey (in this case earthworms in a meadow). Movement after a capture tended to include more turns in one direction, so that birds searched a smaller area more thoroughly than in the same number of hops before the capture. Smith presented birds with 'populations' or cryptically-coloured artificial prey, arranged in regular, random and clumped distributions. Each distribution was presented at a high and a low density. The movement of birds on each distribution and density was then recorded. Smith found that thrushes made larger turns and turned more often in a given direction (rather than alternating left and right) on high-density supplies, but there was little difference in behaviour before and after a find. At low densities, however, they showed area restricted searching on both clumped and random supplies (because of their heterogeneity, random distributions possess a degree of patchiness) but not when 'prey' were regularly distributed.

House sparrows presented with regular, random and clumped distributions of mealworms on a grid showed similar variations in area restricted searching (Table 6.1). Search path tortuousness after a find increased from a minimum on regular distributions to a maximum on clumped while hopping rate increased. In a recent study of lacewing (*Chrysopa carnea*) larvae, Bond found that area restricted searching increased with the period of food deprivation prior to a test. He suggested that larvae estimated food availability in the environment from the food content of their guts.

The implications of the thrush, sparrow and lacewing results are that predators show some ability to modify their searching movements according to their expectation of the environment. The extent to which they show area restricted searching is influenced by the degree of food patchiness.

6.4.2 Return Times

Food supplies vary in their rate of replenishment after depletion by a predator. *Non-renewing* supplies are those which do not replenish or replenish only very slowly. Standing or stored seed supplies are a good example. Once non-renewing supplies are depleted, the predator may as well move off in search of a new area. Predators exploiting *renewing* supplies, on the other hand, have the opportunity to return to a previously depleted area to feed again. The problem, however, is to time the return visit optimally. If the predator returns too soon, replenishment will not be complete; if it returns later than necessary, it risks travelling further or spending more time on its foraging bout than if it had returned earlier. Either way its foraging efficiency for the bout is decreased.

As Charnov and coworkers[53] point out, the utility of returning to a renewed resource is limited by the amount of competition a predator faces. To achieve an optimal return time, the predator must have exclusive use of (and hence reliable information about) the resource. For this reason, we are most likely to find 'return' strategies among territorial (individuals parcel out areas for their exclusive use) and socially aggregating (individuals are all in one place and hence do not deplete different parts of the environment independently) species.

Territorial nectar-feeding birds are good examples of 'return' strategists. Territory size in golden-winged sunbirds (*Nectarinia reichenowi*)[116] and Hawaiian honeycreepers (*Loxops virens*)[154] is geared to the rate of nectar renewal within depleted flowers. Territory owners regulate their patterns of revisits to enhance their net rate of energy intake over a foraging bout. Insectivorous pied wagtails (*Motacilla alba yarrellii*), feeding on debris washed up along a river bank appear to maximise the time between visits to a given site by patrolling their territory in a systematic way and evicting intruders.[76] Similarly, Cody[58] suggests that flocking behaviour in some species may be a means of minimising independent revisits to depleted areas. By forming a cohesive flock, birds can recognise depleted areas more easily because flocks remove more food than solitary individuals from any given area. However, this hypothesis remains to be rigorously tested. The costs and benefits of territoriality and aggregation are discussed more fully in Chapter 9.

6.4.3 Search Image Formation

As we saw in Section 6.2.1, predators can focus their attention on one particular prey type when presented with a mixture. The advantages of this ability are obvious. The predator can concentrate on more profitable prey types and avoid unprofitable or noxious items. There is a problem, however. Prey mixtures in the real world do not remain constant. Different types of prey may vary temporally and spatially in their relative availability. For *specialist* predators, for example, Koalas (*Phascolarctus cinereus*) and everglade kites (*Rostrhamus sociabilis*), which are geared to taking only one or a few prey types, this may have little effect other than to increase the time and range over which they have to forage. Their prey discrimination tasks remain the same. *Generalist* predators, however, obtain their required nutrition from a wide and varying range of prey. Consequently, they may have to change their prey selection quite frequently. There is evidence that changes in selection may involve learning and central perceptual changes.

Functional Responses. A common problem facing generalist predators is the changing density of different prey types. While a predator is feeding on prey type A, which was initially the more available, type B might increase in availability either through depletion of A or because more of B appear in the environment. Other things being equal, the predator should *switch* to the more available prey because of reduced travel costs. However, if switching was simply a matter of tracking changes in prey availability, capture frequency for any given prey type should match its availability. This is often not the case. As Krebs[166] points out, when prey availability first begins to change, the predator's degree of selection for the new most available prey tends to lag behind the change in availability (Figure 6.10a). After a while, however, the rate of predation increases disproportionately to the change in availability. Krebs suggests that this relationship between prey availability and the predator's

Figure 6.10: (a) Sigmoid Type-3 Functional Response for Deer Mice (*Peromyscus*) Taking Cocoons. (b) Type-2 Functional Response for House Sparrows Taking Seed in a Cattleshed. Dashed curve is the Type-2 relationship expected on the basis of Holling's Disc Equation.

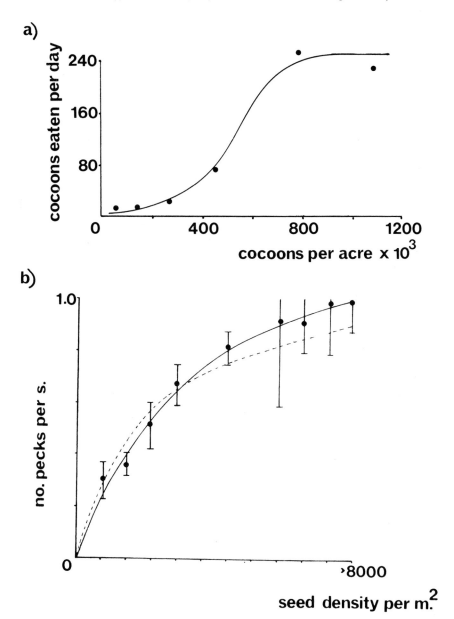

Source: (a) Modified after Holling, C.S. (1959). Some characteristics of simple types of predation. *Can. Ent., 91*: 385–98. (b) From Barnard[13].

attack rate results from an interaction between learning and forgetting. Only when prey availability is high enough will the predator learn about the prey faster than it forgets and hence increase its attack rate. The S-shaped curve in Figure 6.10a characterises predator responses when more than one type of prey is available or when the predator is feeding on cryptic prey which it must first learn to see, and is sometimes known as a *Type 3 functional response*. When only one type of prey is available and feeding rate is limited by the predator's ability to handle each item (instead of by its ability to 'see' prey), a hyperbolic Type 2 functional response generally occurs (Figure 6.10b). Type 1 responses, shown by animals like filter feeders, are less common and describe a linear relationship rising to a plateau.

Search Images. One way in which a predator might improve its switching efficiency is by forming a 'search image' for specific prey. As it encounters different prey types in its environment, the predator might selectively 'learn to see' those which are most profitable. Croze examined search-image formation in the field. Hand-reared and wild crows (*Corvus corone*) were observed on a beach foraging for coloured mussel shells some of which covered pieces of meat. The colours resembled the range found in the pebbles on the beach and all seemed initially to be equally cryptic to the birds. However, after two or three finds under shells of a particular colour, the crows selected only shells of that colour. They had apparently learned to see those shells while the others remained cryptic to them.

By far the most detailed analysis of search image formation is that by Dawkins. Dawkins examined the ability of domestic chicks to detect cryptically-dyed rice grains. Chicks were observed taking grains from a background of stones glued onto hardboard. Grains were dyed green or orange and were stuck on backgrounds of either the same (cryptic prey) or the opposite colour (conspicuous prey).

In her first experiment, Dawkins compared the time taken for naive chicks to find five cryptic and five conspicuous grains at the beginning (1–5 grains eaten) and end (96–100 eaten) of a test. Not surprisingly, chicks took longer to find five cryptic grains at the beginning of a test than they did to find five conspicuous grains. However, by the end of a test, their ability to find cryptic grains had improved dramatically (Figure 6.11). It seems, therefore, that chicks did not take the grains initially because they didn't notice them. To test this, Dawkins compared the amount of time 'cryptic' and 'conspicuous' chicks spent in the *head down* posture (suggesting visual assessment of objects on the ground) and the number of erroneous pecks they made to the background. At the beginning of their tests, 'cryptic' chicks spent about 34 seconds in *head down* before pecking, compared with about 1.3 seconds in 'conspicuous' birds. Furthermore, 'cryptic' chicks pecked at pebbles more often suggesting they had trouble seeing prey even when they were looking straight at the ground.

Dawkins then went on to see how well chicks retained their ability to find

Figure 6.11: The Relative Ease with which Chicks Found Cryptic (Dotted Line) and Conspicuous (Dashed Line) Rice Grains. (a) Birds searching for green (conspicuous) or orange (cryptic) grain on an orange background. (b) Birds searching for green (cryptic) and orange (conspicuous) grain on a green background. Chicks took longer to start finding cryptic grain, but then found them at a high rate implying search image formation.

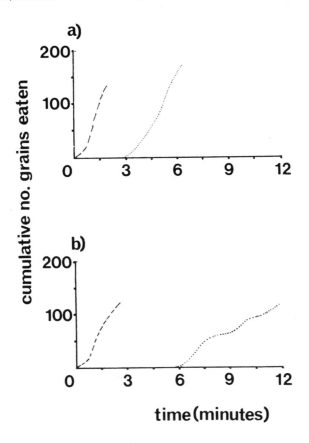

Source: Modified after Dawkins, M.E. (1971). Perceptual changes in chicks: another look at the 'search image' concept. *Anim. Behav.*, *19*: 566-74.

cryptic grain. Chicks were tested as before but with age controlled for, and tests repeated on subsequent days. Again chicks overlooked cryptic grain at the beginning of tests but eventually took them quite rapidly. However, they did not retain their ability to see cryptic grain from one day to the next. They still had trouble seeing grains at the beginning of each test.

Finally Dawkins looked at the effect of experience in eating conspicuous grains on chicks' ability to see cryptic grain. She arranged the apparatus so that a test card could be placed in the centre of the stone background. By arranging different colour combinations of background, test card and grain, Dawkins could investigate the effects of the chicks' experience immediately prior to finding cryptic grain. When released into the arena, chicks usually found between 20 and 100 grains before entering the test square. When the time taken to find the grain on the test card was measured, it was found that chicks discovered cryptic test grains faster (6.7 seconds) when they had just been sampling other cryptic grain than when they had sampled conspicuous grain (66 seconds). This compared with 2.5 seconds and 1.0 second respectively with conspicuous test grains.

Dawkins' results suggest that search-image formation does occur and that it is the result of changes within the CNS rather than peripherally (e.g. head or eye orientation). The formation of search images would provide the predator with a reliable and efficient way of focusing attention on specific prey types.

Summary

(1) Because of the importance of energy and nutrient intake to survivorship, we might expect animals to be highly efficient predators. From the evidence available, this expectation appears to be borne out.
(2) Natural food supplies are commonly distributed in a patchy fashion. Predators appear to maximise the efficiency with which they exploit food patches by gearing their sampling and exploitation behaviour to maximise their net rate of food intake during a foraging bout.
(3) Predators faced with a choice of different prey types, tend to select prey or mixtures of prey which maximise their net rate of food intake over a foraging bout. Sometimes selection may be made on the basis of an approximate 'rule of thumb' which reduces assessment costs.
(4) Some predators are constrained to return to a 'home base' or central place to consume prey. The foraging strategies of such 'central place foragers' are modified according to the costs of the round trip for each foraging bout.
(5) Several behavioural mechanisms appear to help predators optimise their foraging behaviour. The most thoroughly investigated include area restricted searching, return time regulation and the formation of search images.

7 ANTI-PREDATOR BEHAVIOUR

In the last chapter we discussed some of the problems facing a hungry animal as it searches for food. Finding food, however, is only one side of a two-sided coin. Apart from finding its own food, an animal must avoid becoming food itself. As we emphasise several times in this book, an animal's behaviour is usually a compromise between several conflicting selection pressures. The degree to which an animal adapts to one pressure depends on how that adaptation affects the other activities it has to perform. If a particular selection pressure is intense, then we might expect adaptation to proceed a good way towards reducing it before the utility of adaptation in that direction is outweighed by other selection pressures. For most animals, predation constitutes one of the severest everyday pressures. It is not surprising, therefore, that many possess extreme morphological, physiological or behavioural adaptations to reduce their risk of predation. As Harvey and Greenwood[135] point out, however, it is important to emphasise that predation need not be a limiting factor in a prey population for a genetically based anti-predator strategy to evolve. Like the hypothetical herbivorous insect example in the last chapter, any mutant individual who is just a little better at avoiding predation is likely to have a higher inclusive fitness than its conspecifics. The mutant will therefore spread even though it does not necessarily facilitate an increase in population size and density. In addition, anti-predator mechanisms may evolve for the protection of individuals other than the one possessing the mechanism. Because natural selection acts on alleles and not individuals, anti-predator mechanisms may function to protect offspring and other close relatives instead of the individual in question.

The most effective way to avoid predation is clearly to avoid encounters with predators in the first place. Many anti-predator adaptations involve camouflage (crypsis), polymorphism or mimicry of noxious species so that predators fail to distinguish potential prey from objects it does not regard as food. Most of these 'primary' defence mechanisms are morphological and do not directly concern us here (for detailed reviews see Edmunds[93] and Harvey and Greenwood[135]). In some cases, however, crypsis and mimicry depend on behaving in the right way, often to enhance the efficacy of morphological adaptations. The Caribbean fish *Lobotes surinamensis*, for example, lives when juvenile among mangroves (*Rhizophora* and *Avicennia* spp.). Leaves which fall from the mangroves float for a while on the surface of the water before sinking. Juvenile *Lobotes* are roughly the same shape and size as the mangrove leaves and, when resting, float on one side near the water surface with the head tilted slightly downwards. Furthermore, *Lobotes* are also a yellowish-brown colour with diffuse brown spots, exactly like the mangrove leaves. If a leaf is thrown onto the water near a fish, the fish often swims towards it and takes up the appropriate orientation. Similar crypsis-enhancing orientation is found in many other species, particularly those like the peppered moth which rests on a highly patterned substrate (e.g. tree bark) and the various grass and leaf-mimicking phasmids, mantids, grasshoppers and butterflies. It is in so-called 'secondary'[93] defence mechanisms, however, that we find most behavioural anti-predator adaptations.

'Secondary' defence mechanisms are those mechanisms an animal brings into play when it encounters a predator. Having failed to avoid detection or having encountered a predator in the course of its daily activity, the animal needs a back-up anti-predator strategy. Some species rely entirely on avoidance after detection rather than investing in any 'primary' defence mechanisms. Indeed, 'primary' defence mechanisms are likely to be of only limited value anyway. Highly cryptic prey are frequently discovered by chance or systematic searching on the part of predators. Nevertheless, Bauwens and Thoen[22] point out that the cost of a cryptic strategy need not be high and relate an instance in common lizards (*Lacerta vivipara*) where females swop adaptively between crypsis and secondary defence mechanisms depending on their reproductive condition. Not surprisingly, many 'secondary' defence mechanisms involve elaborate escape, avoidance and defence behaviours and it is on these that we shall focus our attention in this chapter.

7.1 Escape Behaviour

Many defence mechanisms involve somehow escaping the predator, either by actually fleeing or dashing for cover, or by distracting the predator or reducing its interest and attention.

7.1.1 Withdrawing to Cover

Many animals live in burrows, crevices or other sheltered places into which they can retreat in the event of danger. Such animals are sometimes known as 'anachoretes' (from the Greek for 'retire'). Most anachoretes must emerge from seclusion at some time to feed or perform other vital activities, but stay within a short distance of their retreat. While retreats like the burrows of rabbits (*Oryctolagus cuniculus*) and lugworms (*Arenicola* spp.) are fixed retreats dug into the substrate, some species use more mobile retreats, or even carry their own retreat with them.

The pearl-fish (*Carapus acus*), which feeds partly on shrimps and other small crustaceans, retreats when disturbed into the anus of a sea cucumber (*Holothuria* spp.). In this instance, the fish is parasitic on the holothurian and feeds on its host's gonads. As an adult the fish occasionally leaves its host to feed facultatively on the crustaceans. Hermit crabs (e.g. *Clibariarius, Diogenes* spp.) acquire empty gastropod shells which they carry about on their backs. Shells are chosen visually and crabs move into a succession of larger shells as they grow. Yet other species, which are not strictly anachoretes, possess their own protective cover. Gastropod and bivalve molluscs, tortoises (e.g. *Testudo* spp.), hedgehogs (*Erinaceus* spp.), echidnas (*Tachyglossas* spp.), armadillos (*Tolypeutes* spp.) and millipedes (*Glomeris* spp.), for instance, can all roll up and present the predator with a tough or spiny outer covering.

As Edmunds points out, however, one problem with retreating to cover is that a predator may follow an animal into cover and corner it. It is perhaps not surprising therefore that many species have evolved ways of blocking the entrance of their retreat after withdrawal. Some hermit crabs (*Diogenes* spp.) have evolved a large chela which occludes the opening of their shell after retreat. The fairy armadillo (*Calamyphorus truncatus*) has horny plates at the posterior end of its body which blocks its burrow, preventing predators entering. The African ground squirrel *Xerus erythopus* blocks up the entrance of its burrow when it retires to sleep. At least in some mammal species, retreat-blocking is used as an extension of parental care to protect young while adults are out foraging.

7.1.2 Flight Behaviour

The response of most animals when suddenly faced with a predator is to flee. Selection has acted in a variety of ways in different species to enhance the efficacy of flight behaviour. Perhaps the most direct adaptation is enhanced flight speed and agility. The familiar coevolution between the long-legged savannah herbivores like zebras and gazelles and running carnivores like lions and cheetahs is a good example. Here the arms race between predator and prey has gone a long way towards perfecting animal speed machines.

Adaptations for speed, however, are likely to require sacrifices in other attributes, so we might expect only some species to adopt a simple fast flight strategy. Another and perhaps evolutionarily cheaper, way of enhancing the effectiveness of flight is to move in an erratic and unpredictable way. Many species, like ptarmigan (*Lagopus mutus*), snipe (*Capella gallinago*) and various lagomorphs, antelopes and gazelles flee from predators in a characteristic zig-zag fashion. Rapid unexpected changes in flight direction make it difficult for a predator to track prey (see also Chapter 9). In some species, like the European hare (*Lepus europaeus*), erratic zig-zag flight is interspersed with bursts of running in a straight line, thus introducing another element of unpredictability. Edmunds suggests that zig-zag flight might be more effective in the presence of predators which are faster than their prey and straight flight more effective against predators which are slower. One observation which supports this suggestion is the recorded tendency for slow-flying black-headed gulls (*Larus ridibundus*) to show frequent changes in flight direction when they spot a peregrine falcon (*Falco peregrinus*) (peregrines are adept at taking flying birds).

'Flash' Flight. A quite different way of enhancing escape by flight is to use so-called *'flash'* behaviour. Here, the alarmed prey animal flees for a short

distance and then 'freezes'. Some predators are unexcited by immobile prey[193] and a startling flash of activity followed by immobility may confuse them. 'Flash' behaviour is used in particular by frogs and orthopteran insects which make conspicuous jumps and then sit immobile. In some species, 'flash' behaviour is enhanced by the display of bright body markings. Good examples are the red and yellow underwing moths (*Catocala nupta* and *Triphaena pronuba*). At rest, both species are a cryptic brown colour. When they fly, however, brightly coloured hindwings are exposed which render the moths highly conspicuous.

Many species of orthopterans, mantids and cicadas also have brightly coloured wings which are exposed only in flight. Similarly, some frogs and lizards have brightly coloured patches or frills which may serve as a 'flash' function when they move quickly. The familiar silvery flashing of small fish when a shoal suddenly breaks up may also have an anti-predator effect. Some species even appear to possess 'flash' sounds. The loud buzzing and clicking noises made by some grasshoppers when they jump may serve to emphasise movement.

Distraction and Diversion Displays. Another common way of dealing with predators is to divert their attention and hence probability of attack away from vital areas. Many fish and insect species have evolved pseudo 'eye' spots or false heads at the posterior end of their body, thus presumably deflecting attack from their true head. Such diversion displays often depend on characteristic movement patterns. Familiar examples are the tail-twitching displays of many lizards and snakes. In some lizards, tail-twitching is accompanied by an ability to break off the tail (*autotomy*) if an attack is carried through. The detached tail wriggles and squirms with a 'life' of its own and may keep a predator occupied while its other half makes an escape. Autotomy of body parts in response to predation is also known in bivalve molluscs, lugworms and arthropods.

Another sort of distraction/diversion display is geared to protecting nests and offspring rather than the individual itself. Various ground-nesting bird species, for example, dotterels (*Charadrius marinellus*) and meadow pipits (*Anthus pratensis*) perform very convincing 'broken-wing' displays when danger threatens (Plate 1). Adult birds meander away from their nest trailing an 'injured' wing on the ground. The predator follows this seemingly easy meal only to have it take off when it has been lured a safe distance from the nest.

'Death-feigning'. Potential prey individuals can sometimes avoid death by feigning death (*thanatosis*) (Plate 2). Many predators will not take carrion or are not stimulated by immobile prey. 'Death-feigning' occurs in a taxonomically wide range of species including beetles, bugs, spiders, stick insects, mantids, grasshoppers, reptiles, birds and mammals. When attacked, some beetle species become immobile with their legs folded close to the body to avoid damage. The American hog-nosed snake (*Heterodon* spp.) feigns death if its initial hissing and jerking displays fail to discourage a predator. The snake turns over, exposing its pale belly and remains motionless. While, in some species, death-feigning may

Plate 1: Broken Wing Display by a Ringed Plove (*Charadrius hiaticula*). Birds adopt a tilted, sagging posture and move slowly away from a predator, often with plaintive vocalisations.

Source: Photograph courtesy of C. Galbraith

Plate 2: Death-feigning in a Grass Snake (*Natrix n. abstreptophora*). Note the ventral side up posture and the open mouth.

Source: Photograph courtesy of Dr P.M.C. Davies.

last only a few seconds after which the animal often attempts to flee, some displays may be continued for a long time and through a good deal of interference. Death-feigning individuals of some ground squirrel species can actually be handled without breaking the bluff. Coyotes and American opossums (*Didelphys* spp.) are other mammals which can show convincing and long-lasting death-feigning displays.

7.2 Defence

If an animal is unable to escape for some reason, it may resort to various defensive strategies. These may range from simple 'startle' responses which momentarily bewilder the predator, to physical retaliation. In this section we shall consider some of the ways in which prey may defend themselves against or even deter predators.

7.2.1 'Startle' Responses

Several species have evolved means of startling predators, presumably to create a 'period of grace' within which they can escape. Many cryptically-coloured animals like rabbits and partridges (*Perdix perdix*) remain motionless until an approaching predator is almost upon them. Then they erupt noisily just in front of the predator and speed away to safety. Anyone who has experienced this will be in no doubt as to its unnerving and disorientating effect on the hitherto oblivious approacher. In this case, the 'startle' response is little more than a variation of straightforward flight. In other cases animals have evolved special and sometimes elaborate displays which appear to be designed to startle or intimidate predators. Such displays or responses are sometimes known as *deimatic* or *dymantic* (from the Greek for 'frighten') behaviour.

The use of startling to gain a period of grace in which to escape has been shown quite nicely in saturnoid and sphingid moths. Here, moths disturbed by an interfering prod or peck perform a vigorous wing-flapping and body-rocking display. Such a display may cause a predator to hesitate for a few seconds during which time the moth has used the display activity to warm its flight muscles for take-off. Other moths possess brightly coloured spots on their underwings which crudely resemble vertebrate eyes. These 'eye' spots generally remain hidden until the animal is alarmed where upon they are suddenly revealed. The startle effect in these cases may be enhanced because the spots resemble the eyes of owls which are predators of the insectivorous birds attacking the moths. Predator mimicry is taken even further in some lepidopterous larvae like the snake caterpillar (*Leucorampha* – Figure 7.1). When disturbed, the snake caterpillar swings its anterior end free of its supporting twig and expands it into a highly convincing snake's head. In some cases 'startle' signals may complement another anti-predator adaptation. Some distasteful arctiid moths, for instance, emit sharp clicking sounds which cause approaching bats to withdraw.

Figure 7.1: The Deimatic Display of the Neotropical Hawkmoth *Leucorampha*.

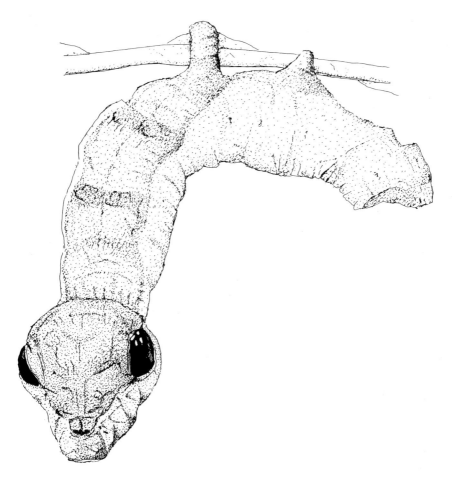

Source: Drawn from a photograph by N. Smythe.

'Startle' responses are also found among vertebrates. Alarmed hog-nosed snakes jerk and writhe about, hissing loudly. Like the rabbits and partridges mentioned earlier, the snakes tend to remain immobile until the unwary approacher is almost on them. Many vertebrate responses involve displays of enhanced body size or 'weaponry' like horns and teeth. Several ungulate species, for instance, display horns or perform kicking actions to display their hooves. African porcupines (Hystricidae) respond to interference by rattling their quills before attacking. American porcupines (Erethizontidae) hiss and erect their spines. In both cases the spines can inflict serious injury if attacks occur.

7.2.2 Chemical Defence

Various species have evolved defence mechanisms based on noxious or distasteful chemicals. In some cases these are associated with special glands which secrete or emit chemicals to dissuade predators from attacking. In others they may have an effect only after the animal has been bitten or otherwise injured by the predator.

Defensive secretory glands are found in a wide range of animals. Gastropod molluscs secrete a variety of substances including various proteinaceous compounds, sulphuric acid and mucopolysaccharides. Several observations suggest that these are defensive. Secretions by species of *Acteon* and *Haminoea* are toxic to at least small zooplanktonic organisms. Many fish are also deterred by acidic secretions and mollusc species with secretory glands are rejected as food by fish. At a more complex level, the opilionid ('harvestman') *Vonones sayi* repels ants with a carefully mixed concoction of body fluid and a quinonoid secretion from carapace glands. The mixture is then 'painted' onto the ant's body with the legs. Ants 'painted' in this way soon give up the attack.

Sometimes noxious chemicals are actively sprayed or ejected at the predator. The whip-scorpion (*Mastigoproctus*) has a well-developed anal gland with a long, thin tail mounted on it. A mixture of acetic and caprylic acid is sprayed from the gland which can be aimed with great accuracy at interfering predators. Perhaps the best-known emission chemical defence mechanism is that of the skunks *Spilogale putorius* and *Mephitis mephitis*. When a skunk is first alarmed it usually performs a deimatic stamping display with the tail held erect. If this fails to dissuade the predator, a pungent and nauseous fluid is squirted from the anal glands.

An entirely different kind of chemical defence is that relying on distasteful, emetic or poisonous qualities of the prey. While distastefulness itself is not a behavioural quality, it is often associated with characteristic and elaborate warning displays. Moreover it may depend on other behavioural features of the prey. While some distasteful species synthesise their own noxious chemicals, others, like the monarch butterfly (*Danaus plexippus*) sequester them from their food. Monarch butterfly larvae are one of the few creatures which can feed with impunity on the highly toxic milkweed (*Asclepias curassavica*) plant. Like many other plants, the milkweed secretes toxic chemicals — in this case cardiac glycosides (which disrupt a number of cellular metabolic processes) to discourage herbivores. Caterpillars which feed on the milkweed store the toxins in their tissues and retain them when they metamorphose. Brower and coworkers[38] showed convincingly that adult toxicity was acquired during larval feeding by raising monarchs on cabbage. Predators (scrub jays (*Aphelacoma*)) fed cabbage-reared monarchs suffered no ill effects, while those fed milkweed-reared butterflies vomited soon afterwards.

An interesting trade-off may occur in natural populations of monarchs between females which choose to lay eggs on toxic plants and those which choose to lay them on harmless plants. Sequestering toxins may well cost the monarch caterpillar something in terms of, say, growth rate compared with non-sequestering larvae on harmless plants. Non-sequesterers may therefore be able to cash in on the presence of sequesterers by gaining the benefits of protection from predation (predators cannot distinguish non-toxic from toxic conspecifics) while not paying the metabolic cost of sequestering. When the frequency of sequesterers in the population is high, therefore, some females can afford to lay their eggs on non-toxic plant species. When it is low, most females would be expected to lay on toxic plants, otherwise predators would learn that monarch caterpillars were palatable. Brower used the term 'automimics' to describe the cheating palatable individuals within monarch populations.

Defences like distastefulness or emetic qualities and the warning colouration that is often associated with them pose another problem for evolutionists. Since their defence mechanism is perceived by the predator only *after* prey has been sampled, it clearly cannot benefit the individuals possessing them. The most likely answer is that such defences have evolved by kin selection[202,319] (see Chapters 3 and 11). Several pieces of evidence bear this out. Many distasteful butterfly species, for instance, are gregarious both as adults and larvae. Their high population 'viscosity' means that individuals are likely to be co-existing with close relatives. Heliconiids in particular have restricted home ranges and range of courtship, roost communally and are long-lived. The widespread association between noxious qualities and warning colouration in animals appears to have arisen because striking colours are more easily remembered by predators then cryptic ones.[117]

7.2.3 Defensive Weaponry

Animals use a variety of morphological structures as defensive weapons. In some cases defence is merely an emergency use for a structure which evolved primarily for some other task like catching food. Many mammals, for instance, defend themselves with their teeth. Wasps sting assailants with a modified ovipositor normally used to paralyse prey. Other weapons used in defence, like the antlers of deer, have evolved for intraspecific display. Oryx (*Oryx* spp.) have long, finely-tapered horns which are usually used in ritualised contests between conspecifics. Occasionally, however, oryx have been known to stab attacking lions with them.

Although retaliatory defence may be a secondary function for a physical structure usually used for something else, some weapons do appear to have evolved for interspecific defence. The pincer-like cerci of earwigs (Insecta: Dermaptera) are a good example. So are the so-called 'gin-traps' found in certain beetle pupae. 'Gin-traps' are toothed jaw-like structures arranged along the rear dorsal surface of the pupa. When closed on the legs of predators or parasites, the serrated 'jaws' soon persuade their owners to try elsewhere. Among vertebrates, hedgehogs, porcupines and tenrecs (*Hemicentetes semi-spinosus*) have evolved formidable spines which, although usually used in passive defence, can in some cases be deployed actively. The spines of various fish

Figure 7.2: Survivorship of Twelve Minnows, Twelve Three-spined Sticklebacks and Twelve Ten-spined Sticklebacks during Predation by a Pike. The three-spined sticklebacks with their long spines survived longest.

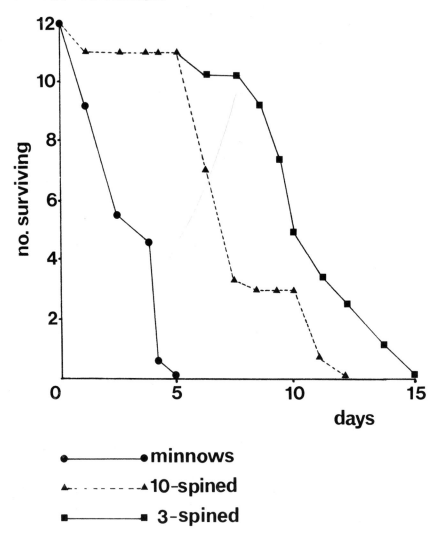

Source: Modified from Hoogland *et al.*[145]

species like sticklebacks and trigger fish (*Balistes* spp.) can also serve an anti-predator function. The spines of these species have a locking device which causes them to jam in the mouths of predators like perch (*Perca fluviatilis*) and Pike (*Esox lucius*). Hoogland and coworkers[145] provided a captive pike with twelve minnows (*Phoxinus phoxinus*), twelve ten-spined sticklebacks

(*Pygosteus pungitius*) and twelve three-spined sticklebacks. Figure 7.2 shows the survivorship curves for the three species over 15 days. The spineless minnows were eaten first, the small-spined *Pygosteus* second and the three-spined stickleback with its large, locking spines, last. Similar comparative survivorship effects have been shown for spiny versus non-spiny morphs of the rotifer *Brachionus calyciflorus* which is preyed upon by other rotifers (*Asplanchna* spp.).

7.2.4 Mobbing

'Mobbing' refers to deliberate predator harassment by prey species. Although it is particularly characteristic of small passerines when they perceive an approaching, but as yet non-attacking, raptor, it also occurs in other bird groups (e.g. gulls), mammals (hyaenas are known to mob lions) and invertebrates. Mobbing may take the form of direct attacks on a predator and often appears to place the performer in considerable danger. At least in birds, mobbing attacks are often associated with characteristic calls. Unlike the alarm calls which we shall discuss later, mobbing calls have frequency characteristics which make them easy to locate. Indeed, they appear to be very effective at recruiting conspecifics and sometimes individuals of other species to join the mobbing action. While mobbing may perform several functions (see below) depending on species and context, it can certainly have a deterrent effect on the predator and can be classed as a rather specialised defensive response.

Although a number of functions have been suggested for mobbing (e.g. announcing to the predator that it has been spotted, deterring the predator to protect self and offspring – see Curio[64] for a comprehensive list), few rigorous investigations have been attempted. An outstanding exception is Curio's[63] study of anti-predator responses in the pied flycatcher (*Ficedula hypoleuca*). Curio was primarily interested in elucidating the stimuli which elicited two distinct anti-predator responses, mobbing and 'snarling' attacks. Over a period of twelve years, various dummy predators were presented to over 2,200 subjects. In the course of his study, Curio noted several adaptive aspects of the flycatcher's responses.

The two main anti-predator responses were used in different contexts. 'Snarling' attacks (swooping flights with a snarling call but at most only slight mobbing) were made close to predators like great-spotted woodpeckers (*Dendrocopus major*), squirrels (*Sciurus* spp.) and other species which were a threat to nests and eggs but not to adults. Mobbing responses (involving prolonged harassment of the predator) were performed at a distance from species like sparrowhawks (*Accipiter nisus*) and shrike (*Lanius* spp.), which were potentially dangerous to adults. Flycatchers therefore graded their responses according to the degree of risk involved. Similarly graded mobbing responses are found in North American squirrels. Squirrels from areas containing non-venomous snakes show strong mobbing responses towards the snakes. Those from areas containing venomous species show much reduced responses. Curio also found that flycatchers from populations where some of the predators were absent (as were certain shrikes from some Spanish populations) did not have an innate predator recognition and hence mobbing response. The stimuli eliciting mobbing therefore appeared to be learned (see later). Interestingly, breeding pairs of flycatchers increased the frequency and vigour of both 'snarling' and mobbing attacks during the nesting and fledgling period. Responses were continued for a short time after the young had fledged, presumably while they were still extremely vulnerable to predators. That such heightened anti-predator responses appeared to be geared to offspring defence was further supported by the low levels of response by adults prior to breeding and by unpaired adults through the breeding season.

Recent experiments by Curio and coworkers[63,66] have shed light on the possible adaptive significance of heightened mobbing responses during the breeding season and added another 'function' to the already long list of proposed 'functions' for mobbing behaviour. Curio's experiments have shown that mobbing may facilitate the cultural transmission of predator recognition. Curio *et al.*[66] found blackbirds (*Turdus merula*) which experienced the mobbing call of a conspecific while in the presence of a harmless honeyeater (*Philemon corniculatus*) (a novel stimulus), subsequently mobbed honeyeaters more strongly when they were encountered on their own. Similar experiments have been tried using a variety of novel objects, including inanimate objects, and different 'tutor' species. In each case subsequent mobbing responses to the object were enhanced.

We mentioned earlier that mobbing calls could have the effect of recruiting both conspecifics and members of other species to the site of mobbing action. Vieth *et al.*[323] investigated the effectiveness of non-conspecific mobbing calls in eliciting mobbing behaviour in blackbirds. Their experiment involved the presentation of a honeyeater to blackbirds while playing a recorded multi-species mobbing chorus. The chorus consisted of chaffinch, great tit, blue tit and nuthatch (*Sitta europaea*) mobbing calls. The chorus appeared to be just as effective as conspecific calls in bringing about cultural transmission of honey-eater recognition. However, it is still not clear whether every *single* call in the chorus is as effective at eliciting mobbing as conspecific calls.

An interesting characteristic of mobbing responses in the wild is the small degree to which they show habituation (see Chapter 4) with repeated 'representations' of a predator. The waning of response in the field appears to be much less than that found in the laboratory. One reason for this might be a failure in laboratory experiments to provide sufficient stimulus specificity. Predators in the wild are constantly changing their position and direction of attack. In a series of laboratory experiments, Shalter[277] found that altering the position of a model predator by as little as 2.5 m rekindled mobbing responsiveness in jungle fowl (*Gallus gallus spadiceus*) subjects which had habituated to the model in a previous location. Shalter[278] also tested his findings in the field by presenting stuffed pygmy owls (*Glaucidium passerinum*) to pied flycatchers. Again, he found that a mobbing response could be evoked in birds which had previously habituated to the owl by presenting the owl in a different location. The sensitivity of the mobbing response to the spatial context of a predator has clear adaptive significance in view of the ever-changing nature of natural predator stimuli.

7.3 Alarm Signals

For obvious reasons, it will benefit palatable prey individuals to spot a predator before the predator spots them. They will be at a clear advantage if they can flee or prepare other defences before, or as soon as possible after, the predator perceives them. As we have seen in Chapter 1, some prey species have evolved special sensory-perceptual mechanisms for the early detection of predators. The noctuid moths studied by Roeder[257], for instance, have evolved a system for detecting the ultrasound produced by approaching predatory bats. When they perceive ultrasonic pulses the moths drop out of the bat's flight path. Perhaps the best known warning displays, however, are those of various social bird and mammal species.

7.3.1 Warning Signals in Mammals

While social mammals may not possess any specially-adapted sensory mechanisms with which to detect the approach of predators, the fact that they live in groups makes it unnecessary for all individuals to be alert the whole time. Because there are many potential detectors acting independently in a group, it is likely that at least one individual will spot a predator even though most are not vigilant at the time. The individual spotting the predator then gives an alarm signal and the group is alerted. The relationship between grouping behaviour, vigilance and anti-predator defence is discussed in Chapter 9 and will not be dealt with *per se* here. In this section we shall be concerned purely with the characteristics and evolution of alarm signalling.

As with mobbing behaviour (indeed, mobbing has been suggested to act as an alarm signal), many possible 'functions' have been suggested for the alarm signals found in social species. The *stotting* or *spronking* behaviour (see below) of some bovids (e.g. Thomson's gazelle (*Gazella thomsoni*)) is a good example. The undoubted effectiveness of stotting in warning other members of the herd has been explained in terms of kin selection. One problem with this, however, is that the costs of stotting would have to be very small so as not to outweigh the low degree of relatedness in nearby animals. Since individuals in bovid herds move about a lot of animals benefiting from the warning are unlikely to be close relatives. An alternative explanation is proposed by Smythe.

As Smythe[294] points out, having detected a predator, the prey animal is in a dilemma. How should it react? If it chooses to ignore the predator, this will only pay-off as long as the predator remains unaware of its presence. The prey must also monitor the predator until it moves away in case it is spotted. Such protracted vigilance may be costly in terms of interfering with other activities the animal must perform. The prey may instead choose to steal away quietly, but in doing so may risk blundering into the predator if it changes its direction of movement. Other possibilities like freezing and remaining cryptic, mobbing and rapid flight also involve problems, either because they waste time or because they may attract the attention not only of the predator in question but also other predators which might be in the vicinity. A possible solution to the dilemma is provided by what, at first sight, seems almost suicidal behaviour. Animals actually approach the predator and then slowly, and with a peculiarly exaggerated gait, (during which they spring up and down with tail and head held erect and legs moved rigidly) run away. In many cases, this stylised, jerky run or stott is associated with the exposure of a conspicuous white patch or other adornment.

Smythe suggests that stotting may be a sort of pursuit 'invitation' to the predator. By encouraging the predator to give chase when the prey is prepared, the latter may outrun the predator so that it gives up and goes in search of less alert prey. As soon as the predator gives up, the prey animal can resume its maintenance activities. The fact that gazelles stott just before and just after full flight, is in keeping with this idea. The stott after flight tests whether the predator really has given up. Zahavi however, suggests that stotting may have precisely the opposite 'function' to that proposed by Smythe. Zahavi's idea is that stotting is a sort of advert of good condition. Stotting animals are 'boasting' of their strength to the predator thereby demonstrating how futile an attack would be.

Pitcher[246] criticises both Smythe's and Zahavi's explanations because stotting antelopes and gazelles are often attacked and because he considers pursuit invitation to be too dangerous under many circumstances of attack. Furthermore, it may be difficult for a predator to find a suitable non-stotting target if it ignores a stotter, and predators which depend on concealed ambush are unlikely to perceive the display at all. Pitcher argues that, if the risk incurred by stotting is not too great, stotting may be a means of reducing the effectiveness of co-operatively hunting predators (like lions) where some individuals remain concealed at the start of an attack. Because, under these conditions, flight in a randomly chosen direction could be lethal, prey individuals will be selected to hesitate for a moment after spotting one predator to check on the possible whereabouts of others. From this, Pitcher argues, it is a small step to evolve a jumping activity during hesitation to see over obstacles or long grass. Stotting may therefore be a means of picking the safest flight path in the face of co-operative predation.

The argument over the 'function' of stotting illustrates the caution that must be exercised in concluding that apparent alarm reactions are geared to conspecifics. They may indeed be. Alternatively or in addition, they may be designed to influence the predator or provide purely selfish benefits.

7.3.2 *Alarm Calls in Birds*

Many bird species, particularly those associating in single or mixed species flocks, produce characteristic calls at times of danger or disturbance. These vary from 'general purpose' distress calls to special alarm calls which are usually given in response to a specific predator.

A striking feature of many avian alarm calls, especially those of small passerines, is their similarity. Most are a high-pitched 'seeet' sound. Marler explained this similarity in terms of the functional requirements of the call. One problem with giving an alarm call is the possibility of directing the attention of the predator to the caller. An optimal alarm call, therefore, would be one which was loud enough to fulfil a warning function but of a quality which was difficult for a predator to pin-point. The most important predators of small passerines are hawks and other raptors. The physical qualities of the 'seeet' call appear to be ideally suited to solving the conflict between warning companions and avoiding capture by aerial predators. As Catchpole[49] points out, birds appear to locate sounds binaurally by using *phase*, *intensity* and *time* differences. Phase differences are only useful at low frequencies because the information they contain becomes ambiguous once the wavelength is less than twice the distance between the two receptors (ears). Intensity differences, however, are more effective at high frequencies because the receiver bird's head forms a 'sound shadow' when the size of the head exceeds the wavelength. Time differences can be used across the frequency range but their effectiveness is improved by repeating, interrupting or modulating the call. The further apart the ears of the listener, the more apparent time differences become. From calculations, it appears that calls with a frequency of around 7 kHz would be too high for using phase differences and too low for an appreciable 'sound shadow' effect. As the sonographs in Figure 7.3 show, this is precisely the frequency range of many small passerine calls when real or dummy hawks fly over. Because they are vulnerable to similar predators, these birds have hit on a similar solution to the problem. Perrins has suggested another property of alarm calls which does not so much make them difficult to detect as *direct* the predator elsewhere, perhaps to another victim. Perrins noted that, to a human listener at least, the alarm calls of some species (e.g. great tits) were apparently ventriloquial.

While human listeners may find the high-pitched alarm calls of small passerines difficult to locate or spatially deceptive, there is now evidence that some raptors do not. Shalter and coworkers have shown that at least barn owls, pygmy owls and goshawks (*Accipter gentilis*) can orientate towards the source of such calls. While the results of these experiments certainly underline the dangers of assuming that auditory perception in animals is the same as in humans, they do not necessarily refute the idea that the calls originally evolved to minimise ease of location. Evolution is a continuing process and, in a predator-prey relationship, both sides evolve and adapt with respect to the other. Perhaps we should not be surprised, therefore, if raptors begin to show signs of getting over the problem of locating alarm callers.

Figure 7.3: Sonographs of the Alarm Calls of Five Passerine Bird Species: (a) Reed Bunting, (b) Blackbird, (c) Great Tit, (d) Blue Tit, (e) Chaffinch.

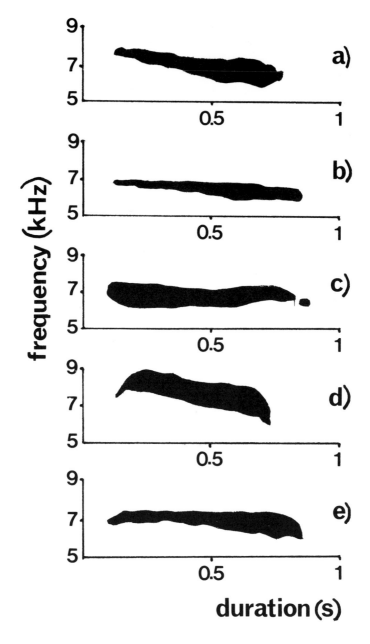

Source: Modified after Marler, P. (1957). Specific distinctiveness in the communication signals of birds. *Behaviour*, *11*: 13–39.

Another functional explanation put forward to explain the evolution of alarm calls depends on kin selection. Since alarm calling is apparently altruistic (it warns companions while perhaps increasing the risk to the caller), it may be that the warning benefits close relatives. While this is unlikely to be true in bird flocks with transient membership, kin selection may well have shaped alarm calls in some mammal species. Intra- and intersexual differences in call frequencies and response times in Beldings' ground squirrel (*Spermophilus beldingi*), for instance, support the notion that calling alerts offspring and close relatives. White-tailed deer (*Odocoileus virginianus*) does, which live in closely-related matrilineal groups, emit characteristic alarm calls when disturbed which are not emitted by bucks in unrelated groups. The evolution of alarm calls will be discussed again in Chapter 10, when an alternative and novel functional explanation is considered in connection with the fundamental nature of communication.

Summary

(1) The ever-present threat of predation has resulted in a variety of anti-predator adaptations in animals.
(2) Behavioural anti-predator responses are generally 'secondary defences' which operate after 'primary defences' (e.g. crypsis) have failed.
(3) Anti-predator behaviour can be divided into the broad categories of escape, defence and early warning.
(4) There are several conflicting ideas concerning the 'function' of apparently warning displays. While they may serve to alert conspecifics and/or heterospecifics, they may also transmit information to predators or be purely 'selfish'.

8 THE ECOLOGY OF REPRODUCTION

Reproducing and passing on copies of its genes to the next generation is biologically the most important act of an animal's life. Surviving to carry out that act and to execute it in the most efficient way is the ultimate goal of the animal's morphological, physiological and behavioural design. When we look at how organisms reproduce, however, we notice that most of them reproduce sexually and that many of their most conspicuous features and activities are directly geared to the sexual act. This is actually a surprising observation. Although the widespread distribution of sexual reproduction is more or less taken for granted, it represents a unique paradox for the evolutionary theorist because, on evolutionary grounds, it is the mode of reproduction we would least expect. While a full discussion of the costs and benefits of sexual versus asexual modes of reproduction is outside the scope of this book, the main points are summarised in Table 8.1 and Figure 8.1.

The main cost of sex is now generally accepted as being the 'fertility' cost of producing males (Figure 8.1c). Depending on the degree of paternal care her offspring can expect (see Ridley[255] for a review), a sexually-reproducing female suffers a reduction in reproductive output of anything from zero to 50 per cent relative to an asexual female. It is this 'cost of males' that the hypothesised short- and long-term benefits of sex (Table 8.1) have to outweigh. Explaining how sex might evolve in the face of asexual competition, however, begs several further questions. If we are to understand sexual behaviour, we must know what makes males and females tick. What is the fundamental distinction between the sexes? What are their evolutionary characteristics? Why are there only two sexes?

8.1 The Evolution of Sexual Behaviour

The question 'what is the difference between the sexes?' is not as easy to answer as it seems. Certainly we have little difficulty identifying males and females among most mammals. Males have a penis, are often larger and may possess various secondary sexual characteristics like facial hair, manes and deep voices. Go further down the phylogenetic scale, or look at any plant species, however, and the distinction is less obvious. The sexes may be identical in outward appearance, even in the characteristics of their external genitalia. Short of dissection, there may be no way of distinguishing between them until mating actually occurs.

Clearly, external morphology is an unreliable guide. What, then, is the basic distinction between males and females? The answer is their gametes. In general,

218

Table 8.1: Costs and Benefits of Sexual versus Asexual Reproduction.

Costs	Benefits
(1) Cost of meiosis (Williams). Sexual females put only half their genotype into the next generation because of meiosis and are therefore at a 50% disadvantage relative to asexual females (Figure 8.1a). Fallacious argument based on individual rather than gene selection (Treisman and Dawkins) (Figure 8.1b).	A. LONG-TERM BENEFITS
	(1) Faster evolution (Muller[221]). Favourable mutations can accumulate in individuals more rapidly through sexual recombination. While this is based on group selection, Fisher and Maynard Smith point to several lines of evidence which suggest that group selection may operate to maintain sex.
(2) Cost of males (Maynard Smith[205]). Sexual females put only half the number of viable reproductives (females) into the next generation that asexual females put in (fully reproductive individuals). They therefore pay a 2:1 cost of producing males (Figure 8.1c). However, depending on the degree of paternal care (Ridley) invested, the cost may be anything from zero to just less than 50%.	(2) 'Muller's ratchet'. Because there is no recombination, asexual organisms may accumulate deleterious mutations which can only be lost by unlikely back-mutations.
	B. SHORT-TERM BENEFITS
(3) Cost of mating (Daly[67]). Mating is unpredictable, consumes time and energy and may incur a risk of injury, disease or death. While Daly considers the cost of mating separately, it will add further to the cost of producing males.	(1) 'Best genotype' advantage (Williams[332]). Sexually-reproducing individuals are more likely to produce the best genotype for colonising a new environment. However, this only works where there is high sibling competition (Maynard Smith). Treisman[312] proposes an extension of this idea where adaptation depends on polygenic units.
	(2) Resource Accrual Theory (Trivers[316]). Sex may be maintained because males are 'free' to spend time and energy gathering and/or defending mates or vital resources. Females can then choose the best. Resource accrual by males may also allow sexual organisms to extend into harsher environments.

Figure 8.1: (a) Williams'[332] 'Cost of Meiosis' in which a Sexually-reproducing Female Apparently Reduces her Future Genetic Representation by Half during Meiosis and Syngamy. (b) Treisman and Dawkins'[313] Gene Selection Model Showing No Net Reduction in the Frequency of a Gene (S) for Sexual Reproduction as a Result of Meiosis. (c) Maynard Smith's[205] 'Cost of Producing Males'. Sexual females compromise their female offsprings' reproductive potential by requiring that they find and mate with males.

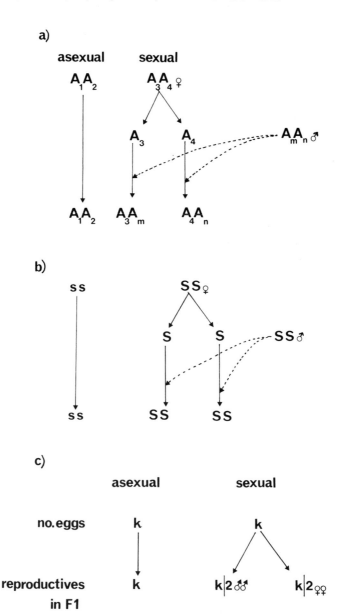

females produce relatively few, large, well-provisioned gametes called *eggs*, whereas males produce prodigious numbers of tiny gametes called *sperm*. The differences between the gametes are critical to an understanding of sexual behaviour. The 'cost of males', referred to above, can be rephrased as the *cost of anisogamy* (the cost of producing different sized gametes). Because they produce many metabolically-inexpensive gametes, males can potentially fertilise a large number of females. Moreover, they can afford to be fairly indiscriminate in their choice of mate, since a few wasted sperm here and there are of no consequence. Females, on the other hand, must guard their few, costly eggs carefully. Mating with a low-quality male might be disastrous in terms of female inclusive fitness. Males can be viewed as in a sense 'parasitic' on the high-quality gametes of females. They may be superfluous to the survival of offspring, yet they account for half of a sexual female's reproductive output. They subsist on the reproductive independence conferred by their small gametes. The question we now have to answer therefore, is not 'why sex?' but 'why anisogamy?'. Isogamous organisms may largely avoid the cost of sex.

8.1.1 Evolution of the Sexes

An ingenious answer has been provided by Parker and coworkers[238]. Anisogamy may be the inevitable evolutionary consequence of size variation in primitive isogametes (isogamy is presumed to be the primitive mode of sexual reproduction). Like any other biological entity, the early isogametes would not have been identical in their morphology. Some would have been a little larger and others a little smaller than the average. An immediate selective advantage can be envisaged for slightly larger gametes. By increasing the amount of provision for the potential zygote, a larger gamete enhances its prospect of surviving to reproduce. We might therefore expect a trend towards larger and larger gametes.

The catch, however, is that the additional provisioning of the larger-than-average gametes can be capitalised on by the smaller-than-average gametes. Fusion between a large and small gamete results in a zygote of equal fitness to one arising from fusion between gametes of average size. While large gametes would do better to fuse with other large gametes, they may have little 'choice' in the matter. By dint of their size, small gametes are more mobile and able to make contact with other gametes faster. Larger gametes are therefore likely to be soaked up by small gametes faster than they can find other large ones. This exploitative relationship between large and small variants sets up a disruptive selection pressure, favouring smaller (and hence more mobile) gametes on the one hand, and large gametes whose size approximates the optimal size for the zygote (to compensate for the lack of provisioning by small gametes) on the other. Intermediate sizes are penalised because they have neither the mobility of the small extreme, nor the provisioning of the large. The net result is the evolution of two highly-dimorphic gametes: sperm and eggs. Through small variations in size, therefore, isogametes may contain the seeds of their own destruction and the evolutionary driving force for two, anisogamous sexes.

Alexander and Borgia take the opposite view to Parker *et al*. They imply that anisogamy might in fact have *preceded* isogamy as the primitive mode of sexual reproduction. They base their argument on the fact that some protists and bacteria reproduce sexually by transferring micronuclei with little or no cytoplasm between cells. The micronuclei could be construed as 'sperm' and the recipient cell as an 'egg'.

8.1.2 Sex Ratios and Mating Systems

Sex Ratios. Since males produce so many expendable gametes, and potentially can mate in a carefree way with as many females as they are able to service, why is the ratio of males to females in natural populations so high? Surely, a small number of males could do the job just as well. This argument, of course, is pitched at the population level and, in that context, is perfectly valid. Populations would get by quite nicely with a very low male:female ratio. The problem is that selection is unlikely to act on populations. What we need to do is examine the utility of different sex ratios from the point of view of the individual. More precisely, we need to ask what the *evolutionarily stable* sex ratio is likely to be.

Although not using the same terminology, it was Fisher[106] who first pointed out that the evolutionarily stable sex ratio was likely to be 1:1. The reason is easy to see. Suppose a female capitalised on the reproductive freedom of males by producing only daughters. Since the daughters would have no trouble finding a mate, the allele for producing only daughters would spread through the population. However, if the proportion of females in the population rises above half, a female which 'decided' to produce only sons would be at an advantage, since her offspring could potentially fertilise a large number of females. The allele for males only would spread and the sex ratio would be pushed back in the opposite direction. Clearly, the net result of this frequency-dependent selection is an evolutionarily stable sex ratio of unity, because deviation in either direction simply sets up a counteracting selection pressure to correct it.

Parental Investment. The above point hinges on the selective pressures facing parents which produce offspring in different sex ratios. It asks what advantage accrues to parents which *invest* differently in sons and daughters. Parents should invest food and other resources equally between the sexes. Conceivably, this could mean producing ratios greater than one if one sex demanded more resources during development than the other. In most cases, however, investment in the two sexes is likely to be equal. From the point of view of offspring sex ratios, the sexes are likely to 'agree' in their relative *parental investment* (PI).[315] Where they may conflict, is in deciding their relative PI for rearing offspring.

By definition, individuals of both sexes will be selected to produce as many surviving offspring as possible. At the same time, each will be selected to minimise its investment in individual offspring, since its net productivity will then be higher. Obviously, the best way to do this is to shift the burden of investment onto the other parent. The opportunity for doing this, however, may not be the same for both parents. Because the female starts by investing more than the male (eggs are more expensive and in shorter supply than sperm), she has more to lose if an offspring dies, or, more strictly, more to invest in the future to bring the next offspring up to the same level of development.[83] This increased commitment makes the female vulnerable to desertion by the male. That is, selection is likely to gear males for copulating with as many different females as possible, but investing little in each one, and females for being cautious about who they mate with and investing a lot in the offspring they produce.

Of course, this only reflects the general direction in which we expect selection to act. Several factors will influence the degree to which males actually capitalise on their potential to desert. Clearly, there is little point in a male deserting if females cannot rear offspring on their own. Depending on the pressure for paternal care, therefore, species may be polygynous (many females per male), monogamous (one female per male) or even polyandrous (many males per female). What are the important factors influencing the type of *mating system* animals adopt?

The Ecology of Mating Systems. The environment in which the animal operates may set a broad constraint on its mating behaviour. Orians[231] and Emlen and Oring[100] suggested that certain environmental factors determine the degree to which mates can be defended or monopolised. The greater the potential for multiple mate monopolisation, the greater the *environmental potential for polygamy* (EPP).[100] Emlen and Oring suggest two main preconditions for the evolution of polygamy:

(1) Multiple mates or resources sufficient to attract multiple mates must be energetically defendable by individuals.
(2) Animals must be able to capitalise on the defendability of mates or resources.

The EPP thus depends on the spatial and temporal characteristics of resource dispersion to the extent that they influence the ability of members of one sex to monopolise more than one member of the other. When important resources are distributed uniformly in space and/or time, there is little opportunity for resource monopolisation. Members of the breeding population would tend towards an even distribution with a low potential for multiple matings. When resources are less evenly distributed, some individuals may be able to control a larger quantity, or better quality, of resource than others, and the EPP would increase. Highly clumped resources create a potential for resource monopolisation by a very few individuals and may result in a high variance in reproductive success.

Figure 8.2: The Likelihood of a Polygamous Mating System Evolving (Areas under Dashed Curves) as a Function of Resource Distribution and the Temporal Availability of Mates.

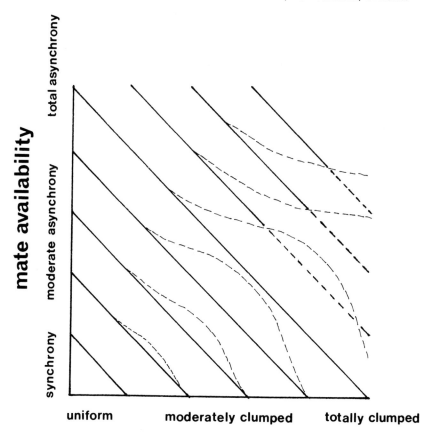

Source: Modified after Emlen and Oring[100].

Of course, the benefit derived from resource defence is strongly influenced by the temporal availability of sexually receptive partners. If females in a population become receptive in unison, individual males have little chance of monopolising a large number. By the time one female has been inseminated other males will have mated with the remainder. With increasing degrees of asynchrony among females, the potential for males to accumulate multiple mates increases. If asynchrony becomes extreme, however, the rate of appearance of new potential mates becomes such that the cost of continued resource defence needed to attract more mates exceeds the additional benefits gained. Continued mate accumulation would no longer be beneficial and tendencies towards polygamy would decline. The joint effects of variation in resource distribution and the temporal availability of mates can be summarised in Figure 8.2.

Of course, even when resources are clumped and mates available, animals may not be in a position to capitalise on them. As we have seen in previous chapters, an individual's response in given situations may be compromised by conflicting needs. Thus, an animal may not be able to devote time to resource or mate defence because of, say, feeding or predation constraints.

When the EPP is high, and animals are able to capitalise on it, polygamy may arise in one of several ways. *Resource defence polygyny (RDP)* may occur when males are able to defend areas which contain resources vital to females. Here, females choose males on the basis of their territory quality rather than their physical characteristics, although the two are likely to be correlated. Female American bullfrogs (*Rana catesbeiana*), for instance, choose territories which are defended by older and larger males. Mortality among developing embryos is lower in these territories because they are subject to less extreme temperature variation (which causes developmental abnormalities) and are infested by fewer predatory leeches. A more extreme form of RDP is found in orange-rumped honeyguide (*Indicator xanthonotus*). Beeswax is an essential part of the birds' diet and males maintain year-round territories at the location of bees' nests. Since these are sparsely distributed on exposed cliffs, a small number of males can monopolise all the food supplies. Courtship centres on the bees' nests and mating success is high for territory owners. One male, for instance, was known to copulate 46 times with at least 18 different females.

Males may sometimes gain access to females by directly defending them against rivals. *Female defence polygyny (FDP)* generally arises when males are able to maintain possession of more than one female. FDP is greatly enhanced if females have an independent tendency to be gregarious, as they do in many ungulate species. During periods of female sexual receptivity, male impala (*Aepyceros melampus*) and water buck (*Kobus defassa*), for example, divide the habitat into defended territories. In this way a few males end up with all the matings when female herds move in. In red deer (*Cervus elephus*), stags compete directly for ownership of female harems.[57] Fighting is fiercest when hinds are most likely to conceive and the pay-off for defending or usurping females is highest (Figure 8.3).

Sometimes males defend neither resources nor females, but instead aggregate and establish rank relationships among themselves. Females then choose between them on the basis of their relative rank status and a few males may again end up with most of the matings. The form this *male dominance polygyny* takes largely depends on the extent to which females synchronise their sexual activity.

So-called *explosive breeding* occurs when all the females come into breeding condition simultaneously and arrive at the male aggregation together. Explosive breeding is common among frogs and toads. Indeed, in populations of the European frog (*Rana temporaria*), mating may occur on only one night in the year.

Figure 8.3: Red Deer Stags Fight More Intensely (Solid Line) at Times When Hinds Are Most Likely to Conceive (Dashed Line).

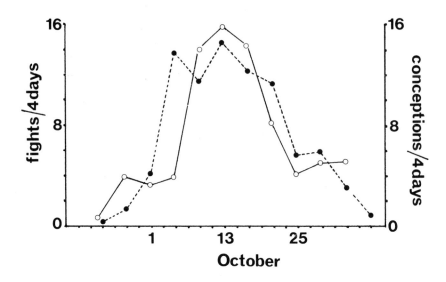

Source: Modified after Clutton-Brock *et al.*[57]

If females are not synchronised in their periods of sexual activity, only a small proportion of them will visit the male aggregation at any one time. As a consequence, competition between males is intense and the resulting mating system known as a *lek*. Since males at a lek control neither resources useful to females, nor the females themselves, females are free to choose males solely on their physical characteristics. Selection in lekking species thus tends to result in the evolution of extreme male mate attraction characteristics and behaviour. Lekking behaviour is discussed in more detail in Chapter 9.

In Emlen and Oring's model, *monogamous* mating systems arise either when the EPP is low or when animals are, for some reason, unable to capitalise on it. The prevalence of monogamy among birds, for instance, probably results from the inability of most species to capitalise on the EPP. Considerable parental care by both sexes is usually needed to raise offspring successfully to maturity. Thus, losses to an individual parent as a result of withholding care from one set of offspring to spend time courting additional mates, are likely to be greater than any benefits arising from polygamy.

Perhaps surprisingly, many monogamous species are sexually dimorphic. Since each individual is potentially assured of a mate, we would expect little preferential choice for particular sexual characteristics. One possibility is that sexual dimorphism in monogamous species is due to differing ecological requirements in the two sexes.[276] However, there is evidence that, at least in arctic skuas, variation in male characteristics is maintained by female preferences even though the mating system is monogamous.[229]

If mate assistance by males is extensive, then there is the evolutionary possibility of *polyandry*. Emancipation gives the female a chance to increase her inclusive fitness through the production of multiple clutches. Female fitness can only be increased, however, to the degree that males are sexually receptive and available to incubate additional clutches. Clearly, males sitting on eggs cannot be sexually active as well so, for females wanting to lay additional clutches, the effective sex ratio will be biased towards a shortage of males. These conditions lead to increased competition between females for access to males, and similar principles to those producing polygyny operate to determine whether or not polyandry will evolve.

An example of *resource defence polyandry* is found in the American jacana (*Jacana spinosa*). Here, females divide up suitable ponds and lagoons into 'superterritories' within which a number of males set up their small breeding territories. Females holding superterritories often have several males incubating clutches simultaneously and readily replace any clutches which are predated. Just like RDP males, resource-defence-polyandrous females are the competing sex and tend to be larger and more aggressive than the males. In the polyandrous spotted sandpiper (*Tringa macularia*), for instance, females are 25 per cent bigger than males and dominate males in aggressive encounters. In the American jacana, females are some 50–70 per cent bigger.

Female access polyandry is also shown by certain bird species. Here females do not defend resources required by males but, through interactions among themselves, limit access to males. The period of defence is often larger than in FDP because females must guard males until a clutch is completed and incubation begins. Female access polyandry is shown by the three species of phalarope (*Phalaropus* spp.). In all three, males and females congregate at bodies of water where they feed, display and copulate. In the red-necked phalarope (*P. lobatus*), the male eventually incubates the eggs alone and cares for the young after hatching. Although there is some debate about the precise form of mating system in phalaropes, it seems clear that certain females can influence the access of others to males during the period of copulation and nest-building.

8.2 Sexual Selection and Mate Choice

During our discussion of mating systems, we have referred several times to selection acting on those characteristics of one sex which can be used by members of the other in choosing a mate. Although not in any way different from natural selection, selection of this kind is generally referred to as *sexual selection*. Sexual selection thus describes the often pronounced directional evolution which is driven by competition within and between sexes.

8.2.1 Sexual Selection

Darwin[69] originally coined the term 'sexual selection' to describe the consequences of two distinct types of sexual competition. He saw it occurring first when males compete with one another for access to a limited number of females (*intra*sexual selection) and, secondly when males compete to attract the attention of females (*inter*sexual or *epigamic* selection). In general, intrasexual selection results in increased strength and elaborate weaponry in the competing sex, while intersexual selection tends to produce exotic ornamentation and behaviour. Although Darwin correctly identified the dual action of sexual selection, it was left to Fisher[106] to provide an evolutionary mechanism for generating and maintaining preferences among the limiting sex.

Fisher maintained that, for sexual selection based on a mating preference in one sex to lead to the directional evolution of characteristics in the other, some advantage must accrue to individuals showing the preference. He suggested that there must be an initial stage in which, say, males possessing a character attractive to some of the females in the population must have some other heritable advantage over other males. Thus an ancestral peacock with a slightly longer than average tail might have been better at flying and escaping predators because the tail allowed greater manoeuvrability in the air. Peacocks with longer tails and peahens preferring them (which then pass long tails onto their sons) would enjoy a high reproductive potential and both traits would spread through the population. Selection would continue to favour longer and longer tails, even though increased length eventually *reduces* flying ability, until the costs of producing and bearing the tail (in terms of energy costs, risk of predation, etc.) outweighed the reproductive advantage of being preferred by females.

The 'Handicap' Principle. An intriguing alternative to Fisher's 'run away' female choice hypothesis has been put forward by Zahavi[344]. Zahavi disputes Fisher's suggestion that dimorphic characteristics in one sex evolve simply because they are attractive to the other. Instead he argues that the choosing sex should select mates on the basis of cues denoting real survival potential. Extreme epigamic features are seen as handicaps which can only be possessed by strong, fit individuals. The possession of a handicap like large horns or an encumbering long tail by a mature male might indicate a good ability to survive. Despite its handicap the male has still managed to find food and avoid predators and so survive to sexual maturity. If she chooses this male, the female stands to pass this high survival potential on to her progeny.

Attractive as the so-called *handicap principle* sounds, it does run into some serious problems as a model for the evolution of epigamic characters. First, as Maynard Smith[204] points out, females which mate with handicapped males are likely to pass the handicap on to their sons. Not all the sons receiving the handicap will have a high survival potential as well. Those that do not will therefore be seriously penalised. Secondly, the handicap would have to be sex-limited so that females choosing handicapped males did not pass the handicap on to their

Figure 8.4: One Reason Why the Handicap Principle May Not Operate is that the Arms Race Producing Differential Handicaps (Open Blocks) May Result in a Reduction of Individual Fitness Budgets to the Base Level Required for Existence (e). Despite outward differences in appearance, therefore, all individuals are of equal competitive ability and the handicap loses its predictive value. Vertical lines represent the fitness budgets of five hypothetical individuals in a population, open blocks represent the amount of fitness budget used up in the production of the handicap and the two plots represent the hypothetical relationship between investment in the handicap and the size of the fitness budget in the early stages of an assessor/cheat arms race (a) and in the late stages of an arms race (b).

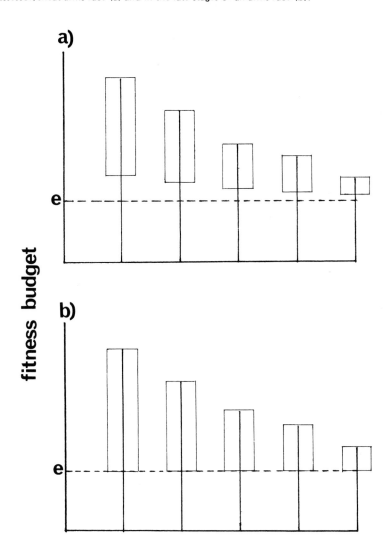

Source: Modified from Barnard and Burk[15].

daughters. Thirdly, an 'arms race' will inevitably arise between truly high-quality males and poor-quality 'mimics' who are likely to arise possessing the handicap (once choice is made on the basis of a marker one like a handicap, rather than fitness *per se*, an enormous selective advantage accrues to 'cheats' who adopt the cue without spending resources on real survival ability). To stay ahead of the cheats, true high-quality males may be driven to evolve larger and larger handicaps until, eventually, the penalty of the handicap cancels out the original fitness advantage (Figure 8.4). Although there is still an asymmetry in handicap size within the population, it is no longer a predictor of survival ability and the handicap should cease to be used by females as a criterion for mate choice. Of course, expensive cues (like large calcium-rich antlers) might correlate with an individual's ability to obtain limited resources from its environment. Since this is often likely to depend on its competitive ability, such *status-limited cues*[15] could be a reliable guide for members of the other sex choosing a mate. However, this is very different from the mechanism envisaged by Zahavi. A 'handicap' might spread through the population if the frequency of females preferring it when it first arises is high (if it is low, the cost of producing the handicap in the first place will militate against it), but the spread would then be due to Fisher's 'sons effect' (see above) rather than to the 'handicap' itself.

8.2.2 Choosing a Mate

Given the fundamental discrepancy in PI between the sexes, and the male's genetic predisposition to desert after copulation, we should expect females to try to minimise their risk of desertion and make sure their expensive eggs are fertilised by good-quality males. One way of doing this would be to mate only with those males whose attributes are most compatible with female reproductive interests; but what attributes might these be? In a recent review, Halliday[129] discussed a number of criteria which females might use:

(a) Sexual Competence. Clearly, male sexual maturity, readiness and competence will affect the efficacy of a mating attempt. In male smooth newts, the rate of performance of courtship sequences correlates positively with the number of spermatophores eventually deposited[128] and may be used as a quality index by females. In monogamous species like the kittiwake (*Rissa tridactyla*), females tend to stay with the same mate until the first poor reproductive season. In some polygamous species, females prefer to mate with males who have already acquired other females, presumably because this indicates desirability.

(b) Good Genes. Females often show preferences for older males (indicating high survival potential) or males that have won aggressive disputes (indicating high competitive ability). Choice may also be based on genetic *complementarity*.[341] Some species tend to mate *assortatively* (matings between similar genetic types), others *disassortatively* (matings between different genetic types). Which is chosen presumably depends on the costs and benefits of maintaining the integrity of the parental genotype.

(c) Parental Ability. In species where paternal care is important for the survival of offspring, choice may be based on the male's potential parental ability. The defence of high-quality feeding territories by males may be one way of indicating an ability to provide for offspring. In some species, elaborate courtship feeding rituals by males may actually force them to invest more directly in the provisioning of potential offspring.[302]

Clearly, then, there are a number of criteria upon which we might reasonably expect a female to base her choice of mate. There is also evidence that females do choose in the ways we have described. However, what we really need to know is whether exercising a choice actually enhances female inclusive fitness, since it is only then that we are in a position to consider its adaptive value. Few experiments and observations have been specifically designed to investigate this. Nevertheless, those that have shown clear reproductive advantages accruing from mate choice. Maynard Smith, for instance, found a preference by female *Drosophila subobscura* for males of an outbred strain rather than those of an inbred strain. When he examined the viability of resulting offspring, he found that females mating with inbred males produced only 25 per cent as many viable offspring as those mating with outbred males. Similarly, Partridge found that female *Drosophila melanogaster* which were allowed to choose mates from a number of males, produced progeny of higher competitive ability than those which were artificially constrained to mate with a male chosen by an experimenter.

Conflict in Mate Choice. From the preceding discussion, we might be forgiven for thinking that females have it more or less their own way in the mate choice stakes. In several species, however, both sexes exercise mate choice. This is particularly true of species in which both sexes show appreciable PI. Even some of the 'typical' sexual behaviours associated with female choice are ambiguous when scrutinised from an evolutionary viewpoint. Protracted courtship sequences, for instance, may be beneficial to females because they indicate a male's commitment to the breeding effort and thus provide an index of his probable future investment (and hence the female's risk of being deserted), but they may also benefit males.

If a breeding female is deserted, there are, broadly speaking, three things she can do. First, she can try to rear any offspring she possesses on her own, but this is likely to pay only if the offspring are quite old and relatively little investment is needed to complete the process. Secondly, if the offspring are very young, perhaps only just conceived, it may pay her to abort or abandon them. Relatively little future investment would then be needed to bring subsequent offspring up to the same level of development. Thirdly, and most beneficial from the female's point of view, she could deceive another male into 'adopting' her offspring and hence lose very little from the desertion. This might be particularly easy if offspring are as yet unborn. If this last strategy is successful, any offspring raised will, of course, be related to the females by 50 per cent. It will share no genes

Table 8.2: Hypothethical Pay-off Matrix for a Range of Male and Female Mating Strategies (see text).

Value of successfully reared child (C) = +15
Cost of rearing (R) = −20
Cost of prolonged courtship (T) = −3

| | | Male strategy | |
		Faithful	Philanderer
		+2(C − ½R − T)	0
	Coy	——————	—
		+2(C − ½R − T)	0
Female strategy			
		+5(C − ½R)	−5(C − R)
	Fast	——————	——————
		+5(C − ½R)	+15(C)

Pay-offs to female strategies above the line, pay-offs to male strategies below.

Source: Based on assumptions and calculations in Dawkins[79].

at all, however, with the duped stepfather. We should therefore expect selection to design males who are extremely wary of the prospect of cuckoldry. Indeed, we should expect males to take very active steps to minimise the possibility. As Dawkins[79] points out, this may well explain instances of spontaneous abortion among certain species. Male mice, for instance, secrete a chemical which, when perceived by pregnant females, causes them to abort (the 'Bruce effect'). In this way, males can be sure any offspring subsequently produced are their own.

A male could also guard against being cuckolded by enforcing a long period of courtship on the female. Before copulating, he could drive away all other males who attempt to mate while also ensuring that the female is not already carrying another male's developing fetus. Elaborate and lengthy courtship rituals may therefore satisfy the evolutionary interests of both sexes. They enable females to be sure they are mating at least with the right species and, more often, with a very particular kind of male; and they provide males with a safeguard against cuckoldry.

At first sight it seems that, if all the females in a population are playing hard to get (*coy* strategy) by imposing a courtship period and all the breeding males are investing heavily in offspring (*faithful* strategy) as a result of female choice, the way is open for more carefree strategists to prosper. In a population of *coy* females and *faithful* males, any females not bothering to impose a courtship period (*fast* strategy) and any males not copulating with females who make them wait but moving off to search for easier conquests (*philanderer* strategy) will be at an advantage through not paying the time costs of courtship. Furthermore, *fast* females will be sure of mating with a reliable, heavily-investing male. The question we are asking, therefore, is: are *coy* and *faithful* strategies evolutionarily

Figure 8.5: Female Yellow-Bellied Marmots Produce Most Young in Monogamous Pairings (Solid Line). Males produce most when there is a ratio of about 2.5 females per male (dashed curve).

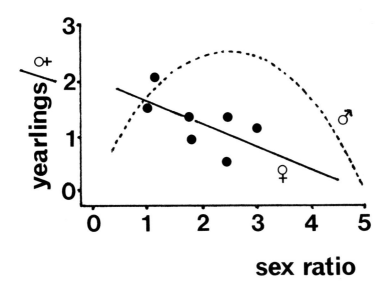

Source: Modified from Downhower and Armitage[92].

stable? To find out, we can carry out a similar game theory analysis to the one described for *hawks* and *doves* in Chapter 3 (Table 8.2).

As with the *hawk/dove* analysis, we can make an educated guess at the relative evolutionary pay-offs accruing to each of the four strategists. In Table 8.2 the figures are those of Dawkins[79]. It is not difficult to see that, although *fast* females will initially spread through a *coy/faithful* population, they create the cause of their own downfall. *Philanderer* males reap an enormous pay-off from *fast* females and spread through the population. *Fast* females, however, are actually *penalised* by mating with *philanderer* males because of their PI costs. So, although the pay-off to *coy* versus *philanderer* is zero, in net terms *coy* is at an advantage to *fast* in a population containing a high frequency of *philanderers*. The frequency of *coy* increases again and the pay-off to *philanderer* is reduced (because successful matings are more difficult to come by). It appears, therefore, that the evolutionarily stable strategy will be a mixture of *coy/fast* females and *faithful/philanderer* males. Using Dawkins' figures, the ratios at equilibrium turn out to be ⅚ *coy* females and ⅝ *faithful* males. As before, this could be achieved by, say ⅚ of the females playing *coy* all the time or all females playing coy ⅚ of the time. However, in a subsequent analysis, Schuster and Sigmund[271] have shown that the ratios are not an ESS but represent the mean of two extremes between which the ratios continuously

Figure 8.6: Solid and Dashed Lines Represent Optimal Mate Size Combinations for Females and Males Respectively in Pairs of Common Toads. Open circles show the ratios observed.

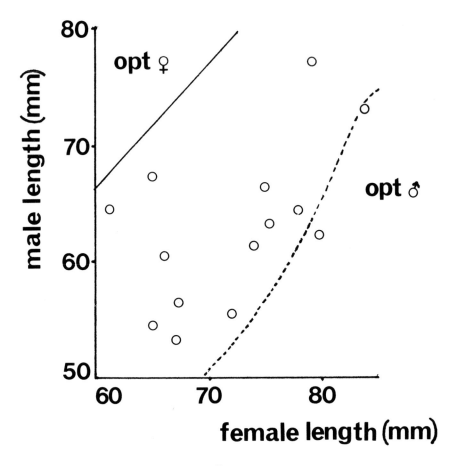

Source: Modified after Davies and Halliday[73].

oscillate. They are therefore what we should on average expect to see, but they are not stable in the same sense as the *hawk:dove* ratio we discussed in Chapter 3.

In some cases, conflict between the sexes with respect to mate choice can be appreciated in more tangible terms. For instance, the sexes may differ in their mating system requirements. In yellow-bellied marmots (*Marmota flaviventris*), males maximise their yearly number of offspring by recruiting a harem of three females.[92] Females, on the other hand, do best in monogamous pairings (Figure 8.5). The most commonly observed mating system in nature is one male to two females, a neat compromise.

Davies and Halliday[73] found a similar compromise in the mate choice requirements of common toads (*Bufo bufo*). Here, two main factors influence the reproductive success of a pair: (1) female size and (2) the similarity in size between a pairing male and female. The bigger the female, the more eggs she lays; the better a pair are matched in size, the greater the proportion of eggs fertilised. Davies and Halliday calculated the optimal mate size for males and females by setting up experimentally-controlled breeding pairs whose size ratios were manipulated. The optimum for females turned out to be the largest male available, but that for males was a female 10-20 mm longer than himself (see dashed and solid lines in Figure 8.6). When pairs mating in the wild were examined, a compromise was found, but one nearer to the male optimum than the female.

Sexual Selection and Mate-searching Strategies. So far, we have produced evidence that sexual selection based on inter- and intrasexual competition actually occurs, but we have said little about how potent an evolutionary force it is likely to be. One penetrating and quantitative investigation of the influence of sexual selection on behaviour has been carried out by Parker[236].

Parker examined the mate-searching behaviour of male dung flies (*Scatophaga stercoraria*). Dung flies mate on and around fresh dung pats and eggs are laid in the pats to develop. Females arrive quickly at a dropping, lay all their mature eggs in one visit and then do not visit a pat again until the next batch of eggs is ready. Sexually-active males arrive at the fresh pat even faster than females, and in much larger numbers. The sex ratio on the pat is around 4-5 males per female and competition between males is fierce. Although females arriving at a pat are likely to have a full sperm supply from the previous oviposition site, they remain sexually active throughout their life and mate many more times than is necessary for fertilisation. Furthermore, sperm competition in the female reproductive tract (see Parker[235] for an excellent review) means that the last male gets most of the fertilisations (because previously stored sperm is displaced).

Searching males can obtain a mating in one of two ways: (1) by catching a female as she arrives on the pat or (2) by taking over a female during copulation or oviposition. The pay-offs are not the same for each strategy. Arriving, copulating and ovipositioning females have characteristic and non-random incidences (both temporally and spatially) on the pat. There is, therefore, a complex set of temporal and spatial equilibria for male searching strategy.

For the first 20 minutes after a pat is deposited, male strategy is determined almost entirely by the availability of newly-arriving females. However, Parker found that males searched in different areas on and around the pat. If searching in different areas is evolutionarily stable, all males must be achieving equal reproductive pay-offs. To test this, Parker compared the observed number of females captured in different zones around the pat with that expected if the number of males searching there was directly proportional to their capture rate. The good agreement shown in Figure 8.7 strongly suggests that male searching

Figure 8.7: Observed (Solid Line) and Expected (Dashed Line — See Text) Number of Females Captured by Male Dung Flies in Different Zones around a Dung Pat. A, pat surface; B, grass periphery to 20 cm; C, 20–40 cm grass zone; D, 40–60 cm grass zone; E, 60–80 cm grass zone.

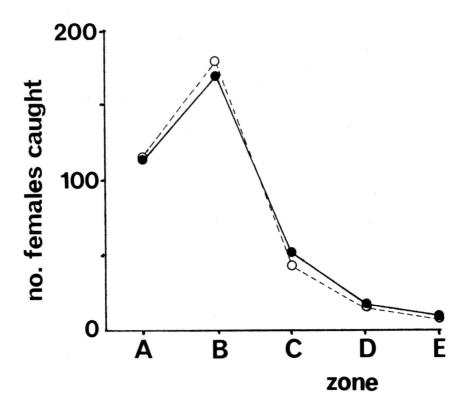

Source: Modified after Parker, G.A. (1974). The reproductive behaviour and the nature of sexual selection in *Scatophaga stercoraria* L. IX. Spatial distribution of fertilization rates and evolution of male search strategy within the reproductive area. *Evolution, 28*: 93–108.

strategy conforms to the mixed ESS we would predict. Davies and Halliday found a similar spatial ESS in male toads searching for females at and away from spawning sites in a pond.

After the first 20 minutes, the equilibrium distribution of males becomes more difficult to calculate, because males can secure matings by take-overs as well. From a knowledge of the temporal and spatial distribution of ovipositions (mainly 50–60 minutes after pat deposition and always on the pat) and copulations (mainly around 40 minutes after deposition and some 1–2 metres from the pat), Parker predicted the ESS for the alternative strategies of searching on the pat and searching in the surrounding grass with time. Figure 8.8a shows the

Figure 8.8: (a) Predicted (See Text) and (b) Observed Percentage of Male Dung Flies Searching in the Grass around a Pat as a Function of Pat Age and the Total Number of Males Present.

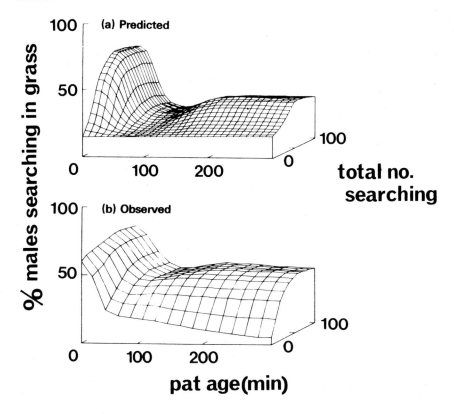

Source: As Figure 8.7.

predicted equilibrium percentage of males searching in the grass as a surface plotted against pat age and male density (to control for competitive effects). The predicted surface assumes that each strategy results in an equal rate of fertilisation per male. Figure 8.8b shows the distribution Parker observed in the field. The observed surface conforms closely to that predicted by the ESS model. The proportion of males in the grass is lowest at low male densities, as we would expect. It also falls sharply with pat age from an initially high value (when most available females are new arrivals) towards the oviposition peak at 50-60 minutes (when oviposition take-overs become important). The surface on the graph then rises again as the oviposition peak passes.

Finally, Parker examined the fertilisation pay-offs to males who stayed for different lengths of time at a pat. If searching for variable lengths of time is an ESS, all males should achieve equal fertilisation rates, despite the fact that no

Figure 8.9: The Temporal Distribution of Male Dung Fly Emigration after Pat Deposition. The observed distribution (solid curve) conforms to that expected (dashed curve) on the assumption that males match their stay time according to female availability (female/male/ minute). The expected curve is corrected to take into account time spent in copulation and oviposition.

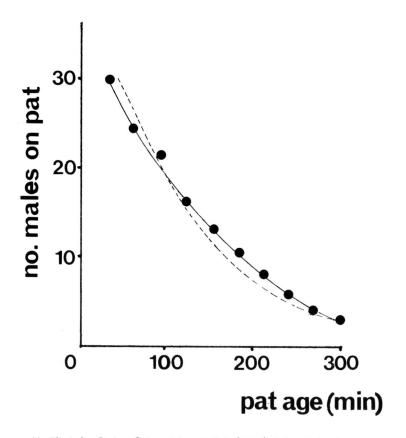

Source: Modified after Parker, G.A. and Stuart, R.A. (1976). Animal behaviour as a strategy optimizer: evolution of resource assessment strategies and optimal emigration thresholds. *Amer. Nat.*, *110*: 1055-76.

two males search for the same time. Parker tested this prediction by comparing the rate of decline in the number of males with that expected if males were leaving according to an 'input matching law'. This 'law' states that the number present at any time should equal the mean rate of captured females (averaged through all reproductive time) multiplied by the rate at which females are arriving. The fit between Parker's observed decline curve and that predicted by the input matching law was very close (Figure 8.9). Male search times therefore approximate well to a mixed ESS.

Summary

(1) The extensive use of sexual modes of reproduction by animals appears, at first sight, to constitute an evolutionary paradox. However, both long-term (mainly group selectionist) and short-term (gene selectionist) evolutionary models help to explain the success of sex. The primary cost of sex is no longer seen as the 50 per cent dilution of the maternal genotype (the 'cost of meiosis') but as the 0–50 per cent cost of producing males (the 'cost of anisogamy').

(2) Anisogamy may be the evolutionary inevitability of natural size variation among primitive isogametes, disruptive selection favouring large and small extremes and eventually leading to the emergence of two distinct sexes.

(3) While males can be viewed as 'parasitic' on the parental investment of females, the two sexes are likely to be maintained in an evolutionarily stable ratio of 1:1 in natural populations.

(4) Despite the evolutionary potential for males to be polygamous, several environmental factors as well as choice criteria exerted by females limit the circumstances under which this potential is actually expressed.

(5) Competition within and between the sexes for access to the preferred type of mate has led to sexually-selected morphological and behavioural characteristics. The importance of sexual selection as an evolutionary force is well illustrated by the mate-searching strategies of male dung flies. Male behaviour appears to have been shaped almost completely by sexual selection.

9 THE ECOLOGY AND ORGANISATION OF SOCIAL BEHAVIOUR

Animals do not operate in isolation within their environment. Their activities are, to a large extent, constrained by the requirements of other individuals sharing the environment. The presence of conspecifics and animals of other species is likely to influence practically every behaviour an animal performs. The efficiency of breeding behaviour, feeding, predator avoidance, even thermal regulation may be affected by the degree of proximity of other animals. In some cases the efficiency of a behaviour is enhanced by close proximity to others; in other cases it is enhanced by ensuring that neighbours are kept at a prescribed distance. In this chapter, we shall look at some of the factors influencing the spacing behaviour and social organisation of animals and the way social behaviour affects the various maintenance activities they have to perform. Since an animal's spatial relationship with its neighbours places important limits on what it can do and where it can do it, it is unlikely that selection will leave spacing patterns to chance. As with the topics discussed in preceding chapters, therefore, we shall examine variations in social organisation from an adaptive viewpoint.

The discussion is divided into two main sections: one deals with the costs and benefits of social *aggregation*, the other examines the reasons for *territorial* behaviour. Many of the examples concern vertebrates and particularly birds because it is these which have provided the best quantitative studies to date.

9.1 The Costs and Benefits of Social Aggregation

Associating with conspecifics or even members of other species may improve an animal's chances of surviving to reproduce in several ways. The blanket term 'associating', however, covers a very wide range of social interactions from the loose, milling herds of savannah herbivores to the highly structured societies of hymenopteran insects. Clearly the costs and benefits of association will vary. Some types of social group will confer similar adaptive advantages on their component individuals, others will have more specialised consequences. Many of the benefits arising from associating in groups, however, fall into two broad categories: those which help an individual avoid predation, and those which help it find food.

9.1.1 Anti-predator Effects

It is not difficult to imagine how associating with other individuals might afford protection from the ever-present threat of predation. There is an obvious truth in the idea of safety in numbers: in a group of 100 individuals, an animal is likely to have only one-hundredth the chance of being caught by a predator than if it was alone. Companions also provide more sets of sense organs with which approaching predators can be detected. In fact there is evidence for at least five different ways in which associating in a group might help an individual avoid predation. The following are based on the divisions of Bertram[28].

Avoiding Detection by Predators. Intuitively, we might expect a group of animals to be more conspicuous than a single individual and hence more vulnerable to detection. However, because groups are likely to be scarcer than single individuals (more of the population is concentrated at fewer points in space), they may be missed more often by predators wandering randomly across their home range. The success rate of a predator which is likely to capture only one individual per group at best may thus be considerably reduced by the tendency for prey to form groups.

Counteracting the scarcity effect, however, is the fact that large numbers of closely-spaced individuals tend to be more conspicuous. Birds in flocks, for instance, sometimes use 'contact calls' which make them audibly conspicuous. Similarly, the characteristic strong odour of many large-scale bird roosts may make associating individuals more vulnerable to olfactory detection. Andersson and Wicklund[2] found that colonially-nesting fieldfares (*Turdus pilaris*) attracted more predators than solitarily-nesting birds, although in this case greater attraction was offset by the fieldfares' effective mobbing response. Treisman[311] and Vine[324] have put forward detailed mathematical models of the advantages of grouping in relation to detectability to predators.

Detecting Predators. Detectability is, of course, a two-sided coin. While aggregating into a group may render prey more or less conspicuous depending on the type of predator concerned, it is also likely to enhance their own ability to detect the predator. The greater the number of detectors in the group, the greater the chances of early detection. Provided an animal can perceive that one of its companions has detected a predator, this 'more pairs of eyes' effect may provide substantial protection. Kenward[158] found that large flocks of wood-pigeons (*Columba palumbus*) responded to an approaching trained goshawk (*Accipiter gentilis*) sooner than small flocks.

Interestingly, the relationship between predator detection and group size may not be a simple one. In flocks of white-fronted geese (*Anser albifrons*), for instance, the number of birds vigilant (i.e. in a characteristic head-up posture) at any given time increases with flock size at a lower rate than that expected by simply multiplying flock size by the vigilance level of a single bird[184] (Figure 9.1). As flock size increases, therefore, the proportion of birds which are vigilant actually goes down. By spending less time looking out for predators, birds in large flocks can spend more time doing other things, particularly finding food. (Recent work currently in press suggests that an alternative reason for the

Figure 9.1: The Proportion of White-fronted Geese Observed Scanning in Flocks of Different Size (Solid Line) Compared with the Number Expected by Assuming that Birds Maintain the Same Level of Vigilance as a Solitary Bird (Dashed Line).

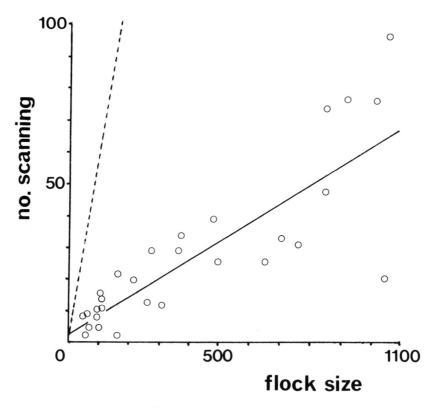

Source: Modified after Lazarus[184].

negative relationship between the number of birds vigilant and flock size is the fact that edge birds tend to be more vigilant but form a smaller proportion of large flocks (J. Lazarus, personal communication).) This 'double benefit' is shown particularly clearly in Powell's[247] experiments with starling flocks. Powell flew model hawks over different sized flocks of captive starlings and calculated the amount of time birds spent scanning for the hawk and feeding. As flock size increased, birds spent less time scanning and more time feeding. We shall say more about the effects of group size on feeding behaviour in Section 9.1.2.

Of course, the improved detection benefit of being in a group may not go on increasing indefinitely. Siegfried and Underhill[286] performed similar experiments to Powell by flying model hawks over dove flocks. Here, however, birds responded most quickly in medium-sized, rather than large flocks (Figure 9.2),

Figure 9.2: Ground Doves in Intermediate-sized Flocks React Soonest to an Approaching 'Predator'. (Reaction time calculated as frames of cine film elapsing between birds responding and a dummy hawk reaching the point where they had been feeding.)

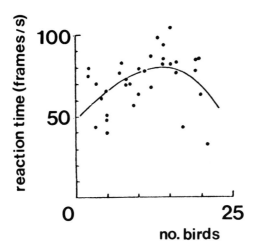

Source: Modified after Siegfried and Underhill[286].

implying that vigilance was somehow impaired when too many birds were present. Caribou (*Rangifer arcticus*) calves also seem to be more easily caught by wolves (*Canis lupus*) when they are in very large herds, perhaps because the wolves' approach is obscured.

Deterring Predators. Grouping by prey may actually deter a predator from attacking. Deterrence may be direct in that the group poses a physical threat to the predator. Musk oxen (*Ovibus moschatus*) form defensive groups with large bulls on the outside when attacked by wolves. The bunching behaviour shown by various bird species in response to spotting a predator may also pose a threat of injury to the predator. As Tinbergen points out, a falcon stooping on a flock of starlings as it bunches together is likely to be seriously injured if it hits any but its target bird.

In a less direct way, predators may be deterred by the lower capture success they might expect with aggregated prey. Neill and Cullen[225] found that cephalopod and fish predators preying on small fish did better when hunting small schools or singletons than when hunting large schools. Furthermore, the predators tended to avoid or ignore schools more than singletons. Page and Whitacre[234] found an optimal flock size in sandpipers (*Calidris* spp.) at which capture success for coastal raptors was minimised. The number of attacks by falcons was lowest for flock sizes yielding the lowest capture rate (Figure 9.3).

Figure 9.3: Coastal Merlins Experienced Reduced Attack Success against Intermediate-sized Flocks of Sandpipers (a) and Tended to Attack Single Birds or Very Large Flocks More Often (b).

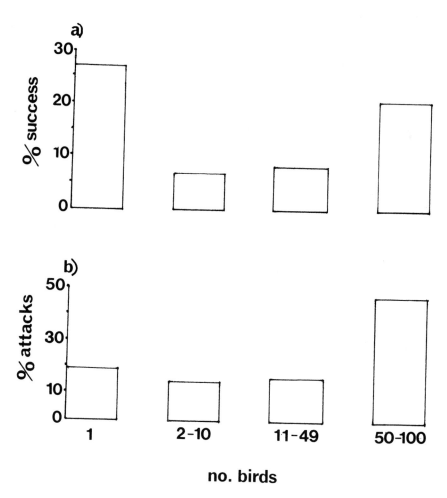

Source: Plotted from data in Page and Whitacre[234].

Predator Confusion. The effect of prey grouping on predator capture success in Neill and Cullen's experiment appeared to be due to predator confusion. When approached, schools broke up and fish scattered in all directions, making it difficult for the predator to track any one of them. Scattering behaviour is, of course, well known in a variety of species. Impala (*Aepyceros melampus*) and other bovid herds, for instance, disperse with rapid erratic jumping movements when a predator appears. To the human observer at least, individual animals are lost in the frantic explosion of activity.

Figure 9.4: Sticklebacks Preferred Singleton *Daphnia* More, the Denser the Swarm next to the Singletons. Singletons were presented in singleton:swarm size ratios of 2:20 and 2:40 and in each case a low (L) and high (H) density swarm of each size was presented. (Density was increased by putting less water in the *Daphnia* tube.)

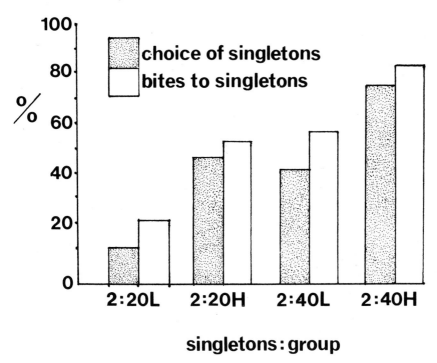

Source: Modified after Milinski[214].

Recently, Milinski and coworkers have examined the confusion effect of grouping in more detail. Milinski[214] provided sticklebacks with a choice between attacking swarms of *Daphnia* and attacking singletons which 'strayed' from the swarm. To control prey movement, singletons and swarms were confined in test tubes and selection by the fish measured as the number of bites or visits to each tube. Milinski found that sticklebacks attacked singletons more frequently the denser the swarm next to it (Figure 9.4). Interestingly, it was the swarms that fish tended to approach first, presumably because they were easier to detect. When they arrived at the swarm, however, they seemed unable to decide which individual to attack. After a few seconds they abandoned the swarm tube and vigorously attacked the singleton.

As we might predict from our discussion of foraging strategies in Chapter 6, fish were less fussy when they were hungry. Hungry fish showed less preference for singletons than those approaching satiation. When deprivation costs were high, therefore, fish tended to go for the first prey they saw (usually large swarms). Similarly, when fish were presented with a model kingfisher (*Alcedo atthis*) (a natural predator), they ceased to be selective for low density swarms.[215] Experience also affected the fishes' choice. If presented with large swarms and allowed to take prey, their subsequent readiness to attack large swarms was markedly increased.

Recently, Treherne and Foster[310] have suggested an alternative explanation for movements apparently designed to confuse a predator. They suggest that movement by the few individuals in a group which detect an approaching predator is communicated to other group members which then have advanced warning of the predator's approach (the so-called *Trafalgar effect*, named after the ship-to-ship signalling system which informed Nelson that the French and Spanish fleet was leaving Cadiz).

Reducing Individual Risk. The final way in which joining a group may help an animal avoid predation is by reducing its individual risk of selection by a predator. In a very simple way, risk is reduced by 'dilution'. Because the predator has a number of potential victims to choose from, the probability of any one individual being selected is the reciprocal of the group size (assuming that all individuals are equally vulnerable). Furthermore, those individuals not caught during an attack can escape while the predator handles its victim. A nice example of the dilution effect has recently been studied in the marine insect *Halobates* by Foster and Treherne[107].

A more complex 'safety in numbers' argument involves the space between a given individual and its nearest neighbour. Hamilton[131] called this unoccupied area around an individual its 'domain of danger'. The 'domain' encloses all those points, and *only* those points, which are closer to the individual in question than to any other. Hamilton suggested that, where a predator emerges at an unpredictable site within a prey group, individuals with the largest 'domain of danger' will have the greatest risk of capture (since they are likely to be nearest to the predator). Selection will therefore favour individuals which minimise the size of their 'domain' by moving towards their nearest neighbour. This 'selfish herd' effect may well explain the tendency of many animals to seek the centre of a group, even to the extent of jumping on top of each other (as commonly happens in herds of sheep).

Lazarus[184] tested Hamilton's predictions with flocks of white-fronted geese. Lazarus assumed that aerial predators are likely to select relatively isolated individuals within flocks (i.e. individuals with a large 'domain of danger') because they would be least able to benefit from alarm signals and would be easier to track. He predicted that the size of a bird's 'domain' would be reflected in its degree of vigilance. Relatively isolated birds should scan more to compensate for their increased vulnerability. When the duration of vigilance bouts (head up) for given individuals was plotted against the size of their 'domain' (measured as the number of birds within nine goose-lengths of the observed bird), a significant negative correlation emerged. When they stopped grazing to scan,

Figure 9.5: House Sparrows Feeding in a Field Tend to Feed for Shorter Bouts and in Larger Flocks the Further They Are from Cover (Hedges).

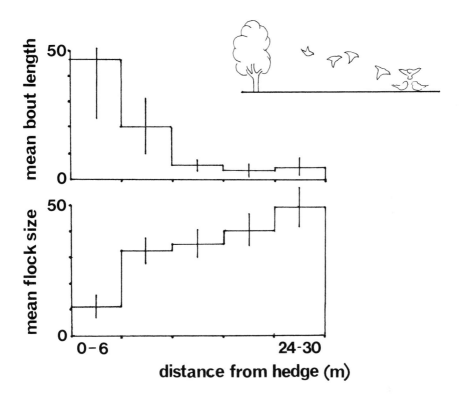

Source: From Barnard, C.J. (1979). Birds of a feather. *New Scientist*, *83*: 818–20.

isolated birds tended to scan for longer.

That animals may regard being in a group as reducing their individual risk of predation is shown in another way. Some intriguing observations on flocks of queleas[185] and ground doves (*Geopelia striata*)[122] have shown that flock size not only influences the distance at which potential predators are detected, but also the way birds respond after detection. The larger the flock the more likely birds are to 'freeze' or orientate rather than take flight or show flight intention movements. In other words, when an approaching object is spotted, birds in large flocks appear to spend time assessing how much of a threat it constitutes before fleeing. This makes sense in that many objects approaching a flock are likely to be harmless and flight at every approach would waste considerable time and energy. Nevertheless it is a characteristic of at least many small passerine flocks that disturbances are frequent and usually in response to innocuous or unidentifiable stimuli.

A major factor which increases an individual's vulnerability, at least to aerial predators, is its distance from cover. In house sparrows, flock size increases and flocks stay out for shorter periods when birds are feeding away from the protection of hedgerows (Figure 9.5). Similarly, Pulliam and Mills found that vesper sparrows (*Pooecetes gramineus*) only formed flocks when they were far from cover in sparsely vegetated habitats.

9.1.2 Feeding Benefits

While joining a group may help an animal to avoid becoming food, it may also help it to *find* food. Like anti-predator effects, however, feeding benefits may arise in a variety of different ways.

More Time to Feed. As we saw in the last section, feeding benefits can arise as a spin-off from being able to spend less time scanning for predators. This 'double benefit' from grouping, however, may be more complex than the observations described earlier suggest. Barnard[13] studied feeding and scanning behaviour in flocks of house sparrows.

The birds in Barnard's study fed in two very different types of habitat. In one, birds were feeding in cattlesheds, taking barley seed from the bedding straw laid down for cattle. In the other, the same birds were feeding in open fields, taking the debris of the autumn harvesting activity and the newly-sown winter crop of barley. An important difference between the habitats lay in the degree of cover (and hence protection from aerial predators) they provided. In the sheds, birds were almost completely sheltered from aerial detection and raptors were seldom seen there. In the open fields, however, they were exposed except when sheltering in the hedges bordering the fields. Both kestrels (*Falco tinnunculus*) and sparrowhawks (*Accipiter nisus*) regularly patrolled the areas of field where the flocks fed. At first sight, flock size appeared to correlate positively with individual feeding rate and negatively with the rate at which individuals scanned. When food availability and distance from cover were taken into account, however, the correlation with flock size disappeared in shed flocks but remained in those feeding in the fields. The allocation of time to feeding and scanning by birds in the sheds was almost completely governed by food availability (flock size just happened to correlate with food availability but did not itself influence time allocation). In the fields, flock size affected time allocation independently of food availability and other enviromental factors. Care must therefore be taken when interpreting grouping behaviour and proposing functions for it.

Finding Better Feeding Areas. A hungry animal may range over a number of potential feeding sites before deciding which to exploit. As we saw in Chapter 6, spending time exploring the environment before deciding where to feed may be essential if the animal is to feed efficiently. Among the cues it uses in deciding where to feed may be the number of other individuals already feeding at a site. Studies have revealed positive correlations between group size and food availability

Figure 9.6: The Size of House Sparrow Feeding Flocks Increased with Millet Seed Density. Open histograms show flock sizes at hedges T, L, V before baiting, and shaded histograms show flock sizes after baiting with different seed densities (light shading = 300 seeds per m², intermediate shading = 600 seeds per m², heavy shading = 1200 seeds per m²); a, b, c represent the three stages of the experiment where seed densities were rotated round the hedges.

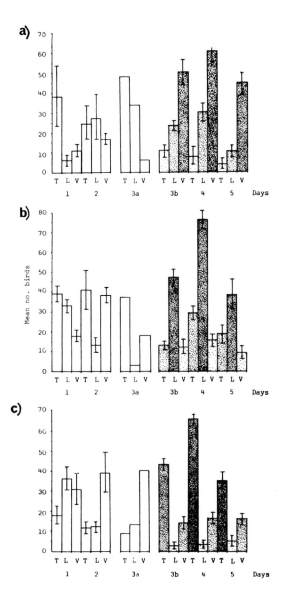

Source: Barnard[11].

for a variety of species, including birds, mammals and insects. While it may be difficult to recognise good feeding areas from the level of food availability alone (e.g. because food takes time to detect), the number of individuals present can provide a simple and reliable indicator.

The use of aggregations in deciding where to start feeding was shown very clearly by Krebs[168] in his study of great blue herons (*Ardea herodias*). Herons fly over large areas of mudflat searching for suitable pools in which to feed. Characteristically, they feed in flocks and flock size correlates positively with individual rate of food intake. Krebs put out life-sized models of herons in varying 'flock' sizes to see whether birds were attracted to pools by the sight of others feeding. Tests showed a strong positive relationship between the number of birds landing at a pool and the number of models present. Furthermore, birds landed close to the models and immediately started foraging.

Barnard[11] carried out a field experiment to see whether house sparrow flocks 'tracked' changing food densities around their habitat. If flock size is to provide an indication of site utility, it is important that it changes with site utility. The experiment consisted of baiting areas near three different hedges in the fields where sparrows were feeding. Counts of the number of birds feeding near each hedge were made before baiting and then each area was baited with a different density of millet seed. The number of birds feeding at each hedge was then counted over the next two hours. The procedure was repeated twice more with the seed densities being rotated around the three areas until all had been tested with each density. The results in Figure 9.6 show very clearly that flock size was positively correlated with seed density and tracked the highest density around the hedges. In this case large flocks built up at high seed densities because individual birds tended to stay longer there.

Providing Local Information about Food. Once a group of animals has built up at a site, further feeding benefits may arise as the result of individuals using information generated by other foragers. When animals feed in sensory contact with one another, information about their foraging success is incidentally transmitted between them. Depending on the type of food supply, individuals can greatly enhance their feeding efficiency by making use of this information. The way in which the information is used varies. In some cases individuals copy the type of behaviour used by a successful forager (social facilitation); in others they direct their attention to the place where that individual is foraging, or to other similar places in the environment (local enhancement). Again, the best studies have been with bird flocks.

In a laboratory study with captive flocks of great tits, Krebs and coworkers[176] found that birds searching for patchily-distributed food hidden in small containers on artificial trees did better in flocks of four than in pairs or singly. The reason appeared to be that, immediately after a bird found food, both the rewarded bird and others in the flock searched on the perch where it had just been successful. Birds in flocks were therefore able to direct their attention to the best feeding sites more quickly. Of course, local enhancement as shown by the tits is only likely to be profitable if the food supply is clumped. Little would be gained from copying neighbours on a dispersed food supply. Indeed, when Krebs *et al.* tested birds on dispersed prey, the incidence of copying was much lower.

Local information benefits are not limited to single-species groups. In a study of captive mixed flocks of chickadees (*Parus* spp.), Krebs[167] found that birds responded to finds by members of other species in exactly the same way as finds by conspecifics. However, when different species were trained separately to search in different parts of their artificial habitat, subsequent copying in mixed flocks resulted in species searching in new types of places. Capitalising on foraging information therefore brought about an extension of each species' 'foraging niche'.

An interesting consequence of the generation of feeding information among groups is that searching effort is open to exploitation by 'parasitic' individuals. Instead of investing time and energy searching for their own food, these individuals usurp food items found by other foragers, most commonly individuals of another species. The best known *kleptoparasites* or *'pirates'* are the gulls and skuas (several species of which 'parasitise' various auks and waders) and the neotropical frigate birds (Aves: Fregatidae) (which 'parasitise' various other seabirds, particularly noddies and boobies (Sulidae)),[37] although they also occur among mammals, insects, and spiders.

Barnard and Sibly[16] have put forward a general frequency-dependent model to predict the number of kleptoparasites (or other 'scrounger' individuals like brood parasites) which can be supported by a given 'host' population. The important predictions of the model are borne out not only by data from inter-specific kleptoparasitic relationships but also from *intra*specific relationships. In captive house-sparrow flocks, for instance, some individuals ('copiers') obtain most of their food by area-copying others which actively search for food ('searchers'). The most stable flocks are those that contain searcher:copier ratios in which both types achieve roughly equal capture rates. The emergence of the 'copier' strategy in sparrow flocks, however, may well depend on the type and distribution of the food supply. A similar intraspecific 'scrounger' strategy has been identified by Rohwer and Ewald[259] in Harris sparrows (*Zonotrichia querula*). Here dominant individuals 'allow' a certain number of subordinates to feed around them as long as the increase in feeding benefit to the dominant as a result of information from the subordinates exceeds the cost of added food depletion by the subordinates. Appropriately Rohwer calls this the 'shepherd' strategy.

Improved feeding efficiency as a result of grouping may also explain the tendency for some species to roost communally. Roosts may act as 'information centres' where individuals who fared badly during the previous day can benefit from those who did well by following them to their presumably more profitable feeding grounds the next day. There is evidence that individuals of some colonially-roosting species, like herons,[168] queleas[185] and heliconiid butterflies[115] do tend to follow neighbours out to feeding sites and that individuals might be able to identify those which have fared better than themselves.

Other Effects on Feeding Efficiency. There is a number of other ways in which being in a group may influence an animal's survival prospects. Other feeding benefits, for instance, include increased efficiency in prey size selection[17] and an increased opportunity to optimise return times to previously depleted but renewing food supplies.[58] Alternatively, grouping behaviour may help individuals *minimise their risk of finding little food* rather than maximise their feeding efficiency. Under certain conditions, minimising risk may be a more pressing requirement. Grouping may allow animals to tackle larger or more dangerous prey, not only because of a numerical advantage but because, as in lions, individuals injured in the process can be supported by the kills of other group members. Large groups of animals may also be able to usurp the prey of other predators when a single animal or a small group would be driven away.

On the deficit side, being in a group may increase the amount of interference an animal experiences in its feeding activity. In boat-tailed grackles (*Cassidix mexicanus*), the proximity of neighbours in a flock restricts the amount of area restricted searching an individual can perform.[291] Often, interference extends to overt aggression with contestants not only paying the cost of increased time and energy wastage but also risking injury (see Chapter 10). Where prey are mobile and can escape, the presence of predators may reduce their availability through disturbance. Inter-individual distance within wader flocks appears to be, at least partly, a consequence of bird density effects on the availability of surface-dwelling invertebrates. Of course, prey disturbance can have the opposite effect on availability. Many interspecific associations are based on 'beating' effects where one species flushes out prey which can be taken by another. Classic examples are the associations between ant (*Insecta: Hymenophera*) and antbird (*Aves: Formicariidae*) species and between domestic cattle and starlings or cattle egrets (*Bubulcus ibis*). In these cases the insects and cattle disturb various species of arthropod prey which the birds take in their wake.

9.1.3 Other Costs and Benefits of Group Living

Grouping behaviour may also directly influence breeding success. Breeding in large colonies where the period of breeding between individuals is synchronised (as in many seabirds for example) effectively saturates the predator population with potential food (e.g. eggs and nestlings). Predators are able to take only a small proportion of the available prey and the risk of loss to individual parents is reduced. A possible penalty of breeding in stable, cohesive groups, however, is the potentially high risk of inbreeding. Unless there are pressing evolutionary reasons for maintaining the integrity of the individual genome (individual level selection — see Chapter 11), a selective disadvantage may accrue to inbreeding parents because of the deleterious genetic consequences of inbreeding. Field evidence suggests that just such a penalty does accrue to inbreeders. Matings between closely-related great tits result in up to 71 per cent higher nestling mortality than unrelated pairings. Similarly, matings between closely-related baboons (*Papio* spp.) produce 40 per cent fewer viable offspring. Of course, as we saw in Chapter 3, the high degree of relatedness within stable groups also creates a potential for the evolution of apparently altruistic acts between individuals, such as food-sharing, guarding behaviour and co-operative breeding and hunting. A high degree of relatedness may therefore enhance group cohesion and provide one mechanism by which initially loose, opportunistic aggregations can evolve into complex and structured societies (see Section 9.1.5).

Group breeders may run a high risk of cuckoldry. Field experiments with colonially-nesting red-winged blackbirds showed that the mates of vasectomised males were sometimes fertilised by other males.[33] Cuckoldry may also occur between females. In groove-billed anis (*Crotophaga sulcirostris*) apparently co-operative nesting turns out to be exploitation of subordinate female incubation effort by dominant females.[321] Dominants wait till last to lay their eggs in a communal nest and then turn out the eggs laid previously by subordinates to ensure it is their own which are incubated most effectively. In some group-breeding species, offspring cannibalism by neighbours constitutes an additional cost.

Finally, in some cases, group formation may improve thermal regulation. Pallid bats (*Antrozotus pallidus*)[317] and dippers (*Cinclus cinclus*)[279] roosting in groups use less energy and maintain their body temperature more effectively than those roosting solitarily.

9.1.4 Costs, Benefits and Group Dynamics

While it helps us to pin-point the main selective pressures favouring aggregation in particular cases, dividing the consequences of grouping behaviour into categories like feeding, breeding, protection, etc., is highly artificial. Groups are ultimately the product of several jointly-acting selective pressures. So far, studies explicitly setting out to investigate the relationship between interacting selective pressures and group build-up and persistence have been few. One that has is Caraco and coworkers'[41,42] study of flock formation in juncos (*Junco phaeonotus*).

By making assumptions about the relative weighting birds should give to each of several mutually exclusive activities (e.g. scanning takes precedence over feeding in all birds, feeding over fighting in dominant birds, fighting over feeding in subordinates and so on) and the way time allocation to each would change with flock size, Caraco was able to predict variation in flock size in relation to temperature, food availability, predation risk and the availability of cover. In field experiments, flocks conformed very closely to prediction in each case, departure perhaps reflecting a trade-off between optima for dominant and sub-ordinate birds.

The organisation and size variation of primate groups has also been analysed in terms of costs and benefits. Using multivariate statistical techniques, Clutton-Brock and Harvey[56] have revealed several interesting relationships between environmental factors and group behaviour. Ground-dwelling, diurnal species, for example, tend to live in larger groups than nocturnal, arboreal species, presumably because the latter are more cryptic and less in need of safety in numbers. One difficulty with this analysis, however, is that it fails to take into account different *types* of primate group. Wrangham[339] argues that group size depends on, among other things, the distribution of costs and benefits of grouping between the sexes. While the key limiting resource for males is receptive females, females are constrained mostly by food availability. Where food supplies are patchy and defensible (like fruit on trees), it may pay females to aggregate in their defence. Where food is more evenly distributed (like leaves), less advantage accrues to grouping females. A series of models with extensive discussion relating to group dynamics in primates can be found in Cohen.[60]

9.1.5 Social Organisation within Groups

During our discussion of the selective pressures favouring aggregation in animals, there were numerous suggestions that individuals may be affected in different ways by being in a group. We cannot tacitly assume that being in a group has the same utility for all concerned. Because individuals respond differently to being in a group, we can often break the group down into sub-groups of behavioural 'types' and imply an internal social organisation. In some cases, the degree of organisation involves little more than a distinction between individuals who fare well in aggressive disputes and those who fare badly. There is little internal structure to the group. In others it involves individuals who fit into well-defined 'roles' or 'castes' which, in turn, fit together into a structured integrated society (see below). The criteria used to describe the pattern of organisation (e.g. dominance relations, family groupings, breeding structure), is likely to vary between species and types of habitat.

Dominance Hierarchies. Many social groups are characterised by sustained aggressive/submissive relationships between individuals. The set of such relationships is often called a *social* or *dominance hierarchy*. The hierarchy may be very simple, involving a single individual ('despot') who is dominant to all others in the group but no rank relationships between subordinates. More often, there is a series of rank relationships such that an 'alpha' individual is dominant to all others in the group, a 'beta' individual is dominant to all except the 'alpha' and so on down to the lowest ranker who is subordinate to everybody. Sometimes, linear dominance hierarchies are complicated by triangular relationships such that A beats B, B beats C, but C beats A, but these are usually unstable. Barnard and Burk[15] distinguish three basic types of hierarchy (Figure 9.7).

Figure 9.7: Model of the Evolution of Different Types of Dominance Hierarchy and Complex Assessment Cues (see text and Chapter 10).

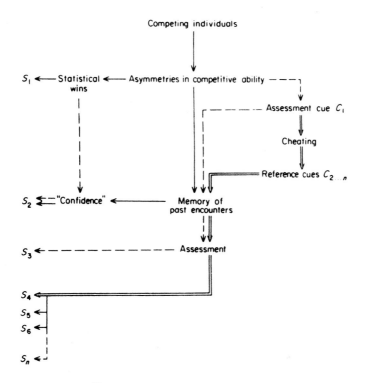

Source: Barnard and Burk[15].

(1) *Statistical hierarchy.* Individuals have no memory of past encounters but fight every time they meet. Because of variation in physical attributes some tend to win and others to lose.

(2) *Confidence hierarchy.* The act of winning a fight increases the vigour with which the victor enters its next encounter, so enhancing its chances of winning again and vice versa. Computer simulation of this 'confidence' effect ultimately produces a linear dominance hierarchy.

(3) *Assessment hierarchy.* Individuals use their memory of past encounters to decide which individuals they will fight and which they will avoid. Fights become calculated risks based on features of opponents (*assessment cues*) which correlate with having won or lost.

Assessment hierarchies are likely to be the commonest type in nature. We would expect selection to favour individuals which are sensitive to any cue which predicts the outcome of a fight because fights are likely to be expensive in time, energy and risk of injury. Indeed, the majority of recorded hierarchies, in insects, amphibians, birds and mammals do appear to be of this type. However, 'confidence' hierarchies have been recorded in crickets (Orthoptera: Gryllidae) and laboratory mice, where the outcome of fights can be predicted more from how individuals fared in their last few fights than from the characteristics of opponents. Because of the costs of fighting and the consequent selection pressures for assessment, purely statistical hierarchies are likely to be rare in nature. However, there is evidence that they may exist at least in some lower invertebrates. Aggression between sea anemones (*Actinia* spp.) appears to conform to a statistical hierarchy where success depends on relative body size. The net effect of a hierarchy, of course, is that a small proportion of the individuals within a group gain priority access to limited resources, like food and mates. Survivorship and reproductive potential may therefore be closely linked with an individual's dominance ranking.

Roles and Castes. Once animals have evolved to operate in social groups, selection may favour more rigid distinctions between individuals than those arising simply through differences in competitive ability. If the group is sufficiently stable, a true 'division of labour' may arise with different individuals carrying out specific 'roles'. A 'role' can be defined as a pattern of behaviour which is characteristic of particular individuals within a group and which occurs repeatedly in different groups belonging to the same species. The behaviour pattern characterising a role affects the behaviour of other individuals in the group in a predictable way, either by the communication of specific signals or by the physical effects of one animal's activity on others. For instance, in some social primate species, certain individuals may spend a lot of time at the periphery of the group scanning for predators. These can reasonably be said to perform a 'scout' or 'vigilant' role. Similarly, other individuals may function in dispute-terminating roles, leadership roles or, in some insect species, colony-guarding roles and so on. 'Role' therefore describes the predominant behavioural tendency of an individual and the impact of this tendency on the group as a whole. The designation of a role is, of course, only helpful as long as it applies to a functionally related group of behaviours, like chasing, biting and stinging by colony guard bees. As with the arbitrary categorisation, there is a risk of creating as many roles as there are behaviours in the animal's repetoire and hence losing whatever explanatory value the concept of 'role' might have.

The division of groups into sub-groups with characteristic roles imposes an internal structure on the group which we can then talk of as a *society*.

In the social insects (termite and hymenopteran species), individual roles are sufficiently well-defined to be regarded as *castes*. Wilson[335] defines a caste as 'a set of individuals, smaller than the society itself, which is limited more or less strictly to one or more roles'. Caste systems may or may not involve specific genetic differences between individuals. In stingless bees (*Melipona* spp.), for instance, queens are determined as polygenic heterozygotes, while, in other species castes appear to be determined mainly by environmental factors. Human castes are hereditary groups consisting of people of similar rank, occupation and social standing. The distinction between castes in human societies appears to be mainly one of social conventions and customs.

The division of a group into different castes, sex or age groups which have different roles in maintaining its integrity is one of five characteristics used by Eisenberg[96] to distinguish societies from casual groups. The other four include (1) a complex communication system, (2) a tendency for individuals to remain in a cohesive group. (3) a more or less permanent group membership (individual composition remains the same from day to day with little or no migration between groups) and (4) a tendency to reject 'outsiders', individuals which are not permanent members of the group.

9.2 The Costs and Benefits of Territoriality

While various selection pressures in the environment favour the formation of more or less close-knit groups, others favour the opposite response, spacing out. Individuals of many species tend to be spaced more widely apart than we would expect by chance (see Davies[72] for a recent review). In some cases, it is possible to account for increased spacing in terms of local microclimatic or mortality effects, especially in sessile organisms like barnacles (*Balanus* spp.). In others, however, spacing is achieved by behavioural means. Animals actively exclude each other from their respective 'home ranges'. When spacing is the result of behaviour, animals can be regarded as defending *territories*, parts of their 'home range' to which they have more or less exclusive access.

As Wilson[335] points out, however, territories are more than simply defended areas. They are dynamic, often changing in size and shape with seasons, population density and the animal's age. Huxley compared them with elastic discs having the resident animal as its centre. In conditions of high population density, the pressure along the disc's boundary is high and it contracts. However, there is a limit beyond which further contraction is not tolerated by the resident. Either the resident aggressively resists further reduction in the size of its territory or the territorial system in the population breaks down. When population density and hence boundary pressure is low, territories expand. Once again however, change does not go on indefinitely; very sparse populations are often characterised by territories whose boundaries are not contiguous (i.e. which have a 'no man's land' or 'buffer zone' between them) or are so diffuse as to be indefinable. North American dunlins (*Calidris alpina*) provide good examples of 'elastic disc' territories (Figure 9.8). Sometimes the pattern of space use within the territory changes with its expansion and contraction. Tree sparrows (*Spizella arborea*), for instance, use the whole of their territory with equal intensity when it is fully contracted, but use only a central area intensively when it is expanded.

Figure 9.8: Elastic Disc Territories in North American Sandpipers. Territories are small in the high population density area at Kolomak, but large with 'buffer zones' in the low density Barrow area.

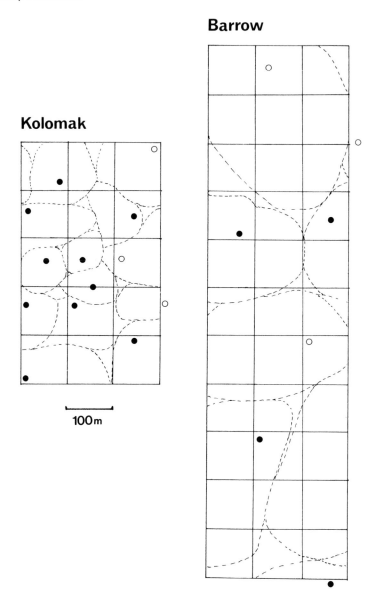

Barrow

Kolomak

100m

Source: Modified after Holmes, R.T. (1970). Differences in population density, territoriality, and food supply of dunlin on arctic and subarctic tundra. In (A. Watson, ed.) *Animal Populations in Relation to their Food Resources*: 303-19. Oxford, Blackwell.

Figure 9.9: Hexagonal Territory Boundaries in Male Mouthbrooders, *Tilapia mossambica.*

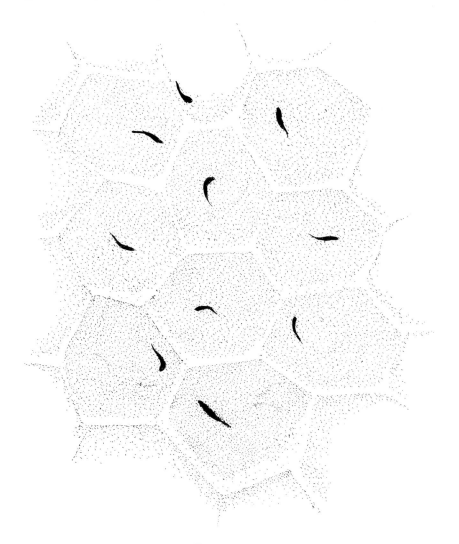

Source: Drawn from a photograph by G. Barlow.

Hexagonal Boundaries. If several elastic discs are allowed to expand in close proximity to one another so that they meet, their originally circular boundaries distort to form hexagons. An important point is that perfectly-matching and contiguous hexagons leave no space between them. This feature would be of immense advantage to territory owners in high-density populations. By allowing their territory boundaries to distort into mutually contiguous hexagons, owners would maintain the maximum amount of space possible within their territories. Careful analysis of the dunlin territory boundaries mentioned above revealed that, at high population densities, these did indeed conform to hexagons. In the mouthbrooder fish (*Tilapia mossambica*), the hexagonal territory boundaries of breeding males are sometimes actually visible as sandy ridges surrounding the owners (Figure 9.9).

9.2.2 The Functions of Territoriality

Like other aspects of behaviour, the adaptive value of territoriality lies in the benefits it confers in increased survivorship and reproductive potential relative to the time, energy and risk costs involved. We should therefore expect animals to invest in territorial defence only when the reproductive benefits outweigh the costs of defence. What sort of adaptive value might territorial behaviour have?

A survey of the immense literature on territoriality reveals two major areas in which it appears to contribute towards individual reproductive success. In many species territorial defence appears to be linked with feeding requirements and food distribution. Here, territories tend to be maintained under specific conditions of food availability. In other instances, territories function in mate attraction and ownership. Members of the mate-choosing sex use characteristics of territories defended by members of the competing sex or features of the latter which correlate with some aspect of territoriality to select a mate.

Territoriality and Feeding Benefits. There are many loosely quantitative pieces of evidence which suggest that territorial behaviour may be linked with food availability. Correlations between individual or group biomass and territory size in many species indicate that the area defended may be decided on the basis of food requirements. Similarly social primate species which include a lot of foliage in their diet tend to have smaller territories than those taking mainly fruit and blossom. Presumably this is because fruit and blossom are scarcer in the environment.

Recently, several studies have attempted to quantify the relationship between territory size and food availability more rigorously. Because territories should only be defended when it is energetically profitable to do so, we might expect animals to be territorial under some conditions but not under others (this is modelled explicitly by Jerram Brown in his 'principle of economic defendability').

Carpenter and MacMillen[45,46] studied changes in territorial behaviour in Hawaiian honeycreepers (*Vestiaria coccinea*). Honeycreepers are small nectar-feeders and territorial defence is based on the need to maintain access to a sufficient number of flowers to ensure survival. By carefully quantifying the costs and benefits of defence, Carpenter and MacMillen were able to predict exactly when honeycreepers should and should not be territorial. They measured benefit in terms of the increased amount of energy available to defenders as a result of preventing other birds taking nectar. Costs were measured as the amount of energy spent defending the territory. Their prediction was very simple. Birds should only be territorial when:

Basic energy + cost of living (V)	Additional energy cost of territorial defence (T)	$<$	Amount of energy gained without defence (P)	$+$	Additional energy gained as a result of defence (Ed)

While this is nothing more than a simple weighing up of the costs and benefits of defence, Carpenter and MacMillen used it to predict two cost:benefit thresholds above and below which honeycreepers should not defend territories. They predicted that defence should begin when P was too low to outweigh V. Here, birds need to reduce the extra depletion of their nectar supply by intruders by actively excluding them and gaining Ed. However, they should cease being territorial if P + Ed together are still too low to outweigh V + T, i.e. when nectar productivity drops to such a level that not even the exclusion of intruders ensures enough food. Here, birds should abandon the area altogether and search for new flowers.

To test their predictions, Carpenter and MacMillen carried out some careful time-budget analyses and converted the time spent by birds in various activities into energy costs. The costs and benefits pertaining to territorial defence and calculated in energy terms are shown in Table 9.1. From their calculations, Carpenter and MacMillen predicted that birds should begin to defend territories when the number of available flowers falls to about 207. They should then remain territorial until the number of flowers falls as low as 60. At this level it will pay them to move to a new area. The predictions were tested with ten birds feeding in an aviary. Figure 9.10 shows that out of the ten birds, nine conformed to the model's predictions.

In the honeycreepers' case, territories yielded short-term benefits in terms of enough energy to see a bird through the day. In some cases, territorial defence is geared to longer-term benefits. Davies studied pied wagtails defending winter feeding territories along the banks of a river. Here, birds were taking organic debris washed up by the river. At the same time as some wagtails were defending territories, others were feeding in flocks on nearby flooded meadows. Interestingly, flock-feeding birds often fed at a higher rate than those maintaining territories. Yet, although territorial birds joined the flocks when food availability on their territories was low, they frequently returned to their territories to maintain defence. The reason appeared to be that the food supplies exploited by the flock were ephemeral. Although they sometimes fared well, flock birds often fared very badly. Territorial birds, however, had access to at least some food all through the winter. Territory owners therefore reaped the long-term benefits of a sure, if not always the best available, food supply. Caraco[41] found a similar long-term benefit to food defence in juncos (*Junco phaeonotus*). Dominant birds in feeding flocks defended local patches of seed when

Table 9.1: Calculation of Territorial Defence Thresholds in Hawaiian Honeycreepers.

Hawaiian honeycreepers should defend territories when:

$V + T < P + Ed$ (see text)

From calculations of energy expenditure from time allocation:

$V = 13.4$ kcal
$T = 2.3$ kcal $\left. \right\}$ 24 hours (time interval over which V and T kcal are required)

Intruders stole approximately 75% of a bird's nectar, so, without defence, birds obtained only a quarter of available nectar.

Ed depends on the effectiveness of territorial defence which is here assumed to be perfect. Under these conditions:

Upper threshold for defence (defence should begin)

$$(P + Ed = \frac{V}{0.25}) = 53.6 \text{ kcal} \equiv 207 \text{ flowers}$$

Lower threshold for defence (defence should cease)

$(P + Ed = V + T) = 15.7$ kcal $\equiv 60$ flowers

Source: From data in Carpenter and MacMillen[45,46].

temperatures were high and feeding priority low. Because the juncos were exploiting a non-renewing seed supply, defence in times of low food requirement ensured adequate provisions when requirement was high and time could not be spent in aggression.

In a continuation of the wagtail study, Davies and Houston[76] noted that intruders were not always evicted from territories by the owner. Like the honeycreepers therefore, wagtails did not always bother to defend their territories. The main cost imposed on owners by intruders appeared to be the unpredictable depletion of feeding sites along the river bank. By patrolling their territories in a regular pattern, owners were able to space their visits to previously depleted sites. This allowed time for food supplies to replenish. Intruders tended to deplete these renewing supplies between visits by the owner and so reduced the owner's feeding rate (effectively by halving their return times to feeding sites — see also Section 6.4.2). On occasions, however, owners benefited from the presence of another bird on their territory because additional birds assisted in expelling further intruders. What we might expect, therefore, is that owners will only exclude intruders if the cost of their additional food depletion exceeds the benefit of assistance in defence.

In general, this appeared to be the case. Owners chased off intruders if their expected feeding rate on their own was greater than that in the presence of an intruder. Intruders were allowed to remain (and were now called 'satellites') when feeding rate was little affected by the 'satellite' and the owner could benefit from the additional defence. The importance of intruder impact on the owner's feeding rate was also reflected in interspecific disputes. Birds tended to chase off those species like grey wagtails (*Motacilla cinerea*) and meadow pipits (*Anthus pratensis*) which overlapped with pied wagtails in their dietary

Figure 9.10: The Observed Degree of Territorial Defence in Honeycreepers in Relation to the Number of Flowers in the Feeding Area. N, no defence; M, medium defence; I, intense defence.

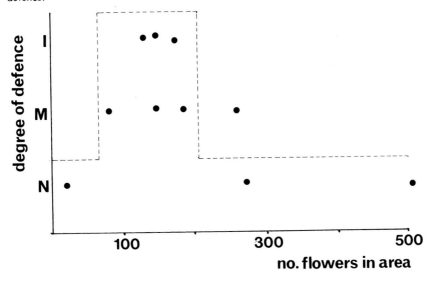

Source: Modified from Carpenter and MacMillen[45].

requirements, but allowed non-overlapping species like linnets (*Acanthis cannabinna*) to remain.

Territoriality and Breeding. Territorial behaviour plays a crucial role in the breeding biology of many species. As we mentioned in the last chapter, the parcelling up of vital environmental resources (like food or nesting sites) by members of one sex can secure their access to members of the other sex. Usually, but not always, it is males who defend limited resources. Females who require the resources are then constrained to venture into male territories where they may be mated. Depending on the distribution of resources in the environment and the ability of males to defend them, females may be able to use some feature of the territory in choosing a mate.

One feature they might use is territory size. In long-billed marsh wrens (*Telmatodytes palustris*), female choice appears to be related to the size and the quality of the food supply within a male's territory. Males which are polygynous (i.e. attract more than one female) have the largest territories, monogamous males have slightly smaller territories and unmated 'bachelor' males have the smallest territories. Interestingly, males with large territories may accumulate a number of females while nearby owners of small territories remain unmated. Territory size also appears to be an important determinant of breeding success in male three-spined sticklebacks, this time because it is negatively correlated with egg predation by intruder males.

In other cases, territory qualities other than size seem to be important. In red-winged blackbirds, the type of vegetation on male territories influences the risk of predation for offspring. Males whose territories contain the most effective type of vegetation attract the most females. In pronghorn (*Antilocapra americana*) it is the males whose territories provide the best foraging material who attract most females.

Of course, territories related to breeding behaviour may be maintained even when vital resources are not defendable. Lekking behaviour (see also Chapter 8) is a good example. Here males aggregate at a suitable area and display communally. During display, each male holds a small territory which, although perhaps only a metre or two across and consisting of little more than bare earth, is vigorously defended against other males. Intriguingly, in all the lekking species studied so far − whether birds, ungulates or even insects − the majority of copulation with females which visit the lek are achieved by only a small proportion of the sexually-receptive males. Successful males are often those occupying certain, usually central, positions on the display ground. Alternatively, as in some dragonfly species (Insecta: Odonata), they may be males which display at the time of day when most females visit the area.

Choosing males from a certain position in the lek may be one way a female can be sure of mating with the best male available. In the Uganda kob (*Kobus kob*), a lekking antelope species, males contest ownership of central 'TG' territories to which females come for mating. Females effectively let inter-male competition select a mate for them. In some cases, however, female choice may be based on the characteristics of males themselves. In black grouse (*Lycrurus tetrix*), for instance, male courtship tactics appear to be the deciding factor. In male ruff (*Philomachus pugnax*) it is plumage characteristics.

Other Functions of Territoriality. Territorial behaviour may enhance an animal's reproductive potential in other ways. Many of these involve reductions in the risk of predation. The defence of roosting positions in many species of bats, starlings, sparrows and pigeons, for instance, may result in successful defenders having priority access to the safest sites or those nearest to rich food supplies. In great tits, increased spacing between nest sites reduces the risk of egg and fledgling predation by rats and weasels. In ectothermic animals, like lizards, aggressive spacing may decide access to the best sunning areas. Forest-dwelling anoles (*Anolis* spp.) fight for access to particular sunny perches in trees.

So-called 'secondary functions' are also sometimes ascribed to territorial behaviour, although to do so is always unsatisfactory because we can never be certain as to the order of importance of jointly-acting selection pressures. In particular, animals may become familiar with parts of their home range and thus be able to escape predators, find food and establish safe resting areas more effectively. Increased familiarity is therefore a beneficial spin-off arising from the maintenance of a territory for initially quite different reasons. Another suggested 'secondary function' is a reduction in the spread of parasites and disease between individuals. Certainly spacing reduces the amount of physical contact between animals, but the non-specificity of many pathogens (in terms of susceptible species) makes transmission prevention an unlikely pressure selecting for territoriality. However, in colonially-nesting birds, like gulls, where epidemics cause high juvenile mortality, spacing may help to reduce disease transmission. By defending territories, males may also be in a better position to maintain established pair bonds. If the territory is large enough to encompass the female's normal range of activity, preventing access by other males is easier. Territories can even be 'functionless' in the usual sense of containing limited resources or acting as a mate-attraction mechanism. Instead they may reflect arbitrary but economical ways of settling otherwise costly disputes (see Chapter 10).

Territorial defence, of course, has a marked effect on species' population dynamics. Populations of territorial species usually contain 'surplus' itinerants or floaters who fail to establish themselves. Removal experiments have shown that there may be as many as three times the number of territorial individuals surviving as floaters in some populations.[299] In many cases these are juveniles who may remain as floaters until they are large enough to contest territories successfully. In Chapter 10, we shall discuss some of the ways in which potentially invading floaters can monitor the occupancy of territories.

Summary

(1) Many factors may favour the aggregation of animals into groups. The most widespread appear to be various anti-predator benefits and increased feeding efficiency. However, grouping should be seen as the product of several integrated selective forces, rather than as serving a single 'function'.

(2) An analysis of the effects of grouping and group size on the behaviour of individuals can lead to an understanding of the dynamics of group build-up and persistence.

(3) Once animals have aggregated, different levels of social organisation may evolve within the group. Individuals may come to perform specific roles, in some cases as genetically-determined castes. Groups with a highly structured internal organisation can be regarded as true societies.

(4) In some cases selection will favour the spacing out of animals within their environment. Spacing maintained behaviourally can be described as territoriality. Two of the most important 'functions' of territoriality appear to be the defence of food supplies and the acquisition of mates. Territoriality can be modelled in terms of the reproductive costs and benefits of defence and may affect species' population dynamics.

10 COMMUNICATION

Many of the most important external stimuli impinging on an animal are those emanating from other animals. Predators, mates, flock companions, parents, offspring and so on transmit signals to which animals respond in particular ways. In many cases these signals are the result of deliberate and complex actions which have evolved specifically to influence the behaviour of other animals. When we consider the relationship between a signal transmitted by one animal and the response it elicits in another, we consider an act of *communication*. Communication can be said to have occurred when an animal performs an act that alters the behaviour of another animal. However, neither the signal nor the response constitute communication in themselves. As Wilson points out, even if one animal signals and the other responds, there has been no communication unless the probability of response is altered from what it would have been in the absence of the signal. While we all feel we know what we mean by communication, it has proved a difficult and still elusive phenomenon to define satisfactorily. Whichever way it is defined and however tortuously precise the definition, exceptions to the rule are always easily found.

Nevertheless, despite the difficulties, it is possible to identify the key components of a communicatory act.

10.1 Signal Characteristics, Displays and Animal 'Language'

Listed below are the important components of a communication system recognised by Sebeok[275]:

(1) The *sender*: an individual which transmits a signal.
(2) The *receiver*: an individual whose probability of behaving in a particular way is altered by the signal.
(3) The *channel*: the medium through which the signal is transmitted (e.g. visual or vocal/auditory channels).
(4) The *signal*: the behaviour (e.g. posture, display, vocalisation) transmitted by the sender.
(5) The *context*: the setting in which the signal is transmitted and received.
(6) *Noise*: background activity in the channel which is irrelevant to the signal being transmitted.
(7) The *code*: the complete set of possible signals and contexts.

Variation in any of these factors can affect the way the receiver's behaviour is altered by the signal it receives. Since it is the signal which is transmitted and therefore 'connects' the sender with the receiver, we shall begin our discussion of communication with a brief look at the nature of animal signals.

10.1.1 Discrete and Graded Signals

Sebeok[274] distinguished two fundamental types of signal: discrete (digital) and graded (analogue). Discrete signals are those that operate in an 'on/off' manner, like the flash sequences of fireflies (*Photinus* spp.). Other typical examples include the adoption of steel-blue and red courtship coloration by male three-spined sticklebacks and the use of stylised preening movements during court-ship in some duck species. The dichotomous nature of discrete signals is emphasised by their tendency to be performed with so-called *typical intensity*[216] (i.e. no matter how weak or how strong the stimulus evoking a signal, the signal is always displayed with the same degree of vigour and complexity). We shall return to the idea of typical intensity later.

Graded signals, on the other hand, can be transmitted in varying degrees of intensity and complexity. Ants, for instance, release quantities of alarm substance which are roughly proportional to the degree to which they have been stimulated. Among vertebrates, birds and mammals can transmit an enormous range of messages simply by varying particular postures or vocalisations. Graded signalling in relation to the degree of stimulation is shown nicely by aggressive displays in rhesus monkeys (*Macaca mulatta*). Low-intensity arousal is manifested as a fixed stare. As the degree of arousal increases, new components are added to the display either singly or in combination. These include opening the mouth, bobbing the head up and down, vocalising, slapping the hand on the ground and lunging forward. If all these components appear in the display, the monkey is likely actually to attack.

10.1.2 Signal Specificity and Economy

In many species, particularly insects and lower vertebrates, signals are highly stereotyped. Each signal elicits only one or a small number of responses and each response can be elicted by only a very few signals. A truly astonishing example of signal specificity is found in moth sex attractant pheromones. The sex phero-mones of two species of *Bryotopha* moth differ by the configuration of only a single carbon atom: they are thus geometric isomers. Field observations have shown that *Bryotopha* males respond solely to the isomer of their own species. Not only that, the male's response is actually *inhibited* if the heterospecific isomer is present.

Among vertebrates, some bird and mammal species are able to recognise individuals on the basis of small variations in visual or vocal signals. Indigo buntings and some other small passerines can distinguish the calls of neighbour-ing territory owners from those of nearby strangers. Discriminating the signal characteristics of particular individuals is also important in maintaining family units in some species. In the guillemot (*Uria aalge*), chicks learn to respond selectively to the calls of their parents within the first few days after hatching. They may even learn certain features of parental calls while still in the egg.

Most signals in animals are transmitted by special *displays*. Displays are specific behaviour patterns which have become specialised (through the evolutionary process of ritualisation) for signal transmission. 'Display' is thus a widely applicable term which covers behaviour patterns as diverse as stridulation in mating male crickets, eyelid flashing in irate baboons and the elaborate 'creeping' and 'retreating' movements of courting male smooth newts. Through ritualisation, displays are thought to have evolved from incidental actions or responses by an animal which just happened to say something about its intentions. Many displays seem to have evolved from conflict behaviours (see Chapter 2). The elaborate and stylised drinking and preening movements of courting duck species,[194] for instance, seems to be derived from displacement activities performed during motivational conflict. The marking of territorial boundaries with urine by various mammal species may have evolved from urination through an autonomic fear response when a rival was encountered on the territory border. The familiar oblique threat posture of black-headed gulls seems to be a stylised ambivalent behaviour containing elements of both attack and retreat responses.

Originally, ritualisation was seen as simply reducing the ambiguity of potentially informative actions. In becoming more stylised and exaggerated, the information content, and therefore the signal property, of the action was enhanced. For instance, through ritualization, threat and appeasement postures have come to be extreme opposites – a threatening dog stands erect, while a submissive dog crouches or rolls on its back. Morris viewed the phenomenon of 'typical intensity' in displays as the product of selection to reduce ambiguity. In Section 10.2, however, we shall consider an entirely different interpretation of ritualisation.

Despite the bewildering array of behaviour patterns which can be regarded as displays, the number of displays which is employed by any one species (apart from humans) is remarkably small. Moynihan[219] has carried out a detailed survey of display repertoires of various vertebrate species. His intriguing results showed that the number of displays used by a species increased only slightly up the phylogenetic scale, from an average of 16.3 in fish, through 21.2 in birds to 24.8 in mammals. The maximum range in Moynihan's sample was from 10 in the bullhead (*Cottus gobius*) to 37 in the rhesus monkey.

This is in sharp contrast to the seemingly limitless potential of human language. The potential for infinite expression in human language is simply illustrated by the fact that we can coin a word for any number between zero and infinity. 'Googol', for instance, designates a 1 followed by 100 zeros. However, as Wilson points out, the sound structure of language is based on about 20–60 different sound types (so-called 'phonemes'). Human language is the result of stringing these sound types into words and sentences with sufficient signal redundancy to make them easily distinguishable. It may be that 60 fundamental sound types is the maximum number the human auditory system can distinguish. Similarly, 40 or so displays may be all the sensory/perceptual systems of other animals can cope with.

The analogy between display repertoires and language leads us into a contentious area in the study of behaviour. It has always been tempting to compare the communicatory complexity of animal signals with that of our own languages. To what extent are animal displays loaded with information? Is the potential information they contain actually used by the receivers? How much complexity is a communication system capable of? These and related questions have been the source of much controversy among workers investigating mechanisms of animal communication.

10.1.3 Animal 'Language'

One communication system which has provoked particularly heated argument is the celebrated foraging dance of the honey bee. The fact that foraging honey bees are able to transmit information about food sources to 'recruits' in the hive has been known for centuries. However, it was not until the painstaking work of von Frisch in 1945 that the details of the bee's communication system began to be unravelled (see von Frisch[109] for a detailed overview). Von Frisch discovered that the type of dance bees performed when they returned to the hive apparently provided information about the distance and direction of the food supply they had just visited. Figure 10.1 shows the relationship between dance form and apparent information. Von Frisch's most startling claim was that the so-called 'waggle dance' (Figure 10.1 b, c) was a sophisticated information transfer mechanism constituting a highly developed language unparalleled among invertebrates. Several experiments bore out von Frisch's claim. When bees were fed at set distances and directions from the hive, the recruits which flew out after their return orientated with remarkable accuracy to the experimental feeding stations.

The Bee 'Language' Controversy. Despite early experimental support for the bee 'language' hypothesis, some later experiments suggested an alternative explanation. These experiments were carried out by Wenner and coworkers and their main conclusion was that recruits located food sources by *olfaction*. Wenner's group did not dispute von Frisch's finding that the 'waggle dance' contains information about the distance and direction of food, merely that recruits make use of this information. Wenner points to several instances where the behaviour of animals contains information decipherable to humans but not used by conspecifics of the signaller. A 'dance' used by blowflies, for instance, conveys information to a human observer about the shape and concentration of a food source, but other flies do not respond to it. Although recruit bees respond to the 'waggle dance', Wenner suggests that it merely stimulates them to search for food. Once they start searching food sources are located by olfaction.

Figure 10.1: The Honey Bee Foraging Dance. (a) The 'round dance' performed on the vertical face of the comb when the food source is within 50 m of the hive. Recruits antennate the dancing forager and are stimulated to search within a 50 m radius of the hive. (b) The 'waggle dance', employed when food is further away than 50 m, incorporates an abdominal waggle phase into the circular dance motion. If the dance is performed on a horizontal surface, the forward direction of the waggle phase acts as a simple pointer to the food source. (c) If the waggle dance is performed inside the hive on the vertical comb face, the waggle phase is orientated with respect to gravity (the vertical), and recruits translate the angle with the vertical into the same angle relative to the sun in order to locate the food. A waggle phase orientated vertically up indicates a bearing directly towards the sun. The waggle phase apparently contains complex information about both the direction and the distance of food sources. Its tempo decreases and its duration and the number of waggles performed increase with the distance of a food source from the hive.

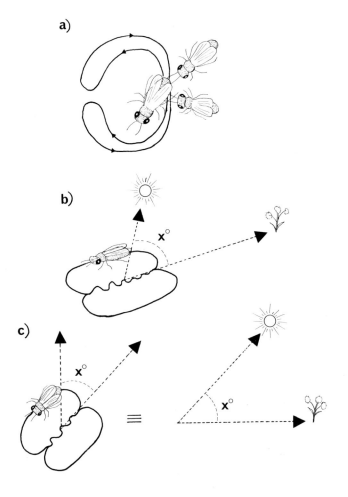

Source: Modified after Curtis, C. (1968). *Biology*. New York, Worth.

To test his olfaction hypothesis, Wenner carried out a series of choice experiments. These were similar in form to von Frisch's except that more note was taken of the olfactory characteristics of food sources and the wind direction during tests. In some experiments, for instance, recruits were allowed to choose between food dishes which had been scented and indicated by dances but were now unscented, and dishes which had never been visited but which were now scented similarly to the original dishes. In these cases, most bees chose novel but scented dishes.

A drawback with this kind of experiment is the high concentrations of scent used. It is now known that artificially high concentrations of scent can stimulate searching behaviour even in the absence of dancing. Nevertheless, other experiments showing the importance of wind direction and the sight and smell of other bees at a food source provide compelling evidence for the use of olfaction in choosing a feeding site. Wenner also pointed out that recruits searching for food dishes often make mistakes. They also tend to take longer to find a dish than might be expected if they had been given precise information about its whereabouts. Such errors are consistent with location by olfactory cues.

The 'information/olfactory cue' dispute continued in the literature for some years (see Gould[121] for an excellent review). The problem was hitting on an experiment which ruled out olfactory cues satisfactorily. Such an experiment, however, was brilliantly designed by Gould[121]. He used the fact that bees respond to a small point of light provided in the hive as if it is the sun. A dancing forager on the vertical comb face thus uses the light in the same way it would the sun if it was dancing on a horizontal surface outside the hive. In other words, it uses the light as a pointer. Hence a 'waggle run' angled at 20° to the left of the light indicates food at 20° to the left of the sun. Gould also made use of the fact that, if a number of ocelli (simple eyes) on the dorsal surface of the head are blacked out, the bee's sensitivity to light is greatly reduced. Bees with blacked ocelli used much stronger light levels to forage and a much stronger artificial light source in the hive to redirect their 'waggle dance'. Gould's experiment depended on having ocelli-blacked foragers dance in the hive where they could not see the light source. Their dances were therefore orientated with respect to gravity, and therefore by extrapolation, the sun. Recruits in the hive, however, could see the light and so interpreted the forager's dance with respect to the light source instead of gravity. When they left the hive therefore, recruits flew at an angle to the sun corresponding to the angle of the forager's 'waggle run' to the light. By changing the position of the light, Gould was able to make predictions about where recruits should go if they were using the erroneous information. As Figure 10.2 shows, most recruits went to the dishes predicted on the basis of the 'wrong information' hypothesis. Despite criticism of the experimental techniques used in the bee 'language' investigation[262] Gould's elegant investigation provides convincing evidence that the 'waggle dance' not only contains information about the direction and distance of food, but that this information is actually used by recruits.

The bee 'language' story shows that relatively simple nervous systems may be

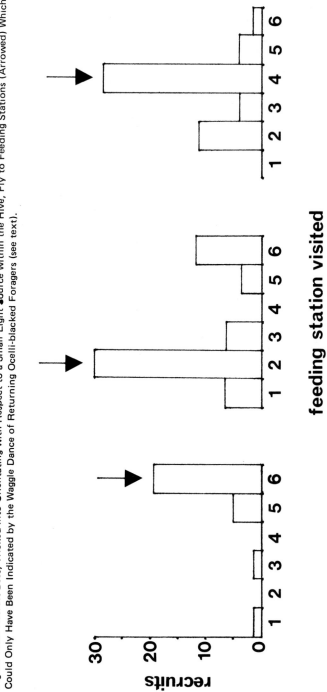

Figure 10.2: Recruit Bees, Tricked into Orientating with Respect to a Small Light Source within the Hive, Fly to Feeding Stations (Arrowed) Which Could Only Have Been Indicated by the Waggle Dance of Returning Ocelli-blacked Foragers (see text).

Source: Modified after Gould[121].

capable of complex, symbolic communication. The story underlines the danger of assuming a simple correlation between the physical complexity of a nervous system and its functional sophistication. It is not only with invertebrates and lower vertebrates that this has tended to be assumed. For a long time, and sometimes still today, one of the key characteristics regarded as separating man from the lower apes is the former's highly complex language. The apparent inability of apes to develop and use a language was held to indicate an enormous evolutionary gulf between the levels of intelligence of ape and man. However, some developments with chimpanzees over the past 20 years have changed all that.

'Language' Learning in Chimpanzees. In the past, there have been several attempts to teach chimpanzees to use spoken human language. In the main, they have been complete failures. Probably the best known was Hayes and Hayes' young chimpanzee 'Viki'. Six years protracted and painstaking work resulted in 'Viki' learning only four sounds which barely approximated English words. The problem is that chimpanzee vocal apparatus and vocal behaviour are very different to our own. The human larynx and buccal cavity are considerably specialised for speech, our main communicating channel. While chimpanzees certainly use a variety of different sounds, they generally only occur at times of high excitement and tend to be very specific. In captivity, undisturbed chimpanzees are often silent. It seems unlikely, therefore, that chimpanzees could be trained to make sophisticated use of vocalisations.

An entirely different approach was taken by Gardner and Gardner[112]. They took an 8–14-month-old chimpanzee and attempted to teach it American sign language (ASL). This is a gesture language widely used in the United States as a means of communication between deaf people. ASL involves manually-produced visual symbols which are analogous to words in spoken language. Just as spoken language can be analysed in terms of phonemes, signs can be analysed as 'cheremes' ASL comprises some 55 cheremes. Nineteen refer to the configurations of the hand(s) making the sign, 12 to the place where the sign is made and 24 to the action of the hand(s). Thus a pointing hand means one thing near the forehead, another near the chin. At a given place, a pointing hand means one thing if it is moved towards the signer, another if it is moved away, yet another if moved vertically and so on.

The Gardners' young chimpanzee 'Washoe' was kept in a room containing a variety of articles normally found in a human dwelling. During her waking hours, Washoe was constantly in the company of humans. Although her human companions took part in her everyday activities — feeding, bathing, dressing, etc. — the only form of 'verbal' communication was by means of ASL. After a time, Washoe acquired several signs which she used in their appropriate context. Furthermore, she paid more attention to and demonstrated that she understood many of the signs her human companions were making.

Many of the signs Washoe learned were acquired by imitation. However, some she invented for herself. In these inventions, the chimpanzee appeared to display impressive cognitive abilities. One particular instance cited by the Gardners is Washoe's request for a bib. Instead of using the sign for a 'napkin' or 'wiper' as she had been taught, she traced the outline of a bib on her chest. Evidence of high cognitive ability in other animals has also been forthcoming. For instance, pigeons trained to recognise human beings in photographs can subsequently do so when presented with only very fragmentary or oddly orientated images, thus showing an impressive conceptual ability. Washoe's capability in this direction was further illustrated by her use of pronouns. When she transferred a sign from herself to another person she frequently accompanied it with gestures indicating 'yours' (usually pointing at possessions of the referent). Although Washoe had clearly not been shown everybody who could be described as 'you' she was able to generalise the pronoun to mean 'other than mine'. Three or four years after the start of training, Washoe was signing simple phrases like 'you me out' or 'you me go out hurry' when passing through a doorway. Comparing Washoe's acquisition of 87 signs similar to those of ASL within three years with Viki's acquisition of four dubious vocalisations, it is clear we need to be more careful in our assumptions about the potential of animal communication. Washoe is not alone in her achievements. More recent experiments,[248] in which a chimpanzee 'Sarah' was taught to communicate using plastic tokens as words, have been just as successful, if not more so, than the ASL experiments with Washoe. As Manning[199] points out, these achievements have revolutionised our estimates of what a 600 ml brain can achieve.

Another distinction which has been drawn between human and non-human communication concerns the possibility of deception.[199] Humans, it has been said, can deceive, but what an animal communicates is a direct reflection of its internal state. More recent ideas, however, suggest that animals are not the unwitting dupes of their nervous systems.[85] Signal systems are likely to evolve which skilfully disguise the real internal state or 'intentions' of the animal. Threat displays, for instance, disguise the animal's urge to flee by making it appear aggressive. Female fireflies (*Photinus* spp.) may use the mating signals of another species to attract heterospecific males which they then eat. Such signals are hardly true reflections of the animal's motivational state. To understand why we can view animal communication as potentially deceitful — indeed why we might *expect* it to be so — we must speculate about the selective pressures shaping it.

10.2 Communication as Exploitation

In a recent stimulating paper, Dawkins and Krebs[85] departed from what they saw as the 'traditional' ethological view of communication and suggested an entirely new way of looking at communication. Hinde[144] has criticised Dawkins and Krebs' resumé of the ethological view of communication and we shall return to his points later. To begin with, we shall briefly examine what Dawkins and Krebs regard as the 'traditional' approach.

Ethologists have characterised the sender and receiver of a communicatory signal as *actor* and *reactor* respectively. Frequently, though by no means always, both parties have been seen as benefiting from signal exchange and evolving to enhance the efficiency of signal transfer. Such a view is still sometimes put forward.[220] The evolutionary improvement in transfer efficiency has been embodied in the concept of *ritualisation*. Through ritualisation, the signal is honed and elaborated and changes from its original form in which it may have served some other 'function' (see Section 10.1.1). Communication is thus seen as a means by which reactors can predict the future behaviour of actors and act appropriately. Moreover, actors are selected to inform reactors of their internal state to make it easy for reactors to predict their future behaviour. Both parties benefit from signals that are as efficient, unambiguous and informative as possible.

When we consider how natural selection is likely to act (see Chapter 3), however, we arrive at a rather different interpretation of communication. In Chapter 3, we saw that animals can be pictured as machines designed to preserve and propagate their genes. As a means to this end, they may manipulate objects — both biotic and abiotic — in the environment. We can regard object manipulation as an extended expression, or part of the *extended phenotype*, of the animal's genes (see Section 3.4.3). Can such a view provide a more satisfactory explanation for communication mechanisms than the so-called traditional view?

10.2.1 Communication as Manipulation

When an animal seeks to manipulate an inanimate object it can do so only by physically moving or handling the object. When the object to be manipulated is another living organism, however, it has an alternative. Instead of spending time and energy in manhandling the organism, it can influence the organism's behaviour to achieve the same results. A male cricket, for instance, does not physically drag a female along the ground to its burrow; it sits at its burrow entrance and sings. The female responds to the song by moving towards the male. From the male's point of view, communication is clearly more efficient than trying to acquire a female by force.

The economy of communicating is by no means a new discovery. Haldane recognised the pronounced efficiency of signalling nearly 50 years ago. Dawkins and Krebs, however, show how signalling efficiency might enable an actor to manipulate the muscle power of a reactor to its own advantage by using the example of a man and a horse-drawn plough. A man's muscles are too weak to pull a plough. By a mixture of sensory stimulation and positive and negative reinforcement, however, the man can manipulate the behaviour of the horse so that it pulls the plough for him. A small tug on the reins results in a large change in the direction of ploughing. In Dawkins and Krebs' view, communication is no longer a co-operative exchange of information between actor and reactor, but an actor manipulating reactors to its own advantage. In many cases, the 'manipulation theory' of communication can only provide a feasible alternative to the more mutualistic ethological approach. Its test as a superior model lies in its ability to provide explanations in areas where ethological theories run into difficulty. One such area concerns the evolution of avian alarm calls (see Chapter 7).

Alarm Calls and Manipulation. As we saw in Chapter 7, several explanations have been put forward concerning the evolution of alarm calls in flock-forming birds. Alarm calls have attracted the attention of evolutionists because, at first sight, they appear to be altruistic. The act of calling seems to increase the survival chances of nearby companions but place the caller in greater immediate danger (by giving away his position).

If members of a flock are closely related, 'altruistic' calling could evolve by kin selection (see Chapters 3 and 11). However, many flocks are transient in membership and it is unlikely that close relatives would tend to remain together long. Trivers[314] suggested that alarm calls could evolve without kin selection if predators are less effective or discouraged when attacking flocks containing callers. Although Trivers' argument could account for the initial spread of calling, selection will cease to favour calling once caller frequency is high enough for most flocks to contain one caller. Most flocks, however, appear to have more than one caller, and there is no reason to suppose that all birds in a flock are not callers. Calling may be able to spread if individual fitness increases with the number of callers in the flock but, for this to be the case, the cost of calling would have to be low. While there are reasons why it might be low (some calls are difficult to locate[200] or even ventriloquial[244] and individual penalty may be reduced by being spread across a number of callers), it would be more satisfactory to be able to explain alarm calls in terms of a positive advantage to the caller. Such an explanation can be found using the manipulation hypothesis.

Charnov and Krebs[52] point out that an individual spotting an approaching predator would be in a good position to avoid being caught by manipulating its flock mates. Having seen the predator, the bird has two pieces of information: (1) there is a predator approaching and (2) the predator's position. By calling, the bird passes on only the first part of this information. Non-callers are told simply that there is a predator and fly to cover without respect to the predator's direction of approach. The caller can then enhance its own safety by flying in to the centre of the flock or on the side opposite to the approaching predator or can benefit from the 'confusion effect' (see Chapter 9).

As stressed before, 'manipulation' is used to mean only that the caller benefits most of all from its action. Clearly, other flock members do not act specifically to enhance the safety of the caller. They respond for their own benefit but in doing so benefit the caller even more.

Hinde[144] criticises Dawkins and Krebs' interpretation of communication on the grounds that it is appropriate only to a limited range of interactions.

'Manipulation' may be appropriate for e.g. prey-warning displays to predators where reaction is likely to be to the reactor's detriment, but it is hardly appropriate to obviously 'co-operative' relationships such as between members of a mating pair. The problem with Hinde's criticism, however, is that it focuses only on the overt nature of the interaction (e.g. whether both parties or only one benefit) rather than on the selective mechanism shaping it. A simple hypothetical example illustrates the point.

Suppose in a mixed-species flock of birds, individuals of species A benefit from mimicking the alarm call of species B. Alarm-call mimicry results in both A and B individuals flying to cover so that A birds become part of a larger airborne flock than if they had flown up on their own. As in Charnov and Krebs' single species model, A individuals may in this way reduce their risk of capture by a predator. The fact that B individuals may also benefit from reacting to the alarm calls of A does not alter the exploitative nature of the selection pressure favouring mimicry by A. At first sight the shared alarm system appears to be mutualistic. Indeed it has been argued that such signal convergence represents a mutually adaptive signal economy measure.[220] In reality it may be a purely exploitative relationship, but one in which self-interested exploitation by the actor incidentally benefits the reactor as well. The 'signal' economy argument also runs into another difficulty in that it focuses on the signal recognition ability of reactors rather than signal characteristics of actors. It is thus difficult to see how it could account for interspecific convergence in signal characteristics.

10.3 Assessment

In the last section, we developed the idea that communicatory acts may not reflect animals helplessly transmitting information about their motivational state. Instead, they may reflect the selective limiting of information transfer which enables animals to exert a self-interested influence over others sharing their environment. The coevolved relationship between actor and reactor is best seen as an arms race (see Chapter 11) of mutually counter-adaptive strategies. Exploitative signalling by the actor leads to counter measures (e.g. change of response to a given signal) by the reactor which in turn leads to signal modification by the actor and so on. In this way complex and apparently co-adapted signal/response relationships may build up over evolutionary time. The cues used during aggressive disputes are a good example.

10.3.1 Symmetrical and Asymmetrical Contests

The signalling conventions which animals often use to settle disputes have long intrigued ethologists. For some time, these conventions were interpreted in group selection terms. Species as a whole benefited if individuals settled disputes without damaging fights. It is only relatively recently that the evolution of conventional displays during contests has been couched in more satisfactory gene selection terms. In Chapter 3, we considered a simple gene selection model which illustrated how conventional ('dove') and escalated ('hawk') strategies might spread through a population. In that model, the *evolutionarily stable* mixture of conventional and escalated strategies depended largely on the injury cost of escalation.

An important assumption of the 'hawk'/'dove' model was that contestants were equal in fighting ability, resource requirement and so on. The only difference between them was the strategy they adopted during disputes. It is therefore reasonable to talk of contests between individuals of equivalent competitive ability and motivational state as 'symmetrical' contests.

'Wars of Attrition' and Other Evolutionary 'Auctions'. In some cases, animals may have no means of escalating effectively (for instance they may not possess any 'weapons' like horns or teeth). Here contests may have to be settled purely by conventional means. It is the time and effort which each contestant puts into display that eventually decides the winner. The crucial decision an animal has to make therefore is how long to persist in display before giving up. Clearly, if all contestants persisted for the same length of time, any individual carrying on for just a little bit longer would always win (although there would be an upper limit to display time set by the costs of display relative to the value of winning). The ESS in such a 'war of attrition'[203] is therefore unlikely to be a pure strategy of display time. Instead, we should expect individuals to display for an unpredictable length of time so that opponents cannot anticipate how long to continue. One way of doing this would be to display with a random distribution of persistence times. Some of the data so far gathered bear this out. Data for the duration of contests between female land iguanas (*Iguana iguana*), for instance, conform roughly to a negative exponential distribution[47] (one type of random distribution).

In a stimulating re-analysis of data from ethological literature, however, Caryl[47] has found that distributions of contest duration for a variety of species differ from the negative exponential. One possible reason for this is that the form the distribution of durations takes may depend on the shape of the *cost function* (see Chapter 2) for displaying[226] (Figure 10.3a). If the cost of display goes up sharply with time, the evolutionarily stable distribution of durations may be almost bimodal. Data from contest durations in Siamese fighting fish (*Betta splendens*) and brownheaded cowbirds (*Molothrus ater*) (Figure 10.3b) reinforce the need to identify the form of the cost function for displaying before interpreting the distribution of persistence times.

While the 'war of attrition' as initially modelled by Maynard Smith[203] may be oversimple for the majority of animal contests, the idea that aggressively motivated displays should give away little about the intentions of the performer certainly holds good. Caryl's re-analysis of threat display data in birds shows quite clearly that components preceding attack predict attack much less reliably than other components predict other behaviours (e.g. escape). The performance of displays at 'typical intensity' (i.e. not varying in form or vigour until one individual gives up) is also well documented in the ethological literature (see also Section 10.1).

The 'war of attrition' is, in a sense, an evolutionary analogue of a commercial auction. Individuals contest a resource which goes to the highest bidder. A similar type of evolutionary game is the *Scotch auction*.[237] Here each contestant makes a fixed bid which is unknown to the other. As in the 'war of attrition' the highest bidder wins but both bids are paid in full. An interesting possibility arising from the 'war of attrition' and 'Scotch auction' is a contest in which an individual makes a reserve bid. A reserve bid is a bid the individual will never exceed, but, if its opponent withdraws before the bid is reached, it may not persist to pay the bid in full. Such reserve bid strategies are likely to be acted on by natural selection to reduce the extent to which the opponent's bid is exceeded. Because of the nature of many types of display behaviour, it may not be immediately apparent to the high-bidding contestant that the low bidder has ceased to contest the resource. During the resulting 'noticing period', the high bidder incurs an increasing 'overshoot' cost (the time and energy costs of persisting longer than necessary). Mutant individuals which have improved perceptual abilities and can reduce the 'noticing period' will thus be favoured. Haigh and Rose[126] explore the evolutionary implications of different types of 'overshoot cost'. One conclusion they come to is that the 'war of attrition' and 'Scotch auction' type of symmetrical contest represents the very simplest model for the evolution of animal display behaviour. Moreover a population of 'Scotch auction' strategists is always open to invasion by mutant 'reserve-bidders' and hence cannot be an ESS. It is therefore likely to be at best a transient phase in the evolution of species' assessment strategy.

For various reasons, strictly symmetrical contests are unlikely to be common in the real world. There are a number of ways in which the outcome of a contest might be biased in favour of one or other contestant. Maynard Smith and Parker[207] divide these into three broad categories: (1) *resource holding potential (RHP)* asymmetries (contestants differ in their competitive ability), (2) *pay-off asymmetries* (contestants differ in their expected benefit from gaining the resource) and (3) *uncorrelated asymmetries* (contestants differ in some way which is unrelated either to RHP or pay-off but which can be used as an arbitrary cue to settle contests — rather like humans settling a dispute by tossing a coin).

RHP Asymmetries. If two individuals differ in RHP (i.e. in their ability to fight and contest a resource), the weaker one should withdraw as soon as it assesses its chances of winning are low. Continued persistence or escalation is unlikely to win the contest. In many cases, certain features of an opponent provide information about its RHP. If so, contestants can use these to indicate how they are likely to fare before getting involved in a fight (see also the discussion of 'assessment' hierarchies in Chapter 9). What kinds of cue might an animal use to assess its opponent's RHP?

Figure 10.3: (a) Three Hypothetical Cost Functions for Display Persistence and their Associated Frequency Distributions of Display Times and 'Hazard' Functions (the Probability that a Display Will End at a Given Time). Hazard functions might give some indication of the cost of continuing to display. (b) (i) Frequency Distribution of Display Time and Hazard Function for Displaying Male Siamese Fighting Fish (*Betta splendens*). (ii) Three Frequency Distributions and Associated Hazard Functions for Displaying Male Cowbirds (*Molothrus ater*) during Territorial Disputes.

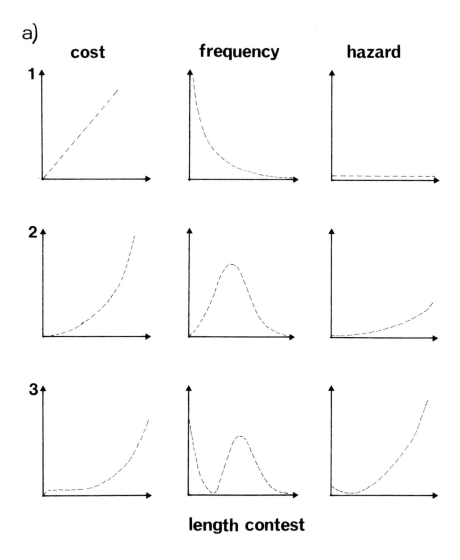

a)

cost   frequency   hazard

length contest

Source: Modified after Norman *et al.*[226]

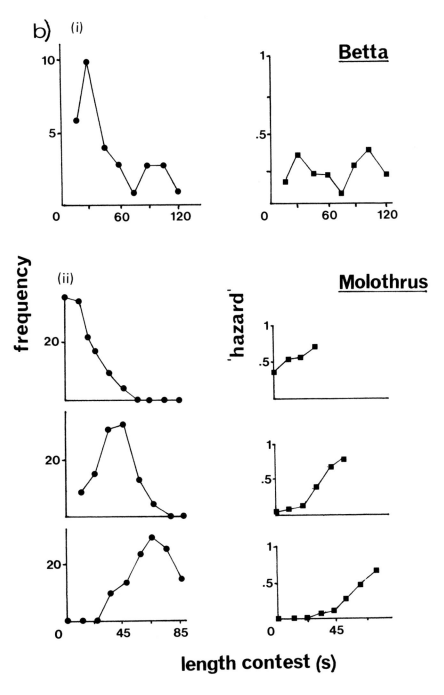

Source: Modified after Caryl[47].

Obviously the best cues to use are those like horn and body size which correlate in some way with RHP. Unlike arbitrary cues, such 'status-limited' cues will be resistant to cheating. Individuals which do not have genuinely high RHP will not be able to mimic high-status cues. Relative body size appears to settle contests directly in hermit crabs (*Clibariarius vitatus*) and less directly via 'trials of strength', in, for instance, water buffalo (*Syncercus caffer*) and foxes (*Vulpes vulpes*). Vocalisations may also impart information about RHP. The pitch of a call depends largely on the size and construction of the animal's resonating chamber. In general the larger the animal, the larger its vocal resonating chamber and the deeper the pitch of its call. Many vocalisations are also energetically expensive to produce. Calls may therefore provide a reliable indication of an opponent's size and strength. Evidence that calls are used as assessment cues comes from red deer and common toads.

Red deer stags usually settle contests over the possession of harems by prolonged roaring 'duels' instead of actually fighting.[57] While one reason for roaring rather than fighting is that fighting involves risk of serious injury, the main reason is probably that roaring is expensive to perform. 'Duelling' opponents roar at a faster and faster rate until, eventually, one gives up and withdraws (Figure 10.4). Roaring at high tempo is extremely exhausting, and even large high-status stags give up 'duelling' with tape-recorded roars when the tempo exceeds a certain rate. In a series of experiments with common toads, Davies and Halliday[74] showed a direct relationship between the pitch of croaking and the probability of a mating male being ousted from the back of a female by an unpaired rival. Rival males attempting to dislodge paired males were deterred from attacking in proportion to the depth of the paired male's croak.

While some animals use assessment cues like body size and call pitch which are relatively simple, others appear to use more complex cues. Barnard and Burk[15] discuss a mechanism by which such complex assessment cues might have evolved. Their model makes the reasonable assumption that potential assessments are tested by 'probing' (persisting in an encounter to see whether an individual's apparent status is real). Because of the advantages gained from high status, low-status cheats are likely to arise which mimic high-status cues provided they are not status-limited. The frequency of probing is likely to be low when there are few cheats in the population and most apparent high-status individuals are genuine. This is because probing is likely to be costly in terms of risk of injury if a genuine high-status individual is challenged. If the frequency of cheats increases in the population, however, the costs of probing go down and the benefits up. The frequency of probing is therefore likely to increase with the frequency of cheats. Barnard and Burk suggest that the resultant increased interference costs to genuine high-status individuals and the risk of challenging high-status individuals will result in a breakdown of the current assessment mechanism. There will be selection to back-up the now unreliable current assessment cue (say vocal pitch) with a more reliable 'reference' cue (say body size). Because reference cues are also potentially open to cheating, the process

Figure 10.4: Roaring Contest between Two Male Red Deer Stags, P and F. F keeps pace
with P until a critical roaring rate is reached, whereupon he gives up.

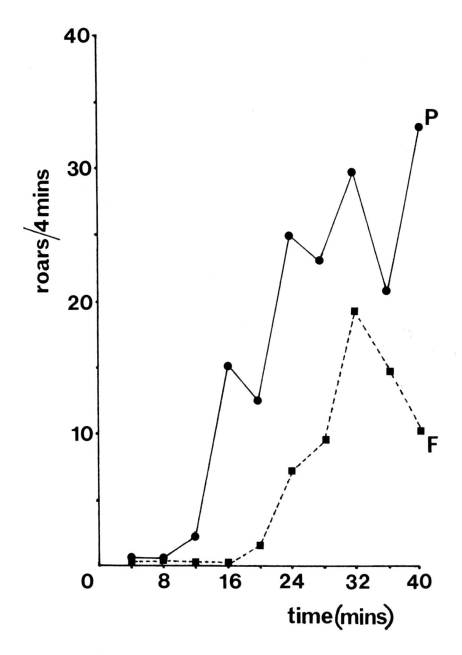

Source: Modified after Dawkins and Krebs[85] from data by Clutton-Brock.

may continue with the addition of several reference cues (see Figure 9.7). With each addition cheating becomes more difficult because multiple cues have to be faked. Eventually costly status-limited cues may arise which halt the 'arms race' between assessors (distinguishing true from false status) and cheats (mimicking high status). Although we can talk of assessors and cheats as if they were qualitatively different types of individual, in reality individuals play both roles.

There is good experimental evidence that animals use the type of 'compound' assessment cue predicted by Barnard and Burk. In Harris' sparrows, for instance, there is a correlation between an individual's status in winter flocks and the size of the black bib of feathers on its breast. Rohwer[258] tried to create cheats by enlarging the bib (apparent status cue) of low-status birds with black dye. Instead of winning more disputes as we might have expected, however, Rohwer's cheats were merely involved in more escalated fights which they ultimately lost. The assessment mechanism was clearly more complex than bib size alone. In a further series of experiments, Rohwer and Rohwer[260] implanted testosterone in low-ranking birds as well as enlarging their bibs. Now birds tended to win escalated contests and consequently increased their status.

In the Harris' sparrow example, the assessment cue appeared to be bib size backed up by taking actual aggressiveness into account. Cheats who failed the 'acid test' of living up to their declared status were penalised. A similar situation was found in common toads by Davies and Halliday. Although croak pitch was a good predictor of the outcome of a dispute, Figure 10.5 shows that this was not the whole story. When paired males were prevented from croaking and tape-recorded croaks played instead, unpaired males were deterred more by large males than small males when both were presented with deep croaks. Assessment in toads therefore appeared to be based on a combination of croak pitch and body size. The reason for also using body size may be that despite a certain dependence on body size, croak pitch is actually cheatable. There is evidence that pitch is affected by temperature. Small toads can produce deeper croaks if temperatures are low. By croaking more during cold days or early in the morning, small toads can sound like large ones. If they are not to be deterred by a small male, therefore, unpaired challengers should check out the call. Similarly, females should also check the apparent information in calls to avoid mating with sub-optimal sized males (see Chapter 8).

Pay-off Asymmetries. Another factor which may influence the outcome of a contest is the relative value of the disputed resource to the contestants. The individual who most requires the resource may be willing to escalate the contest further.

Evidence for such pay-off asymmetries comes from land iguanas (*Iguana iguana*). Female iguanas lay eggs in burrows. In some cases, a female will try to usurp a burrow dug by another female. If disputes arise, we might expect contestants to escalate or persist more if the burrow is deep. Deeper burrows require less work to prepare them for egg-laying. Data collected by Rand and

Figure 10.5: When Male Common Toads, Trying to Oust a Paired Male from the Back of a Female, are Confronted with a Silenced Defender, their Attack Intensity is Determined both by the Pitch of the Pre-recorded Croak they Hear and the Size of the Defending Male.

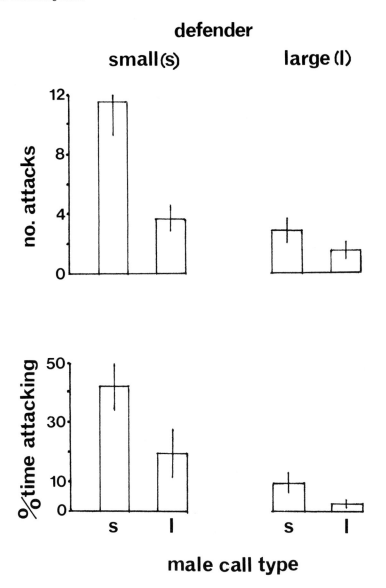

Source: Modified after Davies and Halliday[74].

Rand[252] suggest that both intruder and owner females escalate more when the disputed burrow is deep. Contests over deep burrows tended to involve bites, lunges and chases, while those over shallow burrows merely involved head-swinging and gaping. Interestingly, however, the correlation between degree of escalation and burrow depth was better for owners than intruders. This may be because the owner has better information about and is therefore better at assessing, the value of the burrow. This difference in information may be one reason why owners tend to win disputes.

A similar but more intriguing pay-off asymmetry has been found in digger wasps (*Sphex ichneumoneus*).[84] Like the iguanas, female wasps dig burrows in which to lay eggs. This time only a single egg is laid per burrow and the female provisions the future larva with paralysed katydids (Insecta: Orthoptera) before moving off to start a new burrow. Occasionally (in about 5-15 per cent of cases), a second wasp moves into a female's burrow. In these 'co-occupied' burrows, both females provision the egg chamber and lay an egg. Although both co-occupying wasps provision the burrow full-time, they rarely meet because most of their time is spent hunting. When they do meet, however, they fight.

At first sight, we might not expect any pay-off asymmetry biasing the outcome of fights here because the burrow is apparently worth the same to both contestants and both contestants apparently have the same information about its contents. Dawkins and Brockmann[84], however, found that wasps were not basing their decision about how to fight, on the total number of katydids in the burrow (as we should have expected because this represents the actual utility of the burrow to the winner), but on the number of katydids they had stored themselves. The winner tended to be the female who had contributed most katydids to the burrow (Figure 10.6).

This finding can be interpreted in two ways. First, the wasps may be basing their decision to fight on past investment (the effort spent provisioning the burrow) rather than the utility of the burrow if it is won (the total number of katydids it contains). If they are, then they are committing the so-called 'Concorde fallacy' (fighting so that past investment is not wasted). Secondly, the wasps may not be committing the 'Concorde fallacy' but may be deciding on the basis of a rule of thumb. It may be that wasps are unable (perhaps because of the costs of developing appropriate sensory and neural capacity) to count katydids in the burrow. This is a constraint which affects their decision to fight. One way round the constraint, however, is for a wasp to use its knowledge of its own provisioning trips as an index to how many katydids there are in the burrow. The wasp which brought the most would rate the burrow more highly and fight more vigorously to retain it. Like the shrews using prey size as a convenient guide to profitability discussed in Chapter 6, Dawkins and Brockmann's wasps may have been using the most economical approximation of burrow utility.

A combination of pay-off and RHP asymmetry has recently been found in dung fly (*Scatophaga stercoraria*) contests.[287] Here unpaired males attempt to

Figure 10.6: When Female Digger Wasps Contest a Jointly-owned Burrow, the Winner Tends to be the Female who has Provided the Most Katydids.

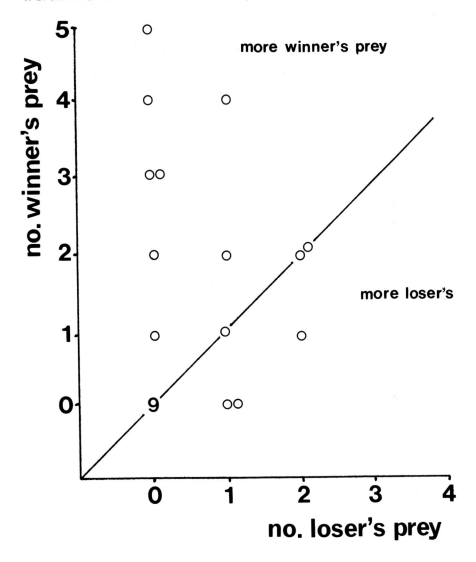

Source: Modified after Dawkins and Brockmann[84].

Plate 3: Fighting between Male Dung Flies (*Scatophaga stercoraria*) Competing for Access to a Female. The attacker male on the left is attempting to push the paired male away from the female.

Source: Photograph courtesy of Dr G.A. Parker and E.J. Brill (Publishers).

usurp females from paired males (Plate 3). The resistance a paired male puts up is negatively related to the time it has been guarding the female. Guarding time is inversely proportional to the number of eggs remaining to be fertilised. The female is thus worth more early on in the guarding phase and paired-male persistence increases accordingly. In addition, paired-male size in relation to the female and to an attacker and the firm grasp he has on the female jointly constitute an RHP asymmetry. Only 20-25 per cent of takeover bids are successful.

Uncorrelated Asymmetries. The third type of asymmetry defined by Maynard Smith and Parker involves neither RHP nor resource value differences between opponents. Because of the potentially high costs of escalated fighting, it may pay contestants to settle a dispute arbitrarily even if there are no important differences between them.

One arbitrary cue which could easily be used to settle a dispute is resource ownership. Although, as we have seen, ownership may bring about RHP or pay-off asymmetries it could be used as a deciding convention in its own right. Contestants could then save valuable time and energy for use in other ways. Because such conventions are not related to any physical or motivational difference between individuals, they are known as uncorrelated asymmetries. A classic example of an uncorrelated asymmetry was found by Davies[71] in the speckled wood butterfly (*Pararge aegeria*).

Davies found that male butterflies defended sunspots on the floor of a wood. Here they courted females which visited the spots. Davies noticed that intruder males always retreated after a brief aerial contest with the owner. Davies's most interesting finding, however, was that the outcome of a dispute could be reversed by allowing an 'intruder' a few seconds' residency. If two males were trained separately to regard the same sunspot as their territory, the aerial disputes which ensued when they were allowed to meet lasted much longer than usual. Davies was unable to find any RHP or pay-off asymmetry which could have been established by only a few seconds' sunspot ownership. Intriguingly, there is evidence from certain ground-dwelling New World spiders and European Lycosid spiders that the opposite convention 'intruder wins' sometimes holds. If animals really are just using an arbitrary convention, 'intruder wins' should be just as likely as 'owner wins'.

Truly uncorrelated asymmetries, however, may be rare in nature. Most environments are likely to be such that resource ownership automatically confers some physical or motivational advantage on the owner (e.g. through improved local knowledge of food supplies). Uncorrelated asymmetries are most likely when contested resources are fairly abundant and the cost of abiding by the convention for intruders (or residents) is low.

Depending on physical and motivational disparities between contestants, therefore, a variety of different types of convention can provide evolutionarily stable alternatives to overt aggression. Arbitrary conventions may even evolve when there are no obvious differences between contestants. Evolutionary games

Figure 10.7: Field Experiment Testing the Hypothesis that Song in Male Great Tits is a Territorial 'Keep Out' Signal. In Experiment 1, territorial males were removed from a wood and areas of the wood designated as experimental (tape-recorded great tit song was played through loudspeakers on a near natural spatio-temporal schedule), control sound (simple musical notes were played through loudspeakers) and control silent areas. The pattern of reoccupation of the three areas (stippled areas) by birds from the surrounding hedgerows was then monitored. Experiment 2 was a repeat of Experiment 1 except that the experimental and control areas were changed around.

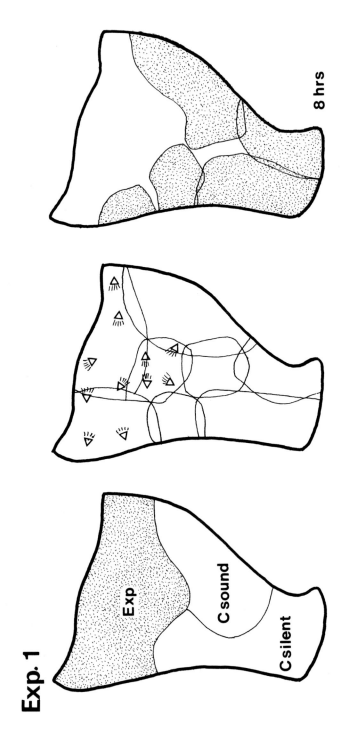

Exp. 1

Exp

C sound

C silent

8 hrs

Exp. 2

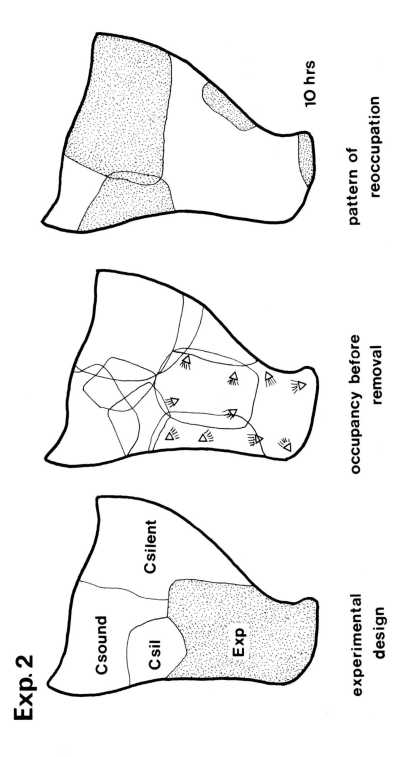

experimental design

occupancy before removal

pattern of reoccupation

10 hrs

theory models of the type discussed above and in Chapter 3 provide a powerful tool for analysing communication between animals. As Haigh and Rose point out, however, games theory models have so far barely scratched the surface of the communication problem. Nevertheless, the results to date offer great encouragement for their further application and extension.

10.4 The Function of Communication

While we know quite a bit about the ontogeny and mechanics of animal communication we are a good deal less certain about the role communicatory acts play in an animal's everyday life. In this section, we shall examine some aspects of the ecology of communication. Because communication is such a vast topic, however, a full discussion of its ecology would constitute a book in itself. We shall therefore examine some recent work on bird song which illustrates very nicely how field experiments and observation can test hypotheses. Bird song is also useful as a model because it is familiar and its ontogeny (see Chapter 4) and mechanics have been studied in some detail.

10.4.1 The Function of Bird Song

It has long been supposed that bird song serves one or both of two functions: territorial advertisement and mate attraction. Until fairly recently, however, the basis for this supposition was little more than a broad correlation between peaks of singing and territorial and/or reproductive activities. Very little *direct* evidence existed for either 'function'. In the last few years, work on a number of species, particularly tits and warblers, has tested the territoriality and attraction hypotheses more rigorously.

Bird Song and Territoriality. One of the best quantitative studies of the territorial 'function' of bird song is that of Krebs[169] on song in male great tits.

In great tits, the peak of singing activity coincides with the establishment of territory. However, territories are not usually established until after males have paired, suggesting that song has little to do with the attraction of mates. Krebs therefore tested the possibility that song functioned as a 'keep out' signal to intruders. In an earlier study, Krebs had shown that, when owners were removed from their territories in preferred woodland habitats, they were rapidly replaced by new birds moving in from surrounding hedgerows (see Chapter 5). Such rapid replacement showed that competition for territory ownership was fierce. Krebs used this to investigate the role of song in territorial defence. Figure 10.7 shows the owner removal experiment carried out by Krebs in a small wood just outside Oxford.

The main conclusion is that, although all areas were eventually fully occupied, the two *control* areas were occupied much more rapidly than the experimental area. To make sure this was not just because birds preferred certain parts of the wood, Krebs repeated the experiment with different areas being designated *experimental* and *control*. The results were the same. Areas in which recorded song was broadcast were the last to be reoccupied.

The loudspeaker experiments showed quite clearly that song in great tits has a role in territorial maintenance. Potential settlers appear to use song as a means of assessing the occupancy of an area. That recorded song was not a perfect 'keep out' signal (experimental territories were eventually reoccupied) may have been due to two things. First, recorded song is a sub-optimal stimulus in that it does not respond immediately to the call of the intruder. Such 'matching' is an important and common feature of territorial song. Secondly, song is only one means of advertisement. If it is not backed up with other signals, e.g. the visual stimulus of the owner itself, intruders may call the 'owner's' bluff and move in.

Territorial advertisement may also underlie an intriguing feature of bird song in some species, its tendency to become elaborated either into a highly complex single song or, more commonly, into a repertoire of relatively simple but different songs. Explanations based on territorial defence involved, until recently, two central ideas. Elaborate song and song repertoires might prevent habituation (see Chapter 4) by intruders. While habituation may indeed occur more easily with a simple signal, it is difficult to see why, in this context, habituation should occur at all, since song effectively warns intruders that they can expect a fight if they intrude further. Song repertoires may also allow matched counter-singing (the use of song types similar to those of an intruder). Counter-singing may signify that the owner has detected a particular intruder and thus constitute a more effective 'keep out' signal. Matching with recorded song types certainly elicits a stronger response in singing birds.

Recently, Krebs[170] has suggested a new and plausible explanation for the evolution of song repertoires based on territorial defence. Krebs pointed out that one way a potential settler could assess the density of residents in an area is by the number of different songs it hears. If a single territory male can give the impression that it is really a number of different males, intruders will be less likely to attempt to settle there. An easy way to do this would be to sing different types of song. Intruders are unlikely to stay long enough to learn the repertoires of each resident because if they stay too long in one habitat, they decrease their chances of finding space in another. This 'Beau Geste' effect (so-called after the famous character in C.W. Wren's novel) has some empirical support. Tits with large repertoires (up to seven songs) tend to have larger territories. Birds also tend to change perches when they change song, a response which would enhance the deceiving effect of a repertoire. Yasukawa[342], using song-playback techniques, tested the 'Beau Geste' hypothesis in red-winged blackbirds. He found that playing repertoires rather than single songs decreased the trespass rate on experimental territories. Recently, McGregor and coworkers have shown substantial reproductive advantages accruing to male great tits with larger repertoires. Figure 10.8 shows that males with larger repertoires put more breeding young into the next generation, although if lifetime reproductive success

Figure 10.8: (a) Male Great Tits with Larger Song Repertoires Put More Breeding Young into the Next Generation. (b) Taking Lifetime Reproductive Success as a Whole, Male Tits with Intermediate-sized Repertoires Did Best.

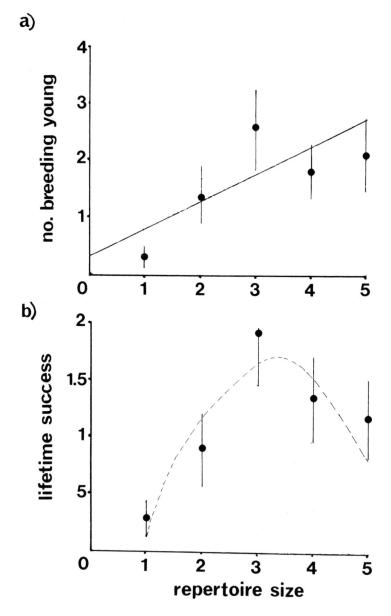

Source: Modified from McGregor, P.K., Krebs, J.R. and Perrins, C.M. (1981). Song repertoires and lifetime reproductive success in the great tit (*Parus major*). *Amer. Nat., 118*: 149–59.

is plotted against repertoire size, it is males with intermediate-sized repertoires which come off best.

Sexual Selection and Bird Song. Darwin originally coined his theory of sexual selection to account for the evolution of complex structures and behaviour patterns which are associated with mating. Complex vocalisations are an obvious candidate for consideration and indeed much work has been done on the relationship between vocal behaviour and breeding.[50,179] In a study of *Acrocephalus* warblers, Catchpole[50] explicitly tested the idea that males with more complex songs were more attractive to females and that polygynous species have evolved more complex songs than monogamous species. Catchpole correlated male repertoire size with a number of variables like date of arrival on the territory and time of pairing. The most convincing relationship was a negative one between repertoire size and pairing date. Males with larger repertoires tended to mate earlier. There was also a weak positive correlation between repertoire size and territory quality. When he examined the relationship between species' repertoire sizes and mating systems, however, Catchpole found that the most complex songs tended to be those of monogamous species. The explanation for this apparent paradox may lie in the relationship between female choice and parental investment (see Chapter 8) in the various *Acrocephalus* species.

Polygynous species are likely to be those in areas of food abundance where deserted females will be able to raise offspring unaided. As we saw in Chapter 8, female choice in situations where males control access to vital resources and hence multiple mates (resource defence polygyny), is likely to be based on quality of territory rather than of the male. Conversely, in poorer areas where polygyny is unlikely to evolve and mating systems tend to be monogamous, choice will probably be for male characteristics. This, Catchpole argues, is the reason why males of the monogamous *Acrocephalus* warblers like the marsh (*A. palustris*) and reed (*A. scirpaseus*) warblers have elaborate songs while the songs of polygynous species like aquatic (*A. paludicles*) and great reed (*A. arundinaceus*) warblers are much simpler. Catchpole's findings do not conflict with the idea that song complexity has evolved through sexual selection.[179] They merely indicate the importance of separating mate choice on the basis of mate quality and choice on the basis of territory quality.

However, an important alternative explanation for varying song complexity, which has nothing to do with either territorial defence or sexual selection, has been put forward by Slater and coworkers.[289] Working with chaffinches, Slater provides evidence that song repertoires may vary through random 'copying errors' during learning. Far from being an adaptive, directionally-selected trend, repertoire change occurs rather like mutation-generated variation in a gene pool. We must therefore be careful when attributing specific functions to variation in behaviour even though adaptive explanations are still the most likely. Furthermore, being able to attribute a particular 'function' to a behaviour does not preclude it from performing other 'functions'. Krebs *et al.*[173] for instance, found that, even though male great tits sung more *after* pairing and appeared to use song in territorial defence, the removal of paired females resulted in a six-fold increase in their level (though not complexity) of singing. Deprived males who sung more also rapidly attracted new females. Great-tit song may, therefore, have a dual 'function' in territory defence and mate attraction.

Environmental Acoustics and Bird Song. Communication mechanisms have to function in an environment. We might therefore expect selection to shape communicatory mechanisms to be maximally efficient in their home environment. A major problem posed by the environment is signal *attenuation*. Because birds live in a wide variety of different habitat types and many species communicate vocally, bird calls and song provide a good opportunity to examine the relationship between environmental and communication characteristics.

In an ideal (for sound transmission) environment, without vegetation or boundaries like the ground, and through a homogeneous, frictionless medium, sound waves diverge in a sphere from their source. Because of this spherical divergence, sound intensity drops at a rate of 6 dB per each doubling of distance owing to the inverse-square law. Natural environments, of course, are far from ideal. Absorption by air, ground, vegetation and other obstacles reduces or redirects sound energy. In a stimulating paper, Morton[217] examined the acoustics of a range of natural environments and related them to the physical characteristics of bird vocalisations used for long-distance communications.

Morton tested sound propagation in neotropical forest interior, forest edge and open grassland habitats to quantify pure tone and random noise band sound transmission. He then recorded the vocalisations of bird species in each habitat and analysed them to determine emphasised sound frequencies (the frequencies used most over the time of recording), frequency range and sound type (whether pure tones or modulated). Morton found that interior forest habitat differed from forest edge and grassland habitats in that a specific range of frequencies (1,585–2,500 Hz) suffered less attenuation than frequencies above and below this range. Sounds recorded from species in interior forest habitat were pure-tone-like in quality and showed emphasised frequencies averaging 2,200 Hz (Figure 10.9). They thus fell within the frequency range expected on the basis of the sound propagation tests.

In forest edge habitat, a wide range of frequencies had very similar attenuation properties. Pure-tone and random-noise band sounds differed little in their attenuation rates. The vocalisations recorded in forest edge birds showed a high degree of variability in emphasised frequency and a higher proportion were composed of both modulated and pure-tone elements than in interior forest and grassland habitats. One interpretation of this high variance is that it reflects a lack of directional selection by environmental acoustics.

Finally, Morton's grassland sound propagation tests showed a positive correlation between sound frequency and degree of attenuation. Contrary to

Figure 10.9: The Distribution of Song Frequencies in Bird Song Varies with their Habitat. Arrows indicate mean frequency for each habitat.

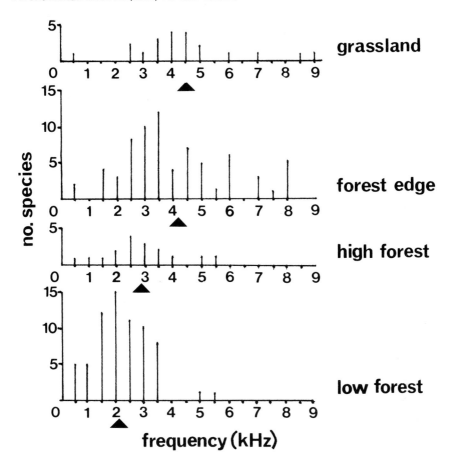

Source: Modified after Morton[217].

expectation, grasslands tend to be poor habitats for sound propagation because of air turbulence and temperature/elevation relationship effects. The rapidly modulated sounds found to be produced by grassland birds may be due to selection *against* communication by signal frequency but *for* communication based on temporal patterning of the sound signal. As long as signals are audible, temporal patterning will be little affected by the factors which might interfere with the transmission of information coded as sound frequency.

Morton's impressive relationships between habitat type and variations in vocal communication by birds certainly suggest a strong selective effect of environmental acoustics. However, the fairly precise predicted ranges (called 'sound windows') obtained by Morton have recently been suggested to be at least partly artefacts of the positioning of his recording apparatus.[256] Work currently underway is seeking to disentangle the products of recording technique and the real selective effects of environmental acoustics. Wiley and Richards have pointed out that Morton's results can be better explained on the basis of signal *degradation* (rather than attenuation) properties of different habitats. By comparison, the corrupting effect of air turbulence is likely to be much greater than the attenuating effects of obstacles. Nevertheless, studies to date suggest quite strongly that we must take account of the constraints operating via the physical qualities of the environment on communication mechanisms when hypothesising about the effects of sexual selection, territory maintenance and other 'higher-order' selective effects.

Summary

(1) 'Communication' describes a process of signal transmission by a 'sender' or 'actor' and signal receipt by a 'receiver' or 'reactor' which results in a change in behaviour by the reactor.
(2) Signals can be discrete or graded and tend to be highly specific and stereotyped. Signals are often transmitted in elaborate displays which enhance the effectiveness of transmission.
(3) The apparent simplicity and economy of animal displays does not necessarily indicate limited simplicity in the nervous system. Comparatively 'simple' nervous systems may be capable of sophisticated symbolic language, while advanced nervous systems may exhibit levels of communication which they can be educated well beyond.
(4) The apparent orchestration of behaviour in two or more individuals during communication may be the product of an exploitative arms race rather than a mutual coevolution to enhance the efficiency of signal transfer.
(5) The selective limiting of information transfer during communication is nowhere better seen than in various assessment behaviours during disputes. The analysis of assessment in an evolutionary context shows how conventional cues may become substituted for actual fighting.
(6) Communication can be approached from a 'functional' point of view. Recent work on bird song shows how different environmental pressures might shape and elaborate a communication mechanism.

11 EVOLUTION AND BEHAVIOUR

Throughout this book we have stressed the adaptive nature of behaviour. Like morphological and physiological characteristics, behaviour has its origins in the animal's genes. It is an evolutionary design feature, equipping the animal for survival and reproduction. It is the product of natural selection between alleles coding for different solutions to internal and external environmental problems. We have seen the shaping force of selection in several aspects of behaviour ranging from motivation and learning to migration, foraging, social behaviour and communication. During the discussion of different topics, a number of key concepts and principles have been mentioned: gene selection, kin selection, individual selection, group selection, evolutionarily stable strategies, optimality theory, arms races and so on. In this chapter we shall examine some of these concepts in more detail. In some cases, for instance the various 'levels' of selection (gene, kin, group, etc.), there is still a good deal of confusion about exactly what concepts mean and how they should be used. Other ideas, like that of biological arms races, are of such ubiquitous importance that they merit further elaboration. It is important to be clear about the nature and application of these ideas if the current evolutionary approach to behaviour is to be understood. We begin, however, by discussing a phenomenon which really underlies all evolutionary changes but which has only been approached quantitatively, and with respect to behaviour, relatively recently, *coevolution*.

11.1 Coevolution

While it is quite apparent that evolutionary change in one component of a community or one party in a set of interacting parties (like predators and prey, hosts and parasites and so on) is likely to cause evolutionary change in other components or parties, the importance of this coevolution as a dynamic process has received close attention only in the last decade or so. Yet, in that time, it has been shown to be a shaping force in the evolution of, for instance, foraging strategies, morphology, host/parasite relationships, life-history strategies, chemical communication and patterns of dispersal. Coevolution is thus fundamental to the study of population dynamics and community structure and stability.

11.1.1 Coevolution and Foraging Behaviour

It is hardly surprising that the interaction between animals and their food organisms should have produced adaptations in the two parties to maximise foraging efficiency on the one hand (see Chapter 6) and minimise predator impact on the other. In some cases, however, the two parties have coevolved so that *both* benefit from the interaction. Some nice examples come from the relationships between various nectar- or pollen-feeding insects and their host plant species.

Species of bumblebee (*Bombus* spp.) are nectar-feeders relying for food on a range of nectar-producing angiosperms. As well as taking nectar, however, the bees also act as the plants' pollen vectors, distributing pollen from the anthers of one flower to the stigmas of others. Recently, Pyke[249],[250] has examined the way bee foraging behaviour and plant morphology are coadapted.

Pyke studied three species of *Bombus* taking nectar from species of *Delphinium, Aconitum* and *Epilobium*. The plant genera were particularly suitable for study because they are pollinated almost exclusively by nectar-feeding bees and they bear flowers in vertical inflorescences which greatly simplifies the range of foraging movements made by the bees. During periods of foraging, bees appear not to perform any other activities (e.g. searching for nest sites or mates) and are not subject to predation. Nectar is also the main source of energy to the bees' colony (growth, survival and reproductive output depend directly on the amount of nectar coming in). We might therefore expect both bees and plants to have evolved maximally efficient ways of 'using' each other to obtain food or reproductive output.

Pyke found that bees showed very characteristic movement patterns on inflorescences. Broadly, these can be summarised as follows:

(1) Almost without exception bees begin feeding at the *bottom* of each visited inflorescence.
(2) Bees showed a strong tendency to move *vertically* up inflorescences.
(3) Bees tended to leave an inflorescence well before reaching the top.
(4) Bees tended to leave many flowers per inflorescence unvisited.

Important properties of the plant species were:

(1) Nectar production (and hence abundance) per flower *decreased* with increasing height up the inflorescence.
(2) Species tended to be *protandrous* and *self-compatible* — that is the stigma did not become receptive until all the anthers had dehisced and shed their pollen, and plants set seeds as often when selfed as when crossed.
(3) Flowers were arranged spirally and their age decreased with height up the inflorescence.

As a result of these characteristics, flowers with receptive stigmas are restricted to the bottom of the inflorescence while those shedding pollen, but with unreceptive stigmas, occur above them.

The tendency for bees to start feeding at the bottom of inflorescences appears to be an adaptation to peak nectar abundance occurring in the lowest flowers. That they tend to move vertically up the inflorescences is, at first sight, less clearly adaptive because, in doing so, they miss out a large number of flowers. A careful analysis of the bees' 'rule' of movement, however, shows that they tend to move next to the nearest flower *which has not previously been visited*. This is usually the flower vertically above the one currently being exploited. This minimises the number of fruitless revisits and the travel costs between flowers and hence maximises the bees' net rate of energy intake within inflorescences. Leaving well before reaching the top of an inflorescence means that the bees do not exploit flowers with low nectar abundance. Since the density of inflorescences in Pyke's study area was high (and hence travel costs between them low), bees would maximise their net rate of energy intake between inflorescences. By leaving inflorescences early when inflorescence density was high, bees were foraging in accordance with the predictions of the marginal value theorem discussed in Chapter 6. In all, therefore, bees appear to optimize their 'rules' of movement within and between inflorescences in relation to the pattern of nectar supply.

The pattern of nectar abundance within inflorescences and the protandrous breeding system of the plants appear to be adaptations promoting outcrossing and minimising pollen wastage. Decreasing nectar abundance with height up the inflorescence ensures that bees start feeding at the bottom and work up. This means that the pollen carried from the previously-visited inflorescence is deposited on flowers with receptive stigmas. As it moves up, the bee collects pollen from the younger flowers near the top of the current inflorescence before moving on to the next. Of the many possible mechanisms that would ensure outcrossing, the one employed results in maximum pollen transfer to *receptive* stigmas for a low cost in nectar production. The spiral arrangement of flowers within inflorescences means that the nearest flower is not usually the next one in the spiral. A bee moving vertically therefore misses flowers. Since a bee cannot carry enough pollen to fertilise all the flowers in an inflorescence (and usually carries enough for only a small number), it is to the plant's advantage if the bee takes nectar from only as many flowers as it can pollinate.

An even more subtle coevolved relationship between nectar-feeding bumble-bees and their host plant has been examined by Whitham.[330] Whitham studied interactions between the bee *Bombus sonorus* and the desert willow *Chilopsis linearis*. He was particularly intrigued by the morphology of nectar production in the willow. Radiating outward from the base of the corolla are five grooves that between them hold (by capillarity) about 1 μl of nectar. When the grooves have filled during nectar production, a further 8 μl of 'excess' nectar accumulates in a pool at the base of the corolla tube. Bees remove this pool nectar at a high rate (about 2 μl/s) but groove nectar can only be obtained more slowly (about 0.3 μl/s).

On the basis of the relative extraction rates for the two types of nectar supply, we should expect bees preferentially to take pool nectar when it is abundant. As the abundance of pool nectar declines so bees should include

Figure 11.1: Amount of Nectar in *Chilopsis* Flowers with Time of Day. Circles show
the mean standing crop of nectar in flowers and squares the mean amount left after
a bee has visited a flower. Bees took proportionately less of the available nectar early on
because they selectively fed on pool nectar (see text).

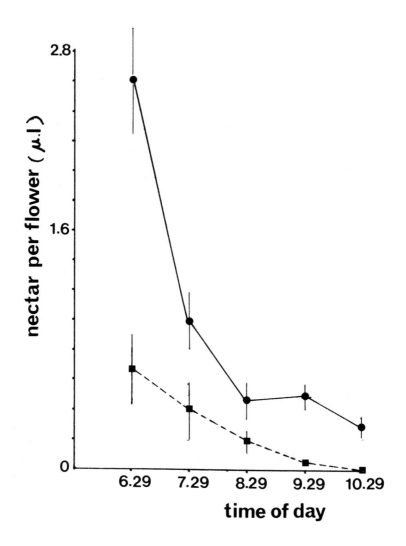

Source: Modified after Whitham[330].

groove nectar in their diet. Figure 11.1 shows that this was the case. Furthermore, bees which followed the 'pool first, groove second' rule had a higher net rate of energy intake than 'constant' bees, which always took both types of nectar.

While we see how bees can optimise their use of the two types of nectar supply, it is not immediately obvious why the willow has evolved them. The reason seems to be an enhanced visitation rate by bees. A plant's fitness (i.e. seed set) is likely to be correlated with the number of bee visits it can attract. If so, it should maximise the number of visits per calorie of nectar production. The result of bees switching from groove to pool nectar when pool nectar is abundant is that they must visit 40 per cent more flowers to obtain the same amount of nectar than if they took all the nectar from each flower. Furthermore, leaving groove nectar means that a *return* visit is profitable. Calculations show that bees which leave groove nectar provide about 1.6 visits per μl nectar produced compared with only 0.8 visits per μl by 'constant' bees. Whitham was also able to calculate that bees should switch from pool-plus-groove to pool-only feeding when nectar availability per flower reached 2 μl. To ensure revisitation, therefore, willows should supply just more than 2 μl of nectar per flower. In fact they supply about 2.4 μl and the 1.1 μl left in the grooves after pool-only depletion is more than enough to attract bees back. Visits only decline when nectar levels fall as low as about 0.4 μl. The relationship between the willows and the bees therefore appears to be a close coevolution based on maximising the energetic efficiency of reproduction and foraging in the two organisms.

11.1.2 Coevolution, Life-history Strategies and Community Structure

The all-embracing operation of coevolution has been painstakingly illustrated by Gilbert[115] in his study of *Heliconius* butterflies and their food plants.

Heliconius is a genus common in low/mid elevation neotropical forest. The larvae feed on the leaves of passion-flowers (*Passiflora* spp.) from which they sequester the chemicals making them distasteful (see Chapter 7). Adult heliconiids have an extraordinarily long lifespan of up to six months. Longevity is facilitated by the butterflies' novel ability to feed on pollen. In fact several lines of evidence suggest that heliconiids have the most behaviourally sophisticated adult phase of any butterfly species. Advanced characteristics include good visual navigation in poor light, good colour discrimination, a circadian memory rhythm, good form perception, gregarious roosting and good learning ability.

Ultimately, however, it is the flowering pattern of the cucurbit vines *Anguria* and *Gurania* which produces pollen that provides the option for extended adult life and reproductive effort in *Heliconius*. Cucurbit flowers are produced in inflorescences on a long peduncle if male and at each node if female. Plants of both sexes are sparsely distributed, females more so than males. However, a single male inflorescence may have a life span of three months to one year. This represents a period of continuous nectar and pollen supply for any butterfly able to locate such a scattered and inconspicuous resource.

The relationship between *Heliconius, Passiflora* and cucurbit vines has resulted in a complex series of coadaptations which affect not only individual behaviour and morphology but ultimately population and community structure as well. The simplest way to discuss them is to consider the costs and benefits of the coevolution for individuals of each genus.

Costs and Benefits to Heliconius. The benefit to the butterflies of their association with *Anguria* and *Gurania* lies in the nitrogen contained in the vines' pollen. A large pollen load contains enough nitrogen for the production of about five eggs. Costs, of course, include time and energy spent foraging, increased risk of predation and the uncertainty about flower availability. If butterflies are foraging optimally, we would expect pollen collection only when the benefits of collection outweigh the costs. The distribution of foraging effort by the two sexes supports this. Gilbert found that 90 per cent of early morning (5.30–6.30 a.m.) foragers were female. Consequently 93 per cent of all large pollen loads were borne by females (early foragers get larger loads) which require nitrogen for egg production. The exception which proved the rule was the occurrence of vigorously-foraging males on days following spermatophore deposition. Nitrogen in the spermatophore contributes directly to egg production and therefore needs to be replaced by males after deposition.

Costs and Benefits to the Cucurbit. The obvious benefit to cucurbit species of *Heliconius* pollen collecting is seed set and hence reproductive output. The cost lies in the production of pollen and nectar. Of course, the brunt of the cost is borne by the male plants which intriguingly seem to produce much more pollen than is necessary to ensure cross-fertilisation. Also, nectar continues to be produced after all the pollen has been removed. Gilbert suggests that this is a botanical example of sexual selection (see Chapter 8). Male genotypes compete with one another for the attention of foraging heliconids. Excess pollen and continued nectar supply is a good way of maintaining interest until the vital opportunity for fertilisation arises.

Consequences of Heliconius/*Cucurbit Coevolution*. The coevolution of pollen feeding in heliconiids and pollen availability in the vines has shifted the burden of reproductive effort from the larval to the adult stage in the butterfly. Any morphological or behavioural trait which improves adult foraging efficiency is therefore likely to be favoured by selection. The fact that most of the 'sophisticated' behavioural characteristics of heliconiids can be interpreted as adaptations for foraging bears this out. For instance, the circadian memory rhythm provides a mechanism for visiting food plants systematically (so-called 'traplining'). There is evidence that the timing of butterfly visits to individual plants coincides with their peak nectar production and pollen release. Gregarious roosts may act as information centres (see Chapter 9) allowing less successful foragers to capitalise on the experience of successful ones. In addition, living in

Figure 11.2: Coevolution between *Heliconius* Butterflies, their Food Plant *Anguria* and their Larval Host *Passiflora* Exerts Effects at Individual, Population and Community Levels.

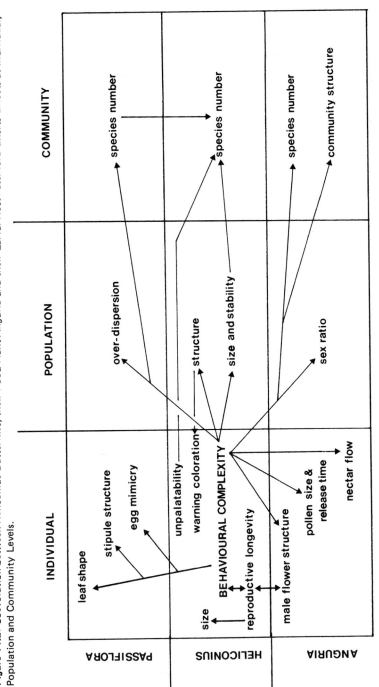

Source: Modified after Gilbert[115].

association with close relatives appears to have provided the opportunity for evolving distastefulness and warning colouration by kin selection.

Gilbert also interprets the strongly male-biased sex ratio in *Anguria* as a sexually-selected consequence of its relationship with heliconiids. Since *Heliconius* is an 'intelligent' forager able to discriminate male from female plants and selectively visit the former, the more male offspring a plant produces, the more likely it is to have its descendants fertilise the sporadic female plants that become available. To get round the butterflies' preference for male plants, female plants appear to have evolved male mimicry. Telescoped inflorescences and pollen-like nodules on the stigma are both male-like features which may encourage pollinating visits from butterflies.

Consequences of the Costs to Passiflora. The apparent importance of vision in heliconiids when selecting a host food plant for larvae might be expected to select for mimicry by Passifloraceae of non-food species. While there is no direct evidence for this, the highly variable leaf shape in the Passifloraceae (in sharp contrast to almost all other tropical forest species) provides strong suggestive evidence. More striking still is the evolution of *Heliconius* egg-like structures in *Passiflora. Heliconius* larvae are to some degree cannibalistic, so egg-laying females avoid plants which already have eggs on them. Interestingly the colour of the egg-mimics is nearer to that of the heliconiid eggs which are about to hatch rather than those which have just been laid.

The interrelationships between the passion-flowers, vines and different life-history stages of heliconiids have therefore resulted in a complex series of coadaptations which exert effects at a variety of ecological levels. These are summarised in Figure 11.2.

Oropendolas, Cowbirds and Insects. Coevolutionary relationships of a different sort have been discovered by Smith[293] among some neo-tropical bird and insect species.

Oropendolas (*Oropendola* spp.) are small black birds with a characteristically pointed bill and long tail. They build hanging, flask-shaped nests in palms and other tall trees. Cowbirds (*Scaphidura*) are common brood parasites which lay their eggs in oropendola nests. Smith's first important observation was that oropendolas in some areas discriminate against cowbird eggs and remove them from their nests whereas those in other areas do not discriminate. Smith then noticed that oropendolas which did discriminate lived in trees which contained active bee or wasp nests. Those which did not had no insect colonies near their nests.

The fourth and crucial component in the relationship was the parasitic bot-fly (*Philornis* spp.). Female bot-flies enter oropendola nests and lay their eggs directly on the chicks. The larvae then feed on the tissues of the chicks and eventually kill them. Bot-flies are a major cause of death in oropendola chicks. By hanging sticky fly-papers near oropendola nests, Smith found that nests in trees containing wasp or bee colonies were not troubled by bot-flies. Those without colonies were inundated with them. Smith's most exciting discovery, however, was that nests in colony-free trees which were parasitised by cow-birds were nine times less likely to suffer bot-fly infestation than unparasitised nests. It turned out that cowbirds are very precocious (develop quickly) com-pared with their hosts. Their eggs need up to seven days less incubation and chicks are covered in protective down when they hatch. Cowbirds' eyes open within 48 hours, contrasting with the six to nine days for oropendolas. While these characteristics have no doubt been selected to enhance exploitation of the oropendolas, they also mean the cowbirds can feed independently on whatever enters the nest. In doing so, the cowbirds not only defend themselves against bot-fly infestation, but incidentally defend the oropendola chicks as well.

The advantage of cowbirds in the nest when no wasp or bee colonies are present is apparently enormous. Smith found that nests with cowbirds had three times the fledging success of those without. In the absence of insect colonies, therefore, oropendolas did not discriminate against cowbird eggs. However, when insects were present and keeping the air space free of bot-flies, cowbird chicks were not able to outweigh their parasitic cost by defending chicks and so egg discrimination among the oropendolas was presumably favoured.

We have limited our discussion of coevolution to worked examples which involve behaviour. As we pointed out, however, coevolutionary effects at the level of the individual have consequences for the structure of whole populations and communities. A number of models and discussions of coevolution at the population/community level can be found in the literature but these are outside the subject of this book. For more information the reader is referred in particular to the papers by Lawlor and Maynard Smith[183] and Roughgarden[265].

11.2 Arms Races

If we think of a predator chasing its prey as, say, a hunting dog might chase a zebra, we can imagine a race going on on two time scales. First there is a race on the immediate time scale with the predator physically chasing the prey. Secondly, there is a race on an evolutionary time scale in which predator and prey adapt and counteradapt with respect to each other. This second, evolu-tionary, type of race is analogous to the progressive improvement of battleships and submarines during a war. It is thus reasonable to talk in terms of a bio-logical *arms race*.

The analogy is not entirely a new one. Cott used the term 'arms race' as long ago as 1940 when countering the argument that some forms of mimicry were too detailed to have been produced by natural selection. Indeed, Darwin[68], while not using the term, correctly attributed an almost universal importance to arms races. Although the basic idea of evolutionary arms races has been about for some time, however, it has recently been elaborated explicitly by Dawkins and Krebs.[86] Dawkins and Krebs' development of the arms-race analogy has led to some new and exciting ways of looking at evolutionary problems. Before we discuss arms races in detail, however, there are two points which should be emphasised.

First, while we talk of an arms race as 'progressing' and their competing parties as 'improving', we do not imply that, for example, predators end up catching more prey. Prey 'improve' too. As Dawkins and Krebs point out, while Recent predators might massacre Eocene prey, Recent prey might outrun Eocene predators. Secondly, the parties that are 'racing' are not individuals but genetic lineages. As we shall see more clearly in the next section, individuals do not evolve.

11.2.1 Classifying Arms Races

Although arms races occur in a wide variety of evolutionary contexts, Dawkins and Krebs have proposed a simple two-way classification which distinguishes the major forms of arms race (Table 11.1). Their classification distinguishes symmetric from asymmetric races and intraspecific from interspecific races.

Table 11.1: A Simple Classification of Biological Arms Races.

	Asymmetric (e.g. attack/defence)	Symmetric (e.g. competition)
Arms analogy:	swords sharper shields thicker swords sharper still	2 megaton bombs by A 3 megaton bombs by B 4 megaton bombs by A
Biological examples:	predator/prey host/parasite assessor/cheat	inter-male competition model/mimic
Distribution:	mainly interspecific	mainly intraspecific

Source: From discussion in Dawkins and Krebs[86].

Asymmetric races are those in which mutual counteradaptations are qualitatively different. An *interspecific* example would be a race between a parasite and its host. A parasite might evolve a mechanism for avoiding detection by its host's immune system (say by incorporating host antigen into its surface tissues as in Schistosomes), while the host evolves keener discriminatory abilities in its immune system. Predator/prey races may also be asymmetric. Many prey species reduce predator impact by evolving thick or armoured skins. In response, predators may evolve stronger, sharper teeth or specialised behaviour patterns to deal with the prey. Selection is therefore acting on different organs or characteristics of the two parties. *Intraspecific* asymmetric races occur between parties with conflicting genetic interests within species. Classic examples are the conflicts between parents and offspring or members of a mated pair. As we saw in Chapter 8, arms races between female choice criteria and male mating strategies can have far-reaching effects on individual morphology and behaviour.

Symmetric arms races are those in which counteradaptations occur in analogous organs or characteristics. The coevolution between cursorial herbivorous mammals and their predators is a good example. Here, both parties have evolved longer legs to enhance running ability. Another symmetric interspecific race occurs between model and mimic in Batesian mimetic relationships. Both model and mimic are attempting to convince predators of their (true or false) noxious qualities. In doing so the model evolves to distinguish itself from the mimic (to reduce the cost of having palatable cheats in the community) and the mimic evolves to 'close the gap' in appearance between itself and the model. The symmetric model/mimic race is, of course, run simultaneously with and 'driven' by the asymmetric race between the predator and both model and mimic. Symmetric *intraspecific* races are the basic stuff of evolution. In a way all arms races are fundamentally symmetric intraspecific races. Adaptations to avoid predation, for instance, have evolved because individuals bearing those adaptations out-compete their conspecifics. Nevertheless, we can draw a distinction between those symmetric intraspecific races which are driven by interspecific selection pressures and those which are driven by purely intraspecific pressures. Anti-predator adaptations are an example of the former. As Dawkins and Krebs point out, if evolution in the predator was somehow 'frozen', counteradaptive change in its prey would eventually cease. An example of the latter type is intermale competition for females. Here directional change is independent of interspecific pressures (except in as much as predation risk might set a limit on it).

11.2.2 Consequences of Arms Races

The question of what arms races might lead to is, of course, absurdly broad. Since arms races occur in any biological context from the evolution of gametes to Batesian mimicry, it is clear they can lead to almost anything. However, there are a number of possible consequences which are of wide importance.

One of these concerns the possibility of *lineage extinction*. It is reasonable to suppose that most selection pressures will favour a 'fixed' optimum or 'ideal' organism for a given environment. However, it is possible that a particularly strong selective pressure might favour change towards a different, 'relative' optimum. Female choice could provide such a pressure. By initially favouring males with, say, longer than average tails (because of their enhanced survival value), females may set up a runaway selection pressure for longer and longer tails. The directional trend continues purely because it is preferred by females who thus produce attractive sons. However, the tendency for longer tails may take males away from the 'fixed' optimum for the environment. The difference between the 'fixed' optimal male and the 'relative' optimal male preferred by females may eventually become so great that males are no longer ecologically viable and the lineage goes extinct. Dawkins and Krebs suggest that the negative relationship between evolutionary rate (as measured, for instance, by the rate of increase in body size) and survivorship during geological time in certain taxa may be just such a product of arms races.

Arms races may also underlie much of the apparent directionality in evolution. Even a casual glance at the history of life reveals immediate trends: simple to complex, small to large, single cell to multicell, invertebrate to vertebrate and so on. Arms races are powerful and ubiquitous selection mechanisms which, almost by definition, are directional in their effect. They may well be part of the key to understanding the apparent 'arrow on time' in evolution.

As well as providing an explanation for directionality in evolution, arms races may also help to explain sudden increases in the rate of evolution of a species. A heated argument has recently developed among geologists and palaeobiologists concerning the tempo and mode of evolutionary change. Some workers[97,297] see evolution as a process of *sudden* directional changes resulting in speciation which are preceded and followed by periods of little or no change (*stasis*). On this view, evolution is seen as a process of *punctuated equilibria*. In some cases, quantum evolutionary 'jumps' are suggested to account for apparently rapid change. The old evolutionary bogey of gaps in the fossil record is seen not as a series of irritating discontinuities, but as evidence for rapid bursts or quantum jumps in *macroevolutionary* (species evolution) history which leave no intermediate forms. This is set against the notion that evolutionary change takes place as a series of small-scale changes eventually compounding to produce a large-scale (speciation) change. This 'gradualist' view is often pitted against 'punctuationism' as a naive and fundamentalist view of evolution. As such it, and its exponents, are largely misrepresented. There is nothing in the orthodox Darwinian (small-scale change) view of evolution which forbids bursts of rapid change. Arms races in conjunction with appropriate genetic variation provide just the mechanism required. Runaway directional selection in response to, for instance, predator pressure, interspecific competition or even intraspecific pressures like sexual selection could bring about a rapid shift to a new optimum with selection acting against forms intermediate between this and the old optimum. Such a change depends on the well-established principles of natural selection and population genetics. On the other hand, it is very difficult to produce a convincing genetic mechanism (except for mechanisms like large-scale chromosomal changes which could work under certain restricted conditions) for quantum jumps.

11.2.3 Who Wins an Arms Race?

Can one coevolving party in an arms race actually *win* the race? In some cases, the answer may be yes. Dawkins and Krebs discuss four situations in which one party might be at an advantage in a race.

Differences in Evolutionary Rates. For one reason or another, one species in an interspecific race may have an inherently higher rate of evolution than the other. Perhaps such a species is involved in a strongly directional intraspecific race.

Figure 11.3: In Terms of the Appearance of Different Families, Mammals (a) Appear to Have Evolved at a Much Higher Rate than Bivalve Molluscs (b). Relative evolutionary rates are likely to be important determinants of the outcome of arms races.

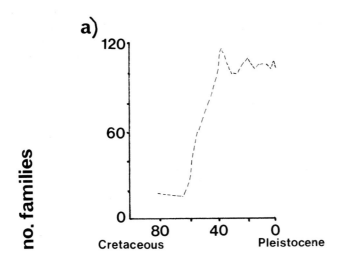

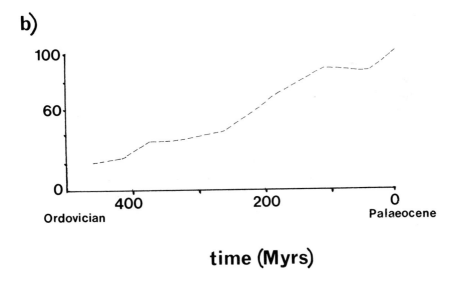

Source: Modified from Stanley[297].

There is suggestive evidence that different taxa may evolve at different rates. Mammals, for instance, appear to have evolved faster than bivalve molluscs (Figure 11.3). In a race between parties with differing rates of evolution, the faster evolving party may be at an advantage. A faster rate of evolution means a long 'period of grace' before an adaptive step is countered by the slower party. An evolutionary rate asymmetry of this sort may underlie the apparently overwhelming advantage of sexual over asexual modes of reproduction (see Chapter 8). However, it is not easy to separate cause and effect when we consider relative evolutionary rates. One party may be at an advantage in a race *because* it evolves faster, or (as we suggested earlier) the race itself may cause one or both parties to evolve more quickly.

Differences in Learning Ability. A similar effect to an asymmetry in evolutionary rates might be produced if one party has a better learning ability. A party with good learning ability may be able to reduce its coevolving partner's 'period of grace' to less than one generation. On the other hand, the 'period of grace' stays the same between generations unless there is cultural transmission.

'Adaptation Budgets'. Adaptations, whether morphological, physiological or behavioural, are likely to be costly. Resources used up in one adaptation are not available for use in another. Dawkins and Krebs use the example of a generalist predator. A predator which is adapted to hunt several prey species is unlikely to be devastatingly effective against any one of them. It is compromised in its effectiveness against each species by having to invest in adaptations to deal with others. A prey species which is able to invest maximally in running away from one or a few predators may be at an advantage in a race with a predator which must invest not only in running ability and adaptations for killing that particular species, but a wide range of other species too. However, generalist predators, by definition, run many races, so may not suffer from small disadvantages in each.

The 'Life-Dinner Principle'. There may be an asymmetry between parties in a race in the penalty of failure. In a predator/prey relationship, the predator runs for its 'dinner', the prey for its life. A predator which fails to kill in a chase is still likely to survive to reproduce. A prey which fails to escape a predator, however, most certainly will not. We might expect the party which has more to lose, and which is thus under the stronger selective pressure, to allocate more of its adaptation budget to improving escape mechanisms. This *Life-Dinner Principle* may therefore give the party under the heaviest selection pressure an advantage in an arms race. The 'Life-Dinner Principle' may be the reason that small gametes (sperm) won out against large gametes (eggs) in the primaeval race to fuse with large gametes (see Chapter 8). Large gametes lost only extra zygote provisions if they fused with a small gamete. Small gametes, however, produced inviable zygotes if they fused with anything other than a large gamete. While certain factors might enable one party to win an arms race, outright victory for one party is by no means the only possible outcome. In some cases, arms races may end in cyclical oscillations between different end points.

Coevolution and arms races have been discussed at length because they are fundamental processes operating across all aspects of evolutionary change. Coevolutionary approaches emphasise the inseparability of change in one organism from changes in others with which it interacts and shares the environment. Nothing evolves in isolation. Change is caused by change and in turn causes change. Arms races are special cases of coevolution which impose strong directional change on coevolving parties. The idea of a biological arms race helps to explain some hitherto difficult problems for evolutionary theory. It shows how naturally-selected traits might drive a lineage extinct instead of 'perfecting' it as earlier views of evolution supposed. It also accounts in terms of orthodox genetics for sudden bursts of rapid evolution and gaps in the fossil record. Helpful as these concepts are, however, we must always be careful to keep sight of the most fundamental concept of all, the unit on which selection acts. Our interpretation of the biological world depends critically on this.

11.3 'Levels' of Selection

A reader sifting through the ethological and behavioural ecological literature will come across a number of different terms describing the supposed mechanism of natural selection for a behaviour pattern. Group selection, kin selection, trait group selection, individual selection, gene selection, to mention the commonest, are variously drawn upon either as explanatory models or as bogeys to be debunked and replaced. As Dawkins[81] points out, however, at whichever 'level' of selection an evolutionary model is formulated it is fundamentally a *gene selectionist* model. All the 'levels' mentioned above operate by altering gene (or more strictly, allele) frequencies within populations. It is the unit on which selection is supposed to act in each case that is a cause of contention.

Problems arise because, although it is allele frequencies which change during evolution, alleles themselves are not free-floating tangible entities in the environment. They reside in individual bodies and are naturally selected through their effects on the individual phenotype. At first sight, therefore, it seems as if it is *individuals* which are the units of selection. After all, it is individuals which have to survive and reproduce in the face of environmental hazards, not genes. This is the *individual selection* view of evolution. Similarly, individuals often do not go around solitarily, but operate in groups. Some workers have seen groups or even populations and species as the basic units of selection (the *group selection* view). We shall discuss the problems of group selection models later. First, we shall examine a sometimes more difficult and confusing distinction, that between models assuming *individuals* as the units of selection and models assuming *alleles* as the units of selection.

11.3.1 Individual Selection

In many cases, the evolutionary interests of an allele and the individual in which it resides are the same. The terms 'gene fitness' and 'individual inclusive fitness' (see Chapter 3) are interchangeable. In some cases, however, they are not and the wrong choice of model can lead to erroneous conclusions. A classic example is the confusion over *probabilistic* versus *certain* coefficients of relatedness (r). The value of r for sexual parents and offspring, for instance, is exactly and always 0.5. Parents and offspring share precisely 50 per-cent of their alleles. Siblings, on the other hand also have an r value of 0.5, but this is only *on average*. Owing to the vicissitudes of the meiotic process, siblings may be related by a little more or a little less than 50 per cent. Some workers have argued that selection recognises the difference between probabilistic and certain degrees of relatedness and acts accordingly. Gibson, for instance, suggested that, given a choice an organism should invest in offspring rather than siblings because selection will favour a genetic certainty rather than a gamble. Fagen supposed that grandparents would be expected to dote on the grandchildren which most resembled them because resemblance would indicate a high proportion of shared genes. In both these cases, the authors have been misled by the ability to define the coefficient of relatedness r in two ways. It can be defined as *the average proportion of one individual's genome which is shared by another* or *the probability that a given allele in one individual will be identical by descent in another*. What selection focuses on is the latter, but what many people think in terms of is the former. The *proportion* definition interprets r from a whole individual's point of view. The *probability* definition sees it strictly in terms of gene frequency. As we know, it is gene frequencies which evolve not individuals. An offspring and a sibling both have a 50 per cent *chance* of sharing a given allele with an individual. From that allele's point of view investment in either yields the same evolutionary pay-off. In Fagen's case, what matters is not the proportion of shared genes coding for physical features, but the probability that grandparent and grandchild share the gene for *doting*. Doting is the adaptive characteristic with which Fagen is concerned. We have already seen another instance where a fallacy has been generated through thinking in terms of individual level selection. Williams' 50 per cent 'cost of meiosis' model of the cost of sexual reproduction (Chapter 8) falls squarely into the 'selection acting on whole genomes' trap. If the expedient of always thinking in terms of *gene selection* is followed, such misinterpretations can be avoided.

11.3.2 Group Selection

Few other concepts in the study of biology have fuelled as much controversy and as many new ideas as that of *group selection*. Although the idea of group selection goes back at least as far as Carr-Saunders in 1922, its most rigorous and explicit application in biology came with Wynne-Edwards'[340] famous book in 1962. Wynne-Edwards saw group selection as the main driving force behind the evolution of social behaviour and population regulation. His main ideas can be summarised as follows:

(1) Animals (especially higher phyla) are adapted to control their own population densities.
(2) Regulatory mechanisms adjust population density to the supply of limited resources.
(3) The mechanisms depend in part on the substitution of conventional 'prizes' (like territories or social status) as proximate objects of competition rather than resources themselves.
(4) Groups of individuals engaged in such conventional competition constitute a society and selection acts on social groups as whole entities rather than on the individual members.

One of the most controversial claims of Wynne-Edwards' theory was that individuals would actually sacrifice their own reproductive potential to ensure the group/population did not exceed the environmental carrying capacity. Selection would therefore favour the social systems which were ecologically most efficient, even if this was against the interests of component individuals.

The criticism which has been heaped on the group selection interpretation of social behaviour is of two main sorts. First, group selection has certainly been responsible for woolly and incorrect thinking in the study of ecology and behaviour. 'Good of the species' arguments were rife in the literature until recently and waved aside more careful thought about genetic mechanisms. Secondly, and more seriously, no convincing mathematical model yet produced can enable group selection to overcome the immediate adaptive advantage accruing to cheating individuals who exploit the social system to their own ends. Selection *between* groups, relying as it does on the slow process of selective group extinction, is no match for the rapidly acting selection *within* groups favouring selfish mutants. For an excellent critical review of group selection see Williams[331].

Despite being open to criticism, however, group selection is not a completely unrealistic model of evolution. It *can* work and can even be demonstrated in the laboratory. Wade[325] carried out a series of experiments with *Tribolium* beetles in which he manipulated laboratory populations so as to stimulate group and individual selective forces affecting the number of adults produced per generation. He found that group selection contributed significantly to the total variance in adult numbers when it was operating both with and against individual selection. The objection is not that group selection *cannot* work but that it is *unlikely* to be a major selective force *under natural conditions*. Among other things, it requires unrealistically low levels of gene flow between groups and mutation rates within groups. Groups would have to be genetically isolated units with little chance of a selfish mutant arising within their own ranks. These requirements are well illustrated in Maynard Smith's 'haystack' model

Figure 11.4: Maynard Smith's 'Haystack' Model of Group Selection. Altruistic mice (genotype aa) can be maintained in the population in the face of competition from selfish individuals (AA) as long as there is little migration between colonies in different haystacks. Heterozygous colonies (Aa) drift towards selfish homozygosity.

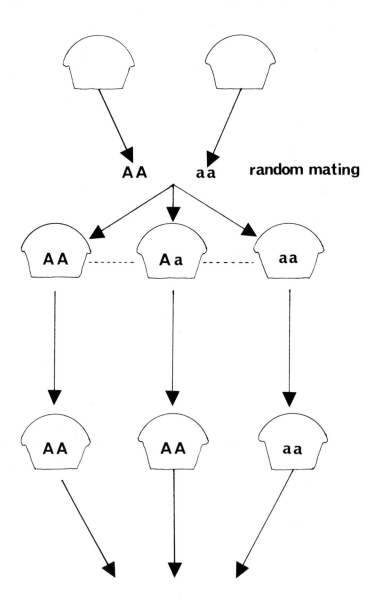

Source: From discussion in Maynard Smith, J. (1964). Group selection and kin selection. *Nature, 201*: 1145-7.

(Figure 11.4). However, data from Wade's experiments suggest that isolation need not be as complete as was at first thought. Although group selection is no longer given the weight previously attributed to it, models based on group selection under more rigorously defined conditions are still forthcoming (see, for example, Wilson[334]) and merit serious consideration.

11.3.3 Kin Selection

The idea of kin selection is a logical consequence of thinking of evolution as changes in allele frequencies within populations. In Chapter 3 we saw that alleles are not the sole possession of one individual. Copies of them are present by common descent in its close relatives. How the individual it helps to build behaves with respect to its relatives is therefore likely to affect an allele's *inclusive fitness*. Adaptations in an individual which are geared towards helping close relatives (e.g. helping to rear siblings, hunting co-operatively with close relatives, being distasteful and deterring predators from attacking relatives in the future) can be said to evolve through kin selection.

However, superficially logical though the concept of kin selection is, several misinterpretations have crept into the literature and discussions. While it is true that the relationship between kin selection and relatedness may not be as clear-cut in every case as we have implied in our simplified discussion (for instance, certain mechanisms acting at chromosome segregation may distort the normal distribution of alleles between gametes — see Boorman and Levitt[31] for a full discussion of genetic models of altruism), this has not been the cause of misinterpretation. Dawkins discusses a number of serious misunderstandings of kin selection. Perhaps chief among them is the assumption that kin selection is a special case of group selection. This error has crept in because 'kin' implies 'family' and 'family' implies 'group'. As should be clear from the above and Chapter 3, however, kin selection acts via shared *genes*, not grouped *individuals*. Another common assumption is that altruistic behaviour must be a virtual certainty between asexually-produced clones if kin selection really operates. What this ignores is the independence of relatedness and the possession of a gene for altruism. If an 'altruism allele' is not present in their genomes, no amount of relatedness will make individuals altruistic. Conversely, altruism *can* evolve between completely unrelated individuals. As long as there is a strong chance of an altruistic act being repaid to the donor in the future, altruistic traits may evolve. Such *reciprocal altruism*[314] has been observed among male baboons co-operating in mating attempts.[233] Curio[65] has also pointed out that there are unexplained sex differences in the degree of apparent altruism shown in certain species. Such differences are difficult to explain by kin selection, but might arise by a 'self-estimation' of reproductive value mechanism. Other common errors based on over-simple or distorted views of relatedness or gene action include the arguments that 'altruistic behaviour is too complex to be coded for by a single gene' (see Chapter 3) and 'all individuals within a species share more than 99 per cent of their genes (for noses, ears, whiskers, etc.) so they should all be mutually altruistic' (based on the assumption that genes for altruism always exist).

It is important to emphasise the errors of thinking underlying these arguments against kin selection if kin selection is not to be thought of as a special and limited form of selective force. Kin selection is a label given to the *natural selection* of traits through the medium of close relatedness. As such it is *one* mechanism which can lead to the evolution of apparent altruism (although we are then, of course, not talking about altruism but of genetic 'selfishness'). Relatedness and altruism, however, do not automatically go hand in hand.

11.4 Evolutionarily Stable Strategies

Time and again throughout this book, we have asked what the evolutionarily stable solution to a behavioural problem might be. While the concept has already been defined and its assumptions outlined, it is important enough to warrant explicit treatment here.

An evolutionarily stable strategy (ESS) is the (assumed genetic) variant of a behaviour which is most likely to be favoured by natural selection in the course of time. We assume that genes coding for feasible alternative solutions (strategies) to a particular problem will arise from time to time, but that the one providing the most efficient solution will eventually spread through the population and become resistant to invasion by alternatives. Within this seemingly simple and logical set of assumptions, however, there are problems.

The first is a conflict with one of the other major themes of the book: the optimality of behaviour in terms of individual reproductive potential. A good working hypothesis when we are seeking an ESS for a behaviour is to assume that the most likely strategy to persist over evolutionary time is the one providing the *optimal* solution to the animal's problem and therefore the best from the individual's point of view. However, as Dawkins lucidly points out in his paper 'Good strategy or evolutionarily stable strategy?', the essential quality of an ESS is not optimality but *uncheatability*. In the familiar *hawk/dove* model discussed in Chapter 3 (Tables 3.1 and 3.2), a 'conspiracy of doves' — an agreement whereby everyone fights conventionally — would be good for each individual but, as we have seen, it is not resistant to invasion by a feasible alternative and so is not an ESS.

A good example of the optimality trap is given by Dawkins. He considers a strategy *retaliator* which under certain conditions in a *hawk/dove/retaliator* population is a pure ESS.[85] Imagine retaliator has evolved in a population of venomous scorpions. If a scorpion is stung by an opponent, it stings back. However, stings are lethal, so it does the *retaliator* no good in terms of survivorship or reproductive potential to sting back. The reason it stings back is that the unconscious, genetically-programmed strategy *retaliator* is an ESS. It is an ESS quite independently of any benefit to individual inclusive fitness. This is another valuable warning about the naivity of assuming that natural selection automatically 'improves' the adaptiveness of organisms.

A problem which arises immediately when we seek ESSs is that of catering for feasible alternatives. The strategy which emerges as an ESS clearly depends on the options it is up against. Similarly, we cannot guarantee that, having thought of a range of strategies, the appropriate genetic variation to code for them will arise in our population. As Dawkins points out, we must rely here on biological intuition and specify strategies which are feasible alternatives given the animal and its environment. Constraints imposed by the sophistication of the animal's nervous system, for instance, may limit its ability to evolve certain strategies. The 'rules of thumb' seen in foraging shrews (Chapter 6) and contesting digger wasps (Chapter 10) are examples where the ESS appears to be only a crude approximation of what we assess as the optimal solution. In shrews this is probably because the ever-changing food supply prohibits the evolution of mechanisms for assessing the profitability of each prey type. In wasps, it is probably due to limited neural sophistication.

A final warning must be sounded about recognising an ESS when we see it. While populations may reach an *equilibrium* for particular strategies where each strategist fares equally well, this is not necessarily an ESS. A good example is the analysis of the evolution of mating strategies in Table 8.2. Here, equilibrium mixtures of *coy* and *fast* females and *faithful* and *philanderer* males can be found, but these are not stable in an evolutionary sense.[271] As with Maynard Smith and Price's initial model of the *retaliator* strategy, a drift away from the equilibrium is not corrected by evolution, but accelerated. The equilibrium is thus unstable, like that of a pencil balanced on its point.

The ESS principle is not confined to an evolutionary time-scale. Baldwin and Meese, for instance, document an interesting 'ESS' arising on a developmental time-scale in pigs. They set up a Skinner box so that a bar pressed at one end released food at the other. When two pigs were put in the box, one, not surprisingly, sat by the food dispenser while the other did all the bar-pressing. Paradoxically, it turned out that it was the dominant of the two pigs who did the pressing. The reason appeared to be that neither pig could get food without one of them pressing the bar, but, if it was the subordinate, he was not able to push the dominant pig aside to get to the feeder. The dominant, however, could press the bar, push the subordinate aside and still obtain food. This *developmentally stable strategy (DSS)* shows exactly the ESS principle at work and its ubiquitous operation wherever the value of an individual's behaviour depends critically on what the other individuals around it are doing.

We can therefore summarise the important points about ESSs as follows:

(1) A strategy is an unconscious programmed behaviour pattern competing during evolution with alternative strategies.

(2) A strategy is an ESS if, when more than a critical proportion of the

population adopts it, it cannot be invaded by any of the range of alternatives.

(3) An ESS is not necessarily an optimal strategy for the individual, but an uncheatable strategy over evolutionary time.

(4) It is a misconception that an ESS is uninvadable by *any conceivable* strategy. Alternatives must be biologically realistic and not outlandish (e.g. a territorial intruder with a machine gun could always destroy a stable 'holder wins' convention).

(5) The ESS concept has been developed mainly in connection with intra-specific aggression, but it is applicable to all aspects of behaviour and has so far proved useful in analyses of the evolution of sex ratios, reciprocal altruism, parent-offspring conflict, mate searching strategies, mate desertion, coevolution, dispersal, deceit and foraging strategies. It is important wherever the value of a behaviour for a particular individual depends on what others are doing and is not limited to an evolutionary time-scale.

Summary

(1) The interaction between an animal and its biotic environment can be usefully examined from a coevolutionary point of view. Coevolution can exert an effect at individual, population and community levels.

(2) Arms races form a broad class of coevolutionary processes which can impose strong directional trends on coevolving parties. They may help to explain apparently maladaptive trends, lineage extinction and changes in evolutionary rates.

(3) Symmetrical and asymmetrical arms races can be distinguished depending on whether counteradaptations are functionally analogous or different.

(4) A number of different 'levels' of natural selection can be envisaged. While all are ultimately gene-selectionist, they view selection as acting on different units. 'Individual selection' and 'gene selection' are often interchangeable, but the former can give rise to fallacies. Group selection can operate but is unlikely under most natural conditions. Kin selection is a logical consequence of thinking of genes as the unit of selection, but has been misinterpreted in several ways.

(5) Evolutionarily stable strategies (ESSs) are not necessarily optimal strategies from the point of view of the individual. The ESS principle is important wherever the value of a behaviour to an individual depends on what other individuals are doing and is not limited to an evolutionary time-scale.

BIBLIOGRAPHY

1. Alcock, J. (1975). *Animal Behaviour: An Evolutionary Approach*. Sunderland, Mass., Sinauer.

2. Andersson, M. and Wicklund, C.C. (1978). Clumping versus spacing out: experiments on nest predation in fieldfares (*Turdus pilaris*). *Anim. Behav.*, *26*: 1207-12.

3. Baerends, G.P. (1970). A model of the functional organization of incubation behaviour. In (G.P. Baerends and R.G. Drent, eds) *The Herring Gull and its Egg. Behav.*, *Suppl. XVIII*: 265-310.

4. ———, Brouwer, R. and Waterbolk, H. Tj. (1955). Ethological studies on *Lebistes reticulatus* (Peters): I. An analysis of the male courtship pattern. *Behaviour*, *8*: 249-334.

5. Baggerman, B. (1962). Some endocrine aspects of fish migration. *Gen. Comp. Endocrinology.*, *Suppl. 1*: 188-205.

6. Baker, M.C. (1974). Genetic structure of two populations of white-crowned sparrows with different song dialects. *Condor*, *76*: 351-6.

7. ——— and Fox, S.F. (1978). Dominance, survival and enzyme polymorphism in dark-eyed juncos, *Junco hyemalis. Evolution*, *32*: 697-711.

8. Baker, R.R. (1978). *The Evolutionary Ecology of Animal Migration*. London, Hodder and Stoughton.

9. ——— (1981). *Human Navigation and the Sixth Sense*. London, Hodder and Stoughton.

10. Barnard, C.J. (1978). Aspects of winter flocking and food fighting in the house sparrow (*Passer domesticus domesticus* L.). DPhil thesis, University of Oxford.

11. ——— (1980). Factors affecting flock size mean and variance in a winter population of house sparrows (*Passer domesticus* L.). *Behaviour*, *74*: 114-27.

12. ——— (1980). Equilibrium flock size and factors affecting arrival and departure in feeding house sparrows. *Anim. Behav.*, *28*: 503-11.

13. ——— (1980). Flock feeding and time budgets in the house sparrow (*Passer domesticus* L.). *Anim. Behav.*, *28*: 293-309.

14. ——— and Brown, C.A.J. (1981). Prey size selection and competition in the common shrew (*Sorex araneus*). *Behav. Ecol. Sociobiol.*, *8*: 239-43.

15. ——— and Burk, T. (1979). Dominance hierarchies and the evolution of 'individual recognition'. *J. Theor. Biol.*, *81*: 65-72.

16. ——— and Sibly, R.M. (1981). Producers and scroungers: a general model and its applications to captive flocks of house sparrows. *Anim. Behav.*, *29*: 543-50.

17. ——— and Stephens, H. (1981). Prey size selection by lapwings in lapwing/gull associations. *Behaviour*, *77*: 1-22.

18. Bastock, M. (1956). A gene mutation which changes a behaviour pattern. *Evolution*, *10*: 421-39.

19. Bateson, P.P.G. (1979). How do sensitive periods arise and what are they for? *Anim. Behav.*, *27*: 470-86.

20. ——— (1980). Optimal outbreeding and the development of sexual preferences in Japanese quail. *Z. Tierpsychol.*, *53*: 231-44.

21. ———, Lotwick, W. and Scott, D.K. (1980). Similarities between the faces of parents and offspring in Bewicks swans and the differences between mates. *J. Zool. Lond.*, *191*: 61-74.

22. Bauwens, D. and Thoen, C. (1981). Escape tactics and vulnerability to predation associated with reproduction in the lizard *Lacerta vivipara. J. Anim. Ecol.*, *50*: 733-44.

23. Bekoff, M. (1974). Social play and play-soliciting by infant canids. *Amer. Zool.*, *14*: 323-40.

24. —— and Byers, J. (1979). A critical reanalysis of the ontogeny and phylogeny of mammalian social and locomotor play: an ethological hornet's nest. In (K. Immelmann, G. Barlow, M. Main and L. Petrinovich eds) *Behavioural Development in Animals and Man*, The Bielefield Interdisciplinary Conference. Cambridge University Press.

25. Bellrose, F.C. (1968). Celestial orientation by wild mallards. *Bird Banding, 29*: 75–90.

26. Benzer, S. (1973). Genetic dissection of behaviour. *Sci. Am., 229*: 24–37.

27. Bertram, B.C.R. (1976). Kin selection in lions and in evolution. In (P.P.G. Bateson and R.A. Hinde, eds) *Growing Points in Ethology*: 281–301. Cambridge University Press.

28. —— (1978). Living in groups: predators and prey. In (J.R. Krebs and N.B. Davies, eds) *Behavioural Ecology: An Evolutionary Approach*: 64–96. Oxford, Blackwell; Sunderland, Mass., Sinauer.

29. Bolles, P. (1970). Species-specific defence reactions and avoidance learning. *Psych. Rev., 77*: 32–48.

30. Bolles, R.C. (1979). *Learning Theory*, 2nd edn. New York, Holt, Rinehart and Winston.

31. Boorman, S.A. and Levitt, P.R. (1980). *The Genetics of Altruism*. New York, Academic Press.

32. Bösiger, E. (1957). Sur l'activite sexuelle des males de plusieur souches de *Drosophila melanogaster*. *C. R. Acad. Sci. (D) (Paris), 244*: 1419–22.

33. Bray, O.E., Kennelly, J.J. and Guarino, J.L. (1975). Fertility of eggs produced on territories of vasectomized redwinged blackbirds. *Wilson Bull., 87*: 187–95.

34. Breland, K. and Breland, M. (1961). The misbehavior of organisms. *Amer. Psychol., 16*: 681–4.

35. Brenner, S. (1973). The genetics of behaviour. *Br. Med. Bull., 29*: 269–71.

36. Brncic, D. and Koref-Santibañez, S. (1964). Mating activity of homo- and heterocaryotypes in *Drosophila pavani*. *Genetics, 49*: 585–91.

37. Brockmann, H.J. and Barnard, C.J. (1979). Kleptoparasitism in birds. *Anim. Behav., 27*: 487–514.

38. Brower, L.P. and Brower, J.V.Z. (1964). Birds, butterflies and plant poisons, a study in ecological chemistry. *Zoologica, 49*: 137–59.

39. Bullock, T.H. (1973). Seeing the world through a new sense: electroreception in fish. *Amer. Sci., 61*: 316–25.

40. Caplan, A. (ed.) (1978). *The Sociobiology Debate: Readings on Ethical and Scientific Issues*. New York, Harper and Row.

41. Caraco, T. (1979). Time budgeting and group size: a test of theory. *Ecology, 60*: 618–27.

42. ——, Martindale, S. and Pulliam, H.R. (1980). Avian flocking in the presence of a predator. *Nature, 285*: 400–1.

43. —— —— and Whitham, T.S. (1980). An empirical demonstration of risk-sensitive foraging preferences. *Anim. Behav., 28*: 820–30.

44. Carlson, A. and Moreno, J. (1981). Central place foraging in the wheatear *Oenanthe oenanthe*: an experimental test. *J. Anim. Ecol., 50*: 917–24.

45. Carpenter, F.L. and MacMillen, R.E. (1976). Threshold model of feeding territoriality and test with a Hawaiian honeycreeper. *Science, 194*: 629–42.

46. —— —— (1976). Energetic cost of feeding territories in a Hawaiian honeycreeper. *Oecologia, 26*: 213–24.

47. Caryl, P.G. (1979). Communication by agonistic displays: what can games theory contribute to ethology? *Behaviour, 68*: 136–69.

48. Catania, A.C. (1963). Concurrent performances: a baseline for the study of reinforcement magnitude. *J. Exp. Anal. Behav., 6*: 299–300.

49. Catchpole, C.K. (1979). *Vocal Communication in Birds*. London, Arnold.

50. —— (1980). Sexual selection and the evolution of complex songs among European warblers of the genus *Acrocephalus*. *Behaviour, 74*: 149–66.

51. Charnov, E.L. (1976). Optimal foraging: the marginal value theorem. *Theor. Pop. Biol.*, *9*: 129–36.

52. ——— and Krebs, J.R. (1975). The evolution of alarm calls: altruism or manipulation? *Amer. Nat.*, *109*: 107–12.

53. ———, Orians, G.H. and Hyatt, K. (1976). The ecological implications of resource depression. *Amer. Nat.*, *110*: 247–59.

54. Chitty, D. (1967). The natural selection of self-regulatory behaviour in animal populations. *Proc. Ecol. Soc. Aust.*, *2*: 51–78.

55. Clark, R.B. (1960). Habituation of the polychaete *Nereis* to sudden stimuli. I. General properties of the habituation process. *Anim. Behav.*, *8*: 92–103.

56. Clutton-Brock, T.H. and Harvey, P.H. (1977). Primate ecology and social organization. *J. Zool. Lond.*, *183*: 1–39.

57. ———, Albon, S.D., Gibson, R.M. and Guinness, F.E. (1979). The logical stag: adaptive aspects of fighting in red deer (*Cervus elephus* L.). *Anim. Behav.*, *27*: 211–25.

58. Cody, M.L. (1971). Finch flocks in the Mohave desert. *Theor. Pop. Biol.*, *2*: 142–8.

59. ——— (1973). Coexistence, coevolution, and convergent evolution in seabird communities. *Ecology*, *54*: 31–44.

60. Cohen, J.E. (1971). *Casual Groups of Monkeys and Men: Stochastic Models of Elemental Social Systems*. Cambridge, Mass., Harvard University Press.

61. Cohen, S. and McFarland, D.J. (1979). Time-sharing as a mechanism for the control of behaviour sequences during the courtship of the three-spined stickleback (*Gasterosteus aculeatus*). *Anim. Behav.*, *27*: 270–83.

62. Cowie, R.J. (1977). Optimal foraging in great tits (*Parus major*). *Nature*, *268*: 137–9.

63. Curio, E. (1975). The functional organization of anti-predator behaviour in the pied flycatcher: a study of avian visual perception. *Anim. Behav.*, *23*: 1–115.

64. ——— (1978). The adaptive significance of avian mobbing. I. Teleonomic hypotheses and predictions. *Z. Tierpsychol.*, *48*: 175–83.

65. ——— (1980). An unknown determinant of a sex-specific altruism. *Z. Tierpsychol.*, *53*: 139–52.

66. ———, Ernst, U. and Vieth, W. (1978). The adaptive significance of avian mobbing. II. Cultural transmission of enemy recognition in blackbirds: effectiveness and some constraints. *Z. Tierpsychol.*, *48*: 184–202.

67. Daly, M. (1978). The cost of mating. *Amer. Nat.*, *112*: 771–4.

68. Darwin, C. (1859). *The Origin of Species*. London, John Murray.

69. ——— (1871). *The Descent of Man and Selection in Relation to Sex*. London, John Murray.

70. Davies, N.B. (1977). Prey selection and the search strategy of the spotted flycatcher (*Muscicapa striata*), a field study on optimal foraging. *Anim. Behav.*, *25*: 1016–33.

71. ——— (1978). Territorial defence in the speckled wood butterfly (*Pararge aegeria*), the resident always wins. *Anim. Behav.*, *26*: 138–47.

72. ——— (1978). Ecological questions about territorial behaviour. In (J.R. Krebs and N.B. Davies, eds) *Behavioural Ecology: An Evolutionary Approach*: 317–50. Oxford, Blackwell; Sunderland, Mass., Sinauer.

73. ——— and Halliday, T.R. (1978). Optimal mate selection in the toad *Bufo bufo*. *Nature*, *269*: 56–8.

74. ——— ——— (1978). Deep croaks and fighting assessment in toads (*Bufo bufo*). *Nature*, *274*: 683–5.

75. ——— ——— (1979). Competitive mate searching in male common toads *Bufo bufo*. *Anim. Behav.*, *27*: 1253–67.

76. ——— and Houston, A.I. (1981). Owners and satellites: the economics of territory defence in the pied wagtail, *Motacilla alba*. *J. Anim. Ecol.*, *50*: 157–80.

77. Davis, M. (1970). Effects of interstimulus interval and variability on startle response habituation in the rat. *J. Comp. Physiol. Psychol.*, *78*: 260–7.

78. Dawkins, R. (1976). Hierarchical organization: a candidate principle for ethology. In (P.P.G. Bateson and R.A. Hinde, eds) *Growing Points in Ethology*: 7–54. Cambridge University Press.

79. ——— (1976). *The Selfish Gene*. Oxford University Press.

80. ——— (1978). Replicator selection and the extended phenotype. *Z. Tierpsychol.*, *47*: 61–76.

81. ——— (1979). Twelve misunderstandings of kin selection. *Z. Tierpsychol.*, *51*: 184–200.

82. ——— (1982). *The Extended Phenotype*. San Francisco, Freeman.

83. ——— and Carlisle, T.R. (1976). Parental investment, mate desertion and a fallacy. *Nature*, *262*: 131–2.

84. ——— and Brockmann, H.J. (1980). Do digger wasps commit the Concorde fallacy? *Anim. Behav.*, *28*: 892–6.

85. ——— and Krebs, J.R. (1978). Animal signals: information or manipulation? In (J.R. Krebs and N.B. Davies, eds) *Behavioural Ecology: An Evolutionary Approach*: 282–309. Oxford, Blackwell; Sunderland, Mass., Sinauer.

86. ——— ——— (1979). Arms races between and within species. *Proc. R. Soc. Lond. B.*, *205*: 489–511.

87. DeCoursey, P.J. (1960). Phase control of activity in a rodent. *Cold Spring Harbor Symposium on Quantitative Biology*, *25*: 49–55.

88. DeFries, J.C., Thomas, E.A., Hegmann, J.P. and Weir, M.W. (1967). Open-field behaviour in mice: analysis of maternal effects by means of ovarian transplantation. *Psychonom. Sci.*, *8*: 207–8.

89. Dewsbury, D.A. (1978). *Comparative Animal Behavior*. New York, McGraw-Hill.

90. Dilger, W. (1962). The behaviour of lovebirds. *Sci. Am.*, *206*: 88–98.

91. Douglass, R.J. (1976). Spatial interactions and microhabitat selection of two locally sympatric voles, *Microtus montanus* and *Microtus pennsylvanicus*. *Ecology*, *57*: 346–52.

92. Downhower, J.F. and Armitage, K.B. (1971). The yellow-bellied marmot and the evolution of polygamy. *Amer. Nat.*, *105*: 355–70.

93. Edmunds, M. (1974). *Defence in Animals*. Harlow, Longman.

94. Ehrman, L. and Parsons, P.A. (1976). *The Genetics of Behavior*. Sunderland, Mass., Sinauer.

95. Eibl-Eibesfeldt, I. (1975). *Ethology: the Biology of Behaviour*, 2nd edn. London, Holt, Rinehart and Winston.

96. Eisenberg, J.F. (1965). The social organization of mammals. *Handbuch der Zoologie*, *8*: 1–91.

97. Eldredge, N. and Gould, S.J. (1972). Punctuated equilibria, an alternative to phyletic gradualism. In (T.J.M. Schopf, ed.) *Models in Paleobiology*: 82–115. San Francisco, Freeman Cooper.

98. Emlen, S.T. (1967). Migratory orientation of the Indigo Bunting, *Passerina cyanea*. *Auk*, *84*: 309–42.

99. ——— (1975). Migration: orientation and navigation. In (D.S. Farner and J.R. King, eds) *Avian Biology*, *Vol. 5*: 129–219. New York, Academic Press.

100. ——— and Oring, L.W. (1977). Ecology, sexual selection and the evolution of mating systems. *Science*, *197*: 215–23.

101. Erichsen, J.T., Krebs, J.R. and Houston, A.I. (1980). Optimal foraging and cryptic prey. *J. Anim. Ecol.*, *49*: 271–6.

102. Fagen, R. (1977). Selection for optimal age-dependent schedules of play behaviour. *Amer. Nat.*, *112*: 395–414.

103. ——— (1981). *Animal Play Behavior*. New York, Oxford.

104. Fantino, E. and Logan, C.A. (1979). *The Experimental Analysis of Behaviour*. San Francisco, Freeman.

105. Farner, D.S. and Lewis, R.A. (1971). Photoperiodism and reproductive cycles in birds.

Photophysiology, 6: 325-70.
106. Fisher, R.A. (1930). *The Genetical Theory of Natural Selection*. Oxford, Clarendon Press.
107. Foster, W.A. and Treherne, J.E. (1981). Evidence for the dilution effect in the selfish herd from fish predation on marine insect. *Nature, 293*: 466-7.
108. Fraenckel, G. and Gunn, D.L. (1964). *The Orientation of Animals*. London, Dover.
109. Frisch, K. von (1971). *Bees: Their Vision, Chemical Senses and Language*. Ithaca, New York, Cornell University Press.
110. Gale, W.F. (1971). An experiment to determine substrate preference of the fingernail clam, *Sphaeriuus transversum* (Say). *Ecology, 52*: 367-70.
111. Garcia, J. and Koelling, R.A. (1966). Relation of cue to consequence in avoidance learning. *Psychonom. Sci., 56*: 801-5.
112. Gardner, R.A. and Gardner, B.T. (1969). Teaching sign language to a chimpanzee. *Science, 165*: 664-72.
113. Gibb, J.A. (1957). Food requirements and other observations on captive tits. *Bird Study, 4*: 207-15.
114. ———— (1960). Populations of tits and goldcrests and their food supply in pine plantations. *Ibis, 102*: 163-208.
115. Gilbert, L.E. (1975). Ecological consequences of a coevolved mutualism between butterflies and plants. In (L.E. Gilbert and P.H. Raven, eds) *Coevolution of Animals and Plants*: 210-40. Austin, Texas, Texas University Press.
116. Gill, F.B. and Wolf, L.L. (1975). Economics of feeding territoriality in the golden-winged sunbird. *Ecology, 56*: 333-45.
117. Gittelman, J.L., Harvey, P.H. and Greenwood, P.J. (1980). The evolution of conspicuous coloration: some experiments in bad taste. *Anim. Behav., 28*: 897-9.
118. Goss-Custard, J.D. (1977). Optimal foraging and the size selection of worms by redshank *Tringa totanus. Anim. Behav., 25*: 10-29.
119. ———— (1977). Predator responses and prey mortality in the redshank *Tringa totanus* (L) and a preferred prey *Corophium volutator* (Pallas). *J. Anim. Ecol., 46*: 21-36.
120. Gottlieb, G. (1961). Developmental age as a baseline for determination of the critical period in imprinting. *J. Comp. Physiol. Psychol., 54*: 422-7.
121. Gould, J.L. (1976). The dance-language controversy. *Q. Rev. Biol., 51*: 211-44.
122. Greig-Smith, P.W. (1981). Responses to disturbance in relation to flock size in foraging groups of barred ground doves, *Geopelia striata. Ibis, 123*: 103-6.
123. Guthrie, D.M. (1980). *Neuroethology: An Introduction*. Oxford, Blackwell.
124. Gwadz, R. (1970). Monofactorial inheritance of early sexual receptivity in the mosquito, *Aedes atropalus. Anim. Behav., 18*: 358-61.
125. Gwinner, E. (1971). A comparative study of circannual rhythms in warblers. In (M. Menaker, ed.) *Biochronometry*: 405-27. Washington, Nat. Acad. Sci.
126. Haigh, J. and Rose, M.R. (1980). Evolutionary game auctions. *J. Theor. Biol., 85*: 381-97.
127. Hailman, J.P. (1969). How an instinct is learned. *Sci. Am., 221*: 98-106.
128. Halliday, T.R. (1976). The libidinous newt. An analysis of variations in the sexual behaviour of the smooth newt, *Triturus vulgaris. Anim. Behav., 24*: 398-414.
129. ———— (1978). Sexual selection and mate choice. In (J.R. Krebs and N.B. Davies, eds) *Behavioural Ecology: An Evolutionary Approach*: 180-213. Oxford, Blackwells; Sunderland, Mass., Sinauer.
130. Hamilton, W.D. (1964). The genetical theory of social behaviour. I, II. *J. Theor. Biol., 7*: 1-52.
131. ———— (1971). Geometry for the selfish herd. *J. Theor. Biol., 31*: 295-311.
132. Harris, G.W. and Levine, S. (1965). Sexual differentiation of the brain and its experimental control. *J. Phsyiol., 181*: 379-400.
133. Harris, V.T. (1952). An experimental study of habitat selection by prairie and forest

races of the deermouse *Peromyscus maniculatus*. *Contrib. Lab. Vert. Biol. Univ. Michigan*, *56*: 1–53.

134. Hart, B.L. (1967). Sexual reflexes and mating behaviour in the male dog. *J. Comp. Physiol. Psychol.*, *64*: 388–99.

135. Harvey, P.H. and Greenwood, P.J. (1978). Anti-predator defence strategies. In (J.R. Krebs and N.B. Davies, eds) *Behavioural Ecology: An Evolutionary Approach*: 129–54. Oxford, Blackwell; Sunderland, Mass., Sinauer.

136. Heiligenberg, W. (1965). The effect of external stimuli on the attack readiness of a cichlid fish. *Z. Vergl. Physiol.*, *49*: 459–64.

137. ——— (1966). The stimulation of territorial singing in house crickets (*Acheta domesticus*). *Z. Vergl. Physiol.*, *53*: 114–29.

138. Heller, H.C. and Poulson, T.L. (1970). Circannian rhythms – II. Endogenous and exogenous factors controlling reproduction and hibernation in chipmunks (*Eutamias*) and ground squirrels (*Spermophilus*). *Comp. Biochem. Physiol.*, *33*: 357–83.

139. Heron, W.T. (1935). The inheritance of maze learning ability in rats. *J. Comp. Psychol.*, *19*: 77–89.

140. Herrnstein, R.J. (1961). Relative and absolute strength of response as a function of frequency of reinforcement. *J. Exp. Anal. Behav.*, *4*: 267–72.

141. ——— and Loveland, D.H. (1975). Maximizing and matching on concurrent ratio schedules. *J. Exp. Anal. Behav.*, *24*: 107–16.

142. Hess, E.H. (1959). Imprinting: an effect of early experience. *Science*, *130*: 133–41.

143. Hinde, R.A. (1970). *Animal Behaviour: A Synthesis of Ethology and Comparative Psychology*, 2nd edn. London, McGraw-Hill.

144. ——— (1981). Animal signals: ethological and games-theory approaches are not incompatible. *Anim. Behav.*, *29*: 535–42.

145. Hoogland, R., Morris, D. and Tinbergen, N. (1957). The spines of sticklebacks (*Gasterosteus* and *Pygosteus*) as a means of defence against predators (*Perca* and *Esox*). *Behaviour*, *10*: 205–36.

146. Hotta, Y. and Benzer, S. (1969). Abnormal electroretinograms in visual mutants of *Drosophila*. *Nature*, *222*: 354–6.

147. ——— ——— (1970). Genetic dissection of *Drosophila* nervous systems by means of mosaics. *Proc. Nat. Acad. Sci. USA*, *67*: 1156–63.

148. Houston, A.I., Halliday, T.R. and McFarland, D.J. (1977). Towards a model of the courtship of the smooth newt, *Triturus vulgaris* with special emphasis on problems of observability in the simultation of behaviour. *Med. and Biol. Eng. and Comput.*, *15*: 49–61.

149. ———, Krebs, J.R. and Erichsen, J.T. (1980). Optimal prey choice and discrimination time in the great tit (*Parus major* L.). *Behav. Ecol. Sociobiol.*, *6*: 169–75.

150. Hubel, D.H. and Wiesel, T.N. (1962). Receptive fields, binocular interaction and functional architecture in the cat's visual cortex. *J. Physiol.*, *160*: 106–54.

151. Iersel, J.J. van and Bol. A.C.A. (1958). Preening of two tern species: a study of displacement activities. *Behaviour*, *13*: 1–88.

152. Janowitz, H.D. and Grossman, M.I. (1949). Some factors affecting the food intake of normal dogs and dogs with esophagotomy and gastric fistulas. *Am. J. Physiol.*, *159*: 143–8.

153. Johns, J.E. (1964). Testosterone-induced nuptial feathers in phalaropes. *Condor*, *66*: 449–55.

154. Kamil, A.C. (1979). Systematic foraging for nectar by Amakihi, *Loxops virens*. *J. Comp. Physiol. Psychol.*, *92*: 288–396.

155. Katz, B. (1966). *Nerve, Muscle and Synapse*. New York, McGraw Hill.

156. Keeton, W.T. (1971). Magnets interfere with pigeon homing. *Proc. Nat. Acad. Sci. USA*, *68*: 102–6.

157. ——— (1979). Avian orientation and navigation: a brief overview. *British Birds*,

72: 451-71.

158. Kenward, R.E. (1978). Hawks and doves: attack success and selection in goshawk flights at woodpigeons. *J. Anim. Ecol.*, *47*: 449-60.

159. Kenyon, K.W. (1969). The sea otter (*Enhydra lutris*) in the Eastern Pacific Ocean. *North Am. Fauna*, *68*: 1-352.

160. King, J.A. (1955). Social behaviour, social organization and population dynamics in a black-tailed prairie dog town in the Black Hills of South Dakota. *Contr. Lab. Vert. Biol. Univ. Mich.*, *67*: 1-123.

161. Knight-Jones, E.W. (1953). Decreased discrimination during settling after prolonged planktonic life in larvae of *Spirorbis boreatis* (Serpulidea). *J. Marine Biol. Ass. UK*, *32*: 337-45.

162. Köhler, W. (1927). *The Mentality of Apes*, 2nd edn. London, Kegan Paul.

163. Kramer, G. (1957). Experiments on bird orientation and their interpretation. *Ibis*, *99*: 196-227.

164. Krebs, C.J., Gaines, M.S., Keller, B.L., Myers, J.H. and Tamarin, R.H. (1973). Population cycles in small rodents. *Science*, *179*: 35-41.

165. Krebs, J.R. (1971). Territory and breeding density in the great tit, *Parus major* L. *Ecology*, *52*: 2-22.

166. —— (1973). Behavioural aspects of predation. In (P.P.G. Bateson and P. Klopfer, eds) *Perspectives in Ethology*, *Vol. I*. New York, Plenum.

167. —— (1973). Social learning and the significance of mixed species flocks of chickadees (*Parus* spp.). *Can. J. Zool.*, *51*: 1275-88.

168. —— (1974). Colonial nesting and social feeding as strategies for exploiting food resources in the great blue heron (*Ardea herodias*). *Behaviour*, *51*: 99-134.

169. —— (1977). Song and territory in the great tit. In (B. Stonehouse and C.M. Perrins, eds) *Evolutionary Ecology*: 47-62. London, MacMillan.

170. —— (1977). The significance of song repertoires. The Beau Geste hypothesis. *Anim. Behav.*, *25*: 475-8.

171. —— (1978). Optimal foraging: decision rules for predators. In (J.R. Krebs and N.B. Davies, eds) *Behavioural Ecology: An Evolutionary Approach*: 23-63. Oxford, Blackwell; Sunderland, Mass., Sinauer.

172. —— and Cowie, R.J. (1976). Foraging strategies in birds. *Ardea*, *64*: 98-116.

173. ——, Avery, M. and Cowie, R.J. (1981). Effect of removal of mate on the singing behaviour of great tits. *Anim. Behav.*, *29*: 635-6.

174. ——, Erichsen, J.T., Webber, M.I. and Charnov, E.L. (1977). Optimal prey selection in the great tit (*Parus major*). *Anim. Behav.*, *25*: 30-8.

175. ——, Kacelnik, A. and Taylor, P.J. (1978). Optimal sampling by foraging birds: an experiment with great tits (*Parus major*). *Nature*, *275*: 27-31.

176. ——, MacRoberts, M.H. and Cullen, J.M. (1972). Flocking and feeding in the great tit *Parus major* – an experimental study. *Ibis*, *114*: 507-30.

177. ——, Ryan, J.C. and Charnov, E.L. (1974). Hunting by expectation or optimal foraging? A study of patch use by chickadees. *Anim. Behav.*, *22*: 953-64.

178. Kreithen, M.L. and Keeton, W.T. (1974). Detection of changes in atmospheric pressure by the homing pigeon *Columba livia*. *J. Comp. Physiol.*, *89*: 73-82.

179. Kroodsma, D.E. (1977). Correlates of song organization among North American wrens. *Am. Nat.*, *111*: 995-1008.

180. Landsberg, J.W. (1976). Posthatch age and developmental age as a baseline for determination of the sensitive period for imprinting. *J. Comp. Physiol. Psychol.*, *90*: 47-52.

181. Larkin, S. and McFarland, D.J. (1978). The cost of changing from one activity to another. *Anim. Behav.*, *26*: 1237-46.

182. Larkin, T.S. and Keeton, W.T. (1978). An apparent lunar rhythm in the day-to-day variations in the initial bearings of homing pigeons. In (K. Schmidt-Koenig and W.T.

Keeton, eds) *Animal Migration, Navigation and Homing*: 92–106. Berlin, Springer-Verlag.

183. Lawlor, L.R. and Maynard Smith, J. (1976). The coevolution and stability of competing species. *Am. Nat.*, *110*: 79–99.

184. Lazarus, J. (1978). Vigilance, flock size and domain of danger size in the white-fronted goose. *Wildfowl*, *29*: 135–45.

185. —— (1979). The early warning function of flocking in birds: an experimental study with captive quelea. *Anim. Behav.*, *27*: 855–65.

186. Lettvin, J.Y., Maturana, H.R., McCulloch, W.S. and Pitts, W.H. (1959). What the frog's eye tells the frog's brain. *Proc. Inst. Rad. Eng.*, *47*: 1940–51.

187. Levick, W.R. (1967). Receptive fields and trigger features in the visual streak of the rabbit's retina. *J. Physiol.*, *188*: 285–307.

188. Levine, L. (1958). Studies on sexual selection in mice. *Amer. Nat.*, *92*: 21–6.

189. Levins, R. (1969). Thermal acclimation and heat resistance in *Drosophila* species. *Amer. Nat.*, *103*: 483–99.

190. Lincoln, G.A., Guinness, F. and Short, R.V. (1972). The way in which testosterone controls the social and sexual behaviour of the red deer stag (*Cervus elephus*). *Hormones and Behaviour*, *3*: 375–96.

191. Lindauer, M. (1961). *Communication among Social Bees*. Cambridge, Mass., Harvard University Press.

192. Loizos, C. (1966). Play in mammals. In (P.A. Jewell and C. Loizos, eds) *Play, Exploration and Territory in Mammals*. London, Academic Press.

193. Lord, R.D. Jr. (1956). A comparative study of the eyes of some Falconiform and Passeriform birds. *Am. Midl. Nat.*, *56*: 325–44.

194. Lorenz, K.Z. (1958). The evolution of behaviour. *Sci. Am.*, *199*: 67–78.

195. MacArthur, R.H. and Pianka, E.R. (1966). On the optimal use of a patchy environment. *Amer. Nat.*, *100*: 603–9.

196. MacKay, T.F.C. and Doyle, R.W. (1978). An ecological genetic analysis of the settling behaviour of a marine polychaete. I. Probability of settlement and gregarious behaviour. *Heredity*, *40*: 1–12.

197. Maier, N.R.F. and Schneirla, T.C. (1935). *Principles of Animal Psychology*. New York, McGraw-Hill.

198. Manning, A. (1961). The effects of artificial selection for mating speed in *Drosophila melanogaster*. *Anim. Behav.*, *9*: 82–92.

199. —— (1979). *An Introduction to Animal Behaviour*, 3rd edn. London, Arnold.

200. Marler, P. (1955). Characteristics of some animal calls. *Nature*, *176*: 6–8.

201. —— (1956). The voice of the chaffinch and its function as a language. *Ibis*, *98*: 231–61.

202. Matthews, E.G. (1977). Signal-based frequency-dependent defense strategies and the evolution of mimicry. *Am. Nat.*, *111*: 213–22.

203. Maynard Smith, J. (1974). The theory of games and the evolution of animal conflicts. *J. Theor. Biol.*, *47*: 209–21.

204. —— (1976). Sexual selection and the handicap principle. *J. Theor. Biol.*, *57*: 239–42.

205. —— (1978). The ecology of sex. In (J.R. Krebs and N.B. Davies, eds) *Behavioural Ecology: An Evolutionary Approach*: 159–79. Oxford, Blackwell; Sunderland, Mass., Sinauer.

206. —— (1978). *The Evolution of Sex*. Cambridge University Press.

207. —— and Parker, G.A. (1976). The logic of asymmetric contests. *Anim. Behav.*, *24*: 159–75.

208. McCleery, R.H. (1978). Optimal behavioural sequences and decision making. In (J.R. Krebs and N.B. Davies, eds) *Behavioural Ecology: An Evolutionary Approach*: 377–410. Oxford, Blackwell; Sunderland, Mass., Sinauer.

209. McCollom, R.E., Siegel, P.B. and Vankrey, H.P. (1971). Responses to androgen in lines of chickens selected for mating behaviour. *Hormones and Behaviour*, *2*: 31–42.

210. McFarland, D.J. (1971). *Feedback Mechanisms in Animal Behaviour*. London, Academic Press.

211. —— (1974). Time-sharing as a behaviour phenomenon. *Advances in the Study of Behaviour*, *Vol. 5*: 201–25. London, Academic Press.

212. —— and Lloyd, I.H. (1973). Time-shared feeding and drinking. *Q. J. Exp. Psychol.*, *25*: 48–61.

213. Miles, C.G. and Jenkins, H.M. (1973). Overshadowing in operant conditioning as a function of discriminability. *Learning and Motivation*, *4*: 11–27.

214. Milinski, M. (1977). Experiments on the selection by predators against spatial oddity of their prey. *Z. Tierpsychol.*, *43*: 311–25.

215. —— and Heller, R. (1978). Influence of a predator on the optimal foraging behaviour of sticklebacks (*Gasterosteus aculeatus* L.). *Nature*, *275*: 642–4.

216. Morris, D. (1957). Typical intensity and its relation to the problem of ritualization. *Behaviour*, *11*: 1–12.

217. Morton, E.S. (1975). Ecological sources of selection on avian sounds. *Am. Nat.*, *109*: 17–34.

218. Mowrer, O.H. (1960). *Learning Theory and Behaviour*. New York, Wiley.

219. Moynihan, M.H. (1970). Control suppression, decay, disappearance and replacement of displays. *J. Theor. Biol.*, *29*: 85–112.

220. —— (1981). A coincidence of mimicries and other misleading coincidences. *Am. Nat.*, *117*: 372–8.

221. Muller, H.J. (1932). Some genetic aspects of sex. *Amer. Nat.*, *66*: 118–38.

222. Muntz, W.R.A. (1964). Vision in frogs. *Sci. Am.*, *210*: 110–19.

223. Murdie, G. and Hassell, M.P. (1973). Food distribution, searching success and predator-prey models. In (M.S. Bartlett and R.W. Hiorns, eds) *The Mathematical Theory of the Dynamics of Biological Populations*. London, Academic Press.

224. Naylor, E. (1958). Tidal and diurnal rhythms of locomotor activity in *Carcinus maenas*. *J. Exp. Biol.*, *35*: 602–10.

225. Neill, S.R. St. J. and Cullen, J.M. (1974). Experiments on whether schooling by their prey affects the hunting behaviour of cephalopod and fish predators. *J. Zool. Lond.*, *172*: 549–69.

226. Norman, R.F., Taylor, P.D. and Robertson, R.J. (1977). Stable equilibrium strategies and penalty functions in a game of attrition. *J. Theor. Biol.*, *69*: 571–8.

227. Nottebohm, F. (1967). The role of sensory feedback in the development of avian vocalizations. *Proc. 14th Int. Ornith. Congr., Oxford*: 265–80. Oxford, Blackwell.

228. —— (1976). Central control of song in the canary. *Serinus canarius*. *J. Comp. Neurol.*, *165*: 457–86.

229. O'Donald, P. (1974). Polymorphisms maintained by sexual selection in monogamous species of birds. *Heredity*, *32*: 1–10.

230. Olds, J. (1961). Differential effects of drives and drugs on self-stimulation at different brain sites. In (D.E. Sheer, ed.) *Electrical Stimulation of the Brain*. Inst. Mental Health, University of Texas.

231. Orians, G.H. (1969). On the evolution of mating systems in birds and mammals. *Amer. Nat.*, *103*: 589–603.

232. —— and Pearson, N.E. (1979). On the theory of central place foraging. In (D.J. Horn, G.R. Stairs and R.D. Mitchell, eds) *Analysis of Ecological Systems*: 155–77. Ohio State University Press.

233. Packer, C. (1977). Reciprocal altruism in *Papio anubis*. *Nature*, *265*: 441–3.

234. Page, G. and Whitacre, D.F. (1975). Raptor predation on wintering shorebirds. *Condor*, *77*: 73–83.

235. Parker, G.A. (1970). Sperm competition and its evolutionary consequences in insects.

Biol. Rev., *45*: 525–68.

236. —— (1978). Searching for mates. In (J.R. Krebs and N.B. Davies, eds) *Behavioural Ecology: An Evolutionary Approach*: 214–44. Oxford, Blackwell; Sunderland, Mass., Sinauer.

237. —— (1979). Sexual selection and sexual conflict. In (M.S. Blum and N.A. Blum, eds) *Sexual Selection and Reproductive Competition in Insects*: 123–66. New York, Academic Press.

238. ——, Baker, R.R. and Smith, V.G.F. (1972). The origin and evolution of gametic dimorphism and the male-female phenomenon. *J. Theor. Biol.*, *36*: 529–53.

239. Partridge, L. (1976). Field and laboratory observations on the foraging and feeding techniques of bluetits (*Parus caeruleus*) and coaltits (*Parus ater*) in relation to their habitats. *Anim. Behav.*, *24*: 534–44.

240. —— (1978). Habitat selection. In (J.R. Krebs and N.B. Davies, eds) *Behavioural Ecology: An Evolutionary Approach*: 351–76. Oxford, Blackwell; Sunderland, Mass., Sinauer.

241. Payne, R. (1962). How the barn owl locates prey by hearing. *Living Bird*, *1*: 151–9.

242. Peeke, H.V.S. and Veno, A. (1973). Stimulus specificity of habituated aggression in three-spined sticklebacks (*Gasterosteus aculeatus*). *Behav. Biol.*, *8*: 427–32.

243. Pengelly, E.T. and Asmundson, S.M. (1969). Free-running periods of endogenous circannian rhythms in the golden-mantled ground squirrel, *Citellus lateralis*. *Comp. Biochem. Physiol.*, *30*: 177–83.

244. Perrins, C.M. (1968). The purpose of high-intensity alarm calls in small passerines. *Ibis*, *110*: 200–1.

245. Pettit, C. (1959). De la nature des stimulations responsables de la sélection sexualle chez *Drosophila melanogaster*. *C. R. Acad. Sci., Paris*, *248*: 3484–5.

246. Pitcher, T. (1979). He who hesitates lives. Is stotting antiambush behaviour? *Am. Nat.*, *113*: 453–6.

247. Powell, G.V.N. (1974). Experimental analysis of the social value of flocking by starlings (*Sturnus vulgaris*) in relation to predation and foraging. *Anim. Behav.*, *22*: 501–5.

248. Premack, D. (1971). Language in chimpanzee? *Science*, *172*: 808–22.

249. Pyke, G.H. (1978). Optimal foraging in bumble bees: pattern of movements between inflorescences. *Theor. Pop. Biol.*, *13*: 79–98.

250. —— (1979). Optimal foraging in bumble bees: rule of movement between flowers within inflorescences. *Anim. Behav.*, *27*: 1167–81.

251. Raisman, G. and Field, P.M. (1973). Sexual dimorphism in the neuropile of the preoptic area of the rat and its dependence on neonatal androgen. *Brain Research*, *54*: 1–29.

252. Rand, A.S. and Rand, W.M. (1976). Agonistic behaviour in nesting iguanas: a stochastic analysis of dispute settlement dominated by the minimisation of energy cost. *Z. Tierpsychol.*, *40*: 279–99.

253. Rescorla, R.A. (1968). Probability of shock in the presence and absence of C.S. in fear conditioning. *J. Comp. Physiol. Psychol.*, *66*: 1–5.

254. —— and Wagner, A.R. (1972). A theory of Pavlovian conditioning: variations in the effectiveness of reinforcement and non-reinforcement. In (A.H. Black and W.F. Prokasy, eds) *Classical Conditioning II: Current Research and Theory*. New York, Appleton-Century-Crofts.

255. Ridley, M. (1978). Paternal care. *Anim. Behav.*, *26*: 904–32.

256. Roberts, J., Kacelnik, A. and Hunter, M.L. Jr. (1979). A model of sound interference in relation to acoustic communication. *Anim. Behav.*, *27*: 1271–3.

257. Roeder, K.D. (1970). Episodes in insect brains. *Amer. Sci.*, *58*: 378–89.

258. Rohwer, S. (1977). Status signalling in Harris' sparrows: some experiments in deception. *Behaviour*, *61*: 107–29.

259. —— and Ewald, P.W. (1981). The cost of dominance and advantage of subordination in a badge signalling system. *Evolution, 35*: 441-54.

260. —— and Rohwer, F.C. (1978). Status signalling in sparrows: experimental deceptions achieved. *Anim. Behav., 26*: 1012-22.

261. Rose, A. and Parsons, P.A. (1970). Behavioural studies in different strains of mice and the problems of heterosis. *Genetics, 41*: 65-87.

262. Rosin, R. (1978). The honey bee 'language' controversy. *J. Theor. Biol., 72*: 589-602.

263. Rothenbuhler, N. (1964). Behaviour genetics of nest cleaning honey bees. I. Responses of four inbred lines to disease-killed brood. *Anim. Behav., 12*: 578-83.

264. —— (1964). Behaviour genetics of nest cleaning honey bees. IV. Responses of F_1 and backcross generations to disease-killed brood. *Am. Zool., 4*: 111-23.

265. Roughgarden, J. (1976). Resource partitioning among competing species: a coevolutionary approach. *Theor. Pop. Biol., 9*: 388-424.

266. Salzen, E.A. (1970). Imprinting and environmental learning. In (L. Aronson, E. Tobach, D. Lehrman and J. Rosenblatt, eds) *Development and the Evolution of Behaviour*. San Francisco, Freeman.

267. Schaller, G.B. (1963). *The Mountain Gorilla*. University of Chicago Press.

268. Schmidt-Koenig, K. (1960). Internal clocks and homing. *Cold Spring Harbor Symp. Quant. Biol., 25*: 389-93.

269. —— (1975). *Migration and Homing in Animals*. New York, Springer-Verlag.

270. Schneirla, T.C. (1933). Motivation and efficiency in ant learning. *J. Comp. Psychol., 15*: 243-66.

271. Schuster, P. and Sigmund, K. (1981). Coyness, philandering and stable strategies. *Anim. Behav., 29*: 186-92.

272. Schutz, F. (1965). Sexuelle Pragung bei Anatiden. *Z. Tierpsychol., 22*: 50-103.

273. —— (1971). Pragung des sexual verhaltens von enten und gansen durch sozialeindrucke wahrend der jugendphase. *J. Neurovisc. Rel. Suppl., 10*: 399-457.

274. Sebeok, T.A. (1962). Coding in the evolution of signalling behaviour. *Behavioural Science, 7*: 430-42.

275. —— (1965). Animal communication. *Science, 147*: 1006-14.

276. Selander, R.K. (1972). Sexual selection and dimorphism in birds. In (B. Campbell, ed.) *Sexual Selection and the Descent of Man*: 180-230. Chicago, Aldine.

277. Shalter, M.D. (1978). Localization of passerine seeet and mobbing calls by goshawks and pygmy owls. *Z. Tierpsychol., 46*: 260-7.

278. —— (1978). Mobbing in the pied flycatcher: effect of experiencing a live owl on responses to a stuffed facsimile. *Z. Tierpsychol., 47*: 173-9.

279. Shaw, G. (1979). Functions of dipper roosts. *Bird Study, 26*: 171-8.

280. Shepard, J.F. (1933). Higher processes in the behaviour of rats. *Proc. Nat. Acad. Sci. USA, 19*: 149-52.

281. Sherrington, C.S. (1906). *The Intergrative Action of the Nervous System*. New York, Schribner's.

282. Shimp, C.P. (1974). Time allocation and response rates. *J. Exp. Anal. Behav., 21*: 491-9.

283. Sibly, R.M. (1975). How incentive and deficit determine feeding tendency. *Anim. Behav., 33*: 437-46.

284. —— and McFarland, D.J. (1974). A state-space approach to motivation. In (D.J. McFarland, ed.) *Motivational Control Systems Analysis*: 213-50. London, Academic Press.

285. —— —— (1976). On the fitness of behaviour sequences. *Am. Nat., 110*: 601-17.

286. Siegfried, W.R. and Underhill, L.G. (1975). Flocking as an anti-predator strategy in doves. *Anim. Behav., 23*: 504-8.

287. Sigurjónsdóttir, H. and Parker, G.A. (1981). Dungfly struggles: evidence for assessment strategy. *Behav. Ecol. Sociobiol., 8*: 219-30.

288. Silcock, M. and Parsons, P.A. (1973). Temperature preference differences between strains of *Mus musculus*, associated variables and ecological implications. *Oecologia*, *12*: 147–60.

289. Slater, P.J.B. and Ince, S.A. (1979). Cultural evolution in chaffinch song. *Behaviour*, *71*: 146–66.

290. Smith, J.N.M. (1974). The food searching behaviour of two European thrushes. I. Description and analysis of search paths. *Behaviour*, *48*: 276–302. II. The adaptiveness of the search patterns. *Behaviour*, *49*: 1–61.

291. —— (1977). Feeding rates, search paths and surveillance for predators in great-tailed grackle flocks. *Can. J. Zool.*, *55*: 891–8.

292. —— and Sweatman, H.P.A. (1974). Food searching behaviour of titmice in patchy environments. *Ecology*, *55*: 1216–32.

293. Smith, N.G. (1980). Some evolutionary, ecological and behavioural correlates of communal nesting by birds with wasps or bees. *Acta XVII Cong. Int. Orn., Berlin 1978*: 1199–205.

294. Smythe, N. (1977). The function of mammalian alarm advertizing: social signals or pursuit invitation. *Am. Nat.*, *111*: 191–4.

295. Southwick, C.H. (1968). Effects of maternal environment on aggressive behaviour of inbred mice. *Comm. Behav. Biol.*, *1*: 129–32.

296. Staddon, J.E.R. (1980). Optimality analyses of operant behaviour and their relation to optimal foraging. In (J.E.R. Staddon, ed.) *Limits to Action. The Allocation of Individual Behaviour*. London, Academic Press.

297. Stanley, S. (1979). *Macroevolution: Pattern and Process*. San Francisco, Freeman.

298. Stevenson-Hinde, J. (1973). Constraints on reinforcement. In (R.A. Hinde and J. Stevenson-Hinde, eds) *Constraints on Learning*. New York, Academic Press.

299. Stewart, R.E. and Aldrich, J.W. (1951). Removal and repopulation of breeding birds in a spruce-fir forest community. *Auk*, *68*: 471–82.

300. Symons, D. (1978). *Play and Aggression: A Study of Rhesus Monkeys*. New York, Columbia University Press.

301. Taylor, L.R. (1961). Aggregation, variance and the mean. *Nature*, *189*: 732–5.

302. Thornhill, R. (1976). Sexual selection and nuptial feeding behaviour in *Bittacus apicalis* (Insecta: Macoptera). *Amer. Nat.*, *110*: 529–48.

303. Thorpe, W.H. (1961). *Bird-song*. Cambridge University Press.

304. —— (1963). *Learning and Instinct in Animals*. London, Methuen.

305. Tinbergen, N. and Perdeck, A.C. (1950). On the stimulus situation releasing the begging response in the newly hatched herring gull chick (*Larus argentatus argentatus* Pont.). *Behaviour*, *3*: 1–39.

306. Toates, F.M. (1975). *Control Theory in Biology and Experimental Psychology*. London, Hutchinson.

307. —— (1980). *Animal Behaviour: A Systems Approach*. Chichester, Wiley.

308. Tolman, E.C. (1924). The inheritance of maze learning in rats. *J. Comp. Psychol.*, *4*: 1–18.

309. —— and Honzik, C.H. (1930). Introduction and removal of reward and maze performance in rats. *Univ. Cal. Publ. Psychol.*, *4*: 257–75.

310. Treherne, J.E. and Foster, W.A. (1981). Group transmission of predator avoidance in a marine insect: the Trafalgar effect. *Anim. Behav.*, *29*: 911–17.

311. Treisman, M. (1975). Predation and the evolution of gregariousness. I. Models for concealment and evasion. II. An economic model for predator-prey interaction. *Anim. Behav.*, *23*: 779–825.

312. —— (1976). The evolution of sexual reproduction: a model which assumes individual selection. *J. Theor. Biol.*, *60*: 421–31.

313. —— and Dawkins, R. (1976). The cost of meiosis: is there any? *J. Theor. Biol.*, *63*: 479–84.

314. Trivers, R.L. (1971). The evolution of reciprocal altruism. *Q. Rev. Biol.*, *46*: 35-7.

315. —— (1972). Parental investment and sexual selection. In (B. Campbell, ed.) *Sexual Selection and the Descent of Man*: 136-79. Chicago, Aldine.

316. —— (1976). Sexual selection and resource-accruing abilities in *Anolis garmani*. *Evolution*, *30*: 253-69.

317. Trune, D.R. and Slobodchikoff, P. (1976). Social effects of roosting on the metabolism of the pallid bat *Antrozotus pallidus*. *J. Mammal.*, *57*: 656-63.

318. Tryon, R.C. (1940). Studies in individual differences in maze ability. VII. The specific components of maze ability and a general theory of psychological components. *J. Comp. Physiol. Psychol.*, *30*: 283-335.

319. Turner, J.R.G. (1971). Studies of Mullerian mimicry and its evolution in burnet moths and heliconiid butterflies. In (R. Creed, ed) *Ecological Genetics and Evolution*: 224-60. Oxford, Blackwell.

320. Usherwood, P.N.R. (1973). *Nervous Systems*. London, Arnold.

321. Vehrencamp, S.L. (1977). Relative fecundity and parental effort in communally nesting anis, *Crotophaga sulcirostris*. *Science*, *197*: 403-5.

322. Vidal, J.-M. (1976). L'empreinte chez les animaux. *La Recherche*, *63*: 24-35.

323. Vieth, N., Curio, E. and Ernst, U. (1980). The adaptive significance of avian mobbing. III. Cultural transmission of enemy recognition in blackbirds: cross-species tutoring and properties of learning. *Anim. Behav.*, *28*: 1217-29.

324. Vine, I. (1971). Risk of visual detection and pursuit by a predator and the selective advantage of flocking behaviour. *J. Theor. Biol.*, *30*: 405-22.

325. Wade, M.J. (1977). An experimental study of group selection. *Evolution*, *31*: 134-53.

326. Ward, J.A. and Barlow, G.W. (1967). The maturation and regulation of glancing off the parents by young orange chromides (*Etroplus maculatus*: Pisces – Cichlidae). *Behaviour*, *29*: 1-56.

327. Ward, P. (1971). The migration patterns of *Quelea quelea* in Africa. *Ibis*, *113*: 275-97.

328. Ward, S. (1973). Chemotaxis by the nematode *Caenorhabditis elegans*: identification of attractants and analysis of the response by use of mutants. *Proc. Nat. Acad. Sci. USA*, *70*: 817-21.

329. Weiss, P. (1941). Does sensory control play a constructive role in the development of motor coordination? *Schweiz. Med. Wschr.*, *71*: 591-5.

330. Whitham, T.G. (1977). Coevolution of foraging in *Bombus* – nectar dispensing in *Chilopsis*: a last dreg theory. *Science*, *197*: 593-6.

331. Williams, G.C. (1966). *Adaptation and Natural Selection*. Princeton University Press.

332. —— (1975). *Sex and Evolution*. Princeton University Press.

333. Willows, A.O.D. and Hoyle, G. (1969). Neuronal network triggering a fixed action pattern. *Science*, *166*: 1549-51.

334. Wilson, D.S. (1980). *The Natural Selection of Populations and Communities*. California, Benjamin/Cummings.

335. Wilson, E.O. (1975). *Sociobiology: The New Synthesis*. Belknap, Harvard.

336. Wiltschko, W. (1968). Uber den Einfluss statischer Magnetfelder auf die Zugorientierung der Rotkehlchen. (*Erithacus rubecula*). *Z. Tierpsychol.*, *25*: 537-58.

337. Wiltschko, R. and Wiltschko, W. (1981). The development of sun compass orientation in young homing pigeons. *Behav. Ecol. Sociobiol.*, *9*: 135-42.

338. Wood-Gush, D.G.M. (1972). Strain differences in response to sub-optimal stimuli in the fowl. *Anim. Behav.*, *20*: 72-6.

339. Wrangham, R.W. (1980). An ecological model of female-bonded primate groups. *Behaviour*, *75*: 262-300.

340. Wynne-Edwards, V.C. (1962). *Animal Dispersion in Relation to Social Behaviour*. Edinburgh, Oliver and Boyd.

341. Yamazaki, K., Boyse, E.A., Miké, V., Thaler, H.T., Mathieson, B., Abbott, J., Boyse, J., Zayas, Z.A. and Thomas, L. (1976). Control of mating preferences in mice by genes in

the major histocompatibility complex. *J. Exp. Med.*, *144*: 1324–35.

342. Yasukawa, K. (1981). Song repertoires in the red-winged blackbird (*Agelaius phoeniceus*): a test of the Beau Geste Hypothesis. *Anim. Behav.*, *29*: 114–25.

343. Young, W.C. (1965). The organization of sexual behaviour by hormonal action during the prenatal and larval periods in vertebrates. In (F.A. Beach, ed.) *Sex and Behaviour*: 89–107. New York, Wiley.

344. Zahavi, A. (1975). Mate selection – a selection for a handicap. *J. Theor. Biol.*, *53*: 205–14.

INDEX